GETTING TO KNOW

ArcGIS®

for DESKTOP

THIRD EDITION

Michael Law | Amy Collins

Esri Press
REDLANDS | CALIFORNIA

Esri Press, 380 New York Street, Redlands, California 92373-8100

Copyright © 2013 Esri
All rights reserved. First edition 2004. Third edition 2013
Printed in the United States of America

17 16 15 14 13   1 2 3 4 5 6 7 8 9 10

Library of Congress Cataloging-in-Publication Data
Law, Michael, 1980–
   Getting to know ArcGIS for desktop / Michael Law, Amy Collins. — 3rd ed.
      p. cm. — (Getting to know ArcGIS Desktop)
      Includes index.
      Earlier edition cataloged under the title.
      ISBN 978-1-58948-308-8 (pbk.)
1. ArcGIS. 2. Geographic information systems. 3. Graphical user interfaces (Computer systems) I. Collins, Amy, 1974– II.
Getting to know ArcGIS for desktop. III. Title.
      G70.212.G489 2013
      910.285'53--dc23 2012032841

Contents

Preface

Getting to Know ArcGIS for Desktop is a workbook for beginners. Its detailed, step-by-step exercises teach you the core functionality of ArcGIS for Desktop software: how to make maps, carry out spatial analysis, and build and edit spatial databases in the context of realistic projects. The exercises are supported by conceptual discussions at the start of each chapter and as needed throughout the book. Abundant color graphics confirm your progress along the way.

Getting to Know ArcGIS for Desktop is a hands-on workbook that's meant to be a practical manual for classroom lab work or on-the-job training. If you have no GIS background, chapter 1 will give you a quick overview. If you have no ArcGIS software experience, chapter 2 will provide a background on Esri GIS software products. The remaining 18 chapters require you to work through software exercises as you read. Each of these chapters contains two to four exercises that focus on a particular GIS task or problem. Many common tasks are covered, including navigating digital maps, finding GIS data online, setting map projections, symbolizing and labeling maps, classifying data, making map layouts, querying maps, analyzing spatial relationships, building spatial databases, editing data, and geocoding addresses.

Most of the exercises can be completed with any type of ArcGIS 10.1 for Desktop license (Basic, Standard, or Advanced), with the exception of chapter 20, which requires a Standard or Advanced license and the ArcGIS 10.1 Spatial Analyst for Desktop extension. (ArcGIS 10.1 for Desktop Advanced, complete with ArcGIS extensions, is provided with this book.)

The book comes with a Maps and Data DVD that contains maps, exercise data, and results. You can download ArcGIS 10.1 for Desktop Advanced software at http://www.esri.com/gtkArcGISforDesktop10-1. The software expires 180 days after installation. You will need the authorization code found on the inside back cover of this book. Appendix C describes the installation process.

How to use this book

Each new exercise in *Getting to Know ArcGIS for Desktop* is a fresh starting point, with the maps and data you need already prepared for you. It is advisable to follow the chapters in order, because tools and functions used often in early chapters may not be described again in later ones. The exercises will work, however, no matter which chapter you start with. Each chapter takes an hour or two to complete.

What's new in this edition

This volume is heavily revised for the third edition. New authors Michael Law and Amy Collins replaced a majority of the datasets for a fresh learning experience. All the exercises were designed for compatibility with ArcGIS 10.1 for Desktop. Although still intended for beginners, the content is no longer limited to the most basic functionality. Some advanced capabilities that are available with Standard and Advanced licenses are discussed. A chapter on raster analysis (chapter 20) requires the Standard or Advanced license and ArcGIS Spatial Analyst extension (available with the included 180-day software). As the chapters progress, instructions and graphics become noticeably more streamlined, assuming increased knowledge and recall by readers as they move forward.

Several new features have been added to the book to enhance the learning experience: learning objectives, chapter summaries, questions for critical thinking, software tips, independent-challenge tasks ("On your own"), a glossary of key terms, and navigation tabs.

Each chapter now begins with learning objectives, which are specific skills you can expect to gain by the time you have completed the chapter. The summary at the end of each chapter reinforces the lessons learned. Questions for the reader are interspersed throughout the book to stimulate learning. Software tips illuminate the content with helpful information and alternative ways to perform software tasks, and "On your own" boxes encourage you to recall what you've learned by performing tasks without step-by-step guidance. Key terms, which are highlighted in the text in a different color and compiled at the end of each chapter, are included in a brand-new glossary at the back of the book. This edition has also added several new sidebars, which offer an expanded discussion of challenging concepts. Finally, a new navigation system, which appears in the page margins, lets you know what chapter and exercise you're working in.

We hope you enjoy *Getting to Know ArcGIS for Desktop.*

Michael Law
Toronto, Ontario

Amy Collins
Napa, California

Acknowledgments

The authors thank the following individuals and organizations for contributing data and images to this book:

- The World Bank provided the air pollution data used in chapters 3 and 4. http://data.worldbank.org
- World Wildlife Fund provided the terrestrial ecosystem data featured in chapter 7. http://www.worldwildlife.org
- The Florida datasets (chapters 7 and 9) were obtained from the Florida Geographic Data Library (FGDL). http://www.fgdl.org/metadataexplorer
- The shipwreck points used in chapters 7 and 9 are based on data obtained from the Automated Wreck and Obstruction Information System (AWOIS) used by the Office of Coast Survey (OSC), a division of the National Oceanic and Atmospheric Administration (NOAA). http://www.nauticalcharts.noaa.gov/hsd/awois.html
- The National Archive of Criminal Justice Data, the Inter-university Consortium for Political and Social Research, and the US Department of Justice, Federal Bureau of Investigation provided crime report data featured in chapter 8. http://www.icpsr.umich.edu/icpsrweb/NACJD
- Natural Earth provided the raster imagery featured in chapters 7, 9, and 10. http://www.naturalearthdata.com
- Department of Environment and Resource Management, Queensland, Australia, provided bioregion and bio-subregion data featured in chapter 10. http://www.derm.qld.gov.au
- Oswald Eichberger, senior director of Consumer Intelligence and Client Services of Generation5 (a Motricity company), provided consumer market data featured in chapter 15, with consultation from Aaron Heppleston, data management specialist of Geomedia Inc. http://www.generation5.ca, http://www.geomedia.ca
- City of Toronto and Open Data provided street data featured in chapter 15. http://www.toronto.ca/open
- The food desert spreadsheet used in chapter 16 was obtained from the US Department of Agriculture, Economic Research Service. http://www.ers.usda.gov/data/fooddesert

- Tim Pudoff, GIS manager of Sonoma County, provided data and input for the Russian River flood zone exercises (chapter 17) and the clinic suitability study (chapter 20). Thanks also to JoAnn Borri of the Sonoma County Department of Health Services. https://gis.sonoma-county.org
- The Oregon Department of Forestry provided fire risk data featured in chapters 18 and 19. http://www.oregon.gov/ODF
- Heidi Mosher, of the City of Manhattan, Kansas, and Robert Nall, of Riley County, Kansas, provided the layers used in chapters 11, 12, and 13.

All other datasets (cities, countries, states, rivers, and census data) were obtained from Esri Data & Maps 2010.

Thanks to everyone at Esri Press.

Although much of this volume has been updated with new material for the third edition, its structure was solid. We are deeply indebted to the original author team who made *Getting to Know ArcGIS* a best seller for many years—Tim Ormsby, Eileen Napoleon, Rob Burke, Carolyn Groessl, and Laura Bowden. No small amount of the original text remains in this edition, and we truly hope their straightforward teaching style and original intent—making sophisticated software accessible to beginners—remain intact.

Getting to know

GIS

Chapter 1
Introducing GIS

Learning objectives

- *Learn what geographic information systems are used for*
- *Learn how GIS layers work*
- *Differentiate between GIS features and surfaces*
- *Obtain preliminary knowledge of feature locations, scale, and attributes*
- *Obtain preliminary knowledge of spatial relationships between features*

FOR A LONG TIME, PEOPLE HAVE STUDIED THE WORLD USING CONVENTIONAL models such as maps and globes. In the last 40 years or so, it has become possible to put these models into computers—with more sophisticated models inside smaller and smaller computers every year. These computer models, along with the tools for analyzing them, make up a geographic information system (GIS).

In a GIS, you can study not only this map or that map, but every map there is. With the right data, you can see whatever you want to see—land features, elevation, weather and climate zones, forests, political boundaries, population density, per capita income, land use, energy consumption, mineral resources, and a thousand other things—in whatever part of the world interests you.

Most importantly, you can map where things are. GIS lets you search for places that have the features you're interested in, find information about these places, and discover new patterns in the data or observe existing ones.

For example, a population density map shows the distribution of population in specific mapping areas (census tracts, counties, states, countries). This type of information plays an important role when the mapping areas vary greatly in size—a total population of 200,000 means different things in an 8-square-mile county versus an 80-square-mile county. Population density in turn provides a context for other maps, which may show demographics, economic statistics, or housing prices. Such maps are commonly seen in the newspaper or on the web.

Some organizations map quantities of things to find locations that meet their criteria and to inform their decision making. A real estate company might keep a GIS database of properties to map property values and generate sales data. A utility company might have thousands of repairs a year to complete and require GIS to coordinate, map, and manage the work efficiently.

GIS can also be used to analyze what's happening around a specific area. An environmental agency, for example, might map wildlife species habitats and take inventory of vegetation across a region to better understand the relationship between them. A water conservation agency might analyze areas adjacent to stream watersheds to gauge how natural features, seasonal water flow, and existing habitats change.

Finally, organizations use GIS to map how and where things evolve over a period of time to learn about subject movement and behavior. Public health officials, for example, might track an influenza outbreak geographically to try to determine its origin, find disease clusters, and identify areas of high risk. GIS can be used to evaluate current situations and anticipate future conditions. Hurricane tracking and modeling, for example, can provide early warning for evacuation schedules that can help save people's lives.

Basic GIS principles

The following sections will introduce you to some basic principles of GIS. These concepts will become clearer and more understandable as you work with the software in chapters 3–20.

A GIS map contains layers

On a paper map, you can't peel cities away from countries or countries away from the ocean, but on a GIS map, you can. A GIS map is made up of layers, or collections of geographic objects that are alike and defined by a visual symbol. To make a map, you can add as many layers as you want. In the map shown, the transmission lines are layered on top of the electricity generation circles, which are then layered on top of the coal basins, and finally layered on top of the terrain and ocean layers. This creates an important layer hierarchy on the map.

This map of North America shows state boundaries, lakes, rivers, terrain, and the ocean. It also shows the amount of electricity generation, transmission lines, and locations of coal basins. The map is accompanied by a legend that shows some of its layers. Map courtesy of Kansas Geological Survey at the University of Kansas.

Layers may contain features or surfaces

In the map shown above, the coal basins layer includes many different basins, and the transmission lines layer contains many different lines. The same is true of the states and lakes layers—each layer contains many of the same types of things. Each geographic object in a layer—each state, lake, transmission line, or coal basin—is called a feature.

Not all layers contain features. The ocean layer is not a collection of geographic objects the way the other layers are. It is considered a single, continuous expanse that changes from one location to another according to the depth of the water. Similarly, the terrain layer shows the elevation of the landscape. A geographic expanse of these kinds is called a surface.

This topographic map shows the terrain of a landmass and the seafloor. Map made with Natural Earth.

Features have shape and size

Geographic objects have an endless variety of shapes. They all, however, can be represented as one of three geometric forms—a polygon, a line, or a point.

Polygons represent objects that are large enough to have boundaries, such as countries, lakes, and tracts of land. Lines represent objects that are relatively too narrow to be polygons, such as rivers, roads, and pipelines. Points are used for objects that are relatively too small to be polygons, such as cities, schools, and fire hydrants. (The same object may be represented by a polygon in one layer and a line or a point in a different layer, depending on how large it is represented.)

Polygons, lines, and points, collectively, are called **vector** data.

Earthquake epicenters and seismic hazards in Indonesia. Polygons are used to represent landmasses, points to represent earthquakes, and lines to represent subduction zones. Map courtesy of US Geological Survey.

Surfaces have numeric values rather than shapes

Unlike countries or rivers, natural phenomena, such as elevation, slope, temperature, rainfall, and wind speed, have no distinct shape. What they have instead are measurable values for any location on the earth's surface. (Wherever you go, for instance, you are either at sea level or a number of meters above or below it.) Geographic phenomena such as this are easier to represent as surfaces than as features.

The most common kind of surface is a *raster*, a matrix of identically sized square cells. Each cell represents a unit of surface area—for example, 10 square meters—and contains a measured or estimated value for that location.

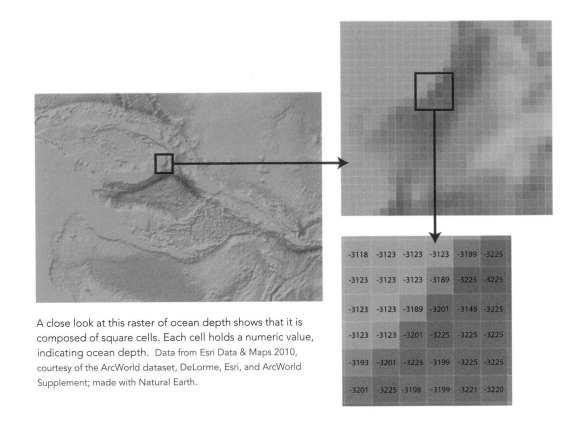

A close look at this raster of ocean depth shows that it is composed of square cells. Each cell holds a numeric value, indicating ocean depth. Data from Esri Data & Maps 2010, courtesy of the ArcWorld dataset, DeLorme, Esri, and ArcWorld Supplement; made with Natural Earth.

The world is not divided neatly into features and surfaces. Many things can be looked at either way—as a feature or as a surface. For example, polygons are often used to mark the boundaries of different vegetation types in a region, but this implies that the change from one type to another is more abrupt than it probably is since one type of vegetation often blends into another. Vegetation can be represented not only as a polygon, but also as a raster surface, with each cell value standing for a type of vegetation.

Features have locations

If you were asked to find Berlin, Germany, on a map of the world, it probably wouldn't take you very long. But suppose Berlin wasn't shown on the map. Could you make a pencil mark exactly where it should be?

Now suppose you could lay a grid or *graticule* over the world map and you knew that Berlin was a certain number of marks up from and to the right of a given starting point. In that case, it would be easy to put your pencil on the right spot. A grid of this kind is called a coordinate system, and it's what a GIS uses to put features in their proper place on a map.

On this map, a coordinate system has an x-axis and a y-axis. The intersection of the axes is called the origin. Feature locations are specified by their distance from the origin in meters, feet, or a similar unit of measure—in this case, lines of latitude and longitude. Data from Esri Data & Maps 2010, courtesy of ArcWorld, DeLorme, Esri, and ArcWorld Supplement; made with Natural Earth.

The location of a point feature on a map is defined by a pair of coordinates. Berlin has coordinates of 52°30'N (latitude) and 13°25'E (longitude). A straight line needs two pairs of coordinates—one at the beginning of the line and one at the end. If the line bends, like a river, there must be a pair of coordinates at every location where the line changes direction. The same holds true of a polygon, which is simply a line that returns to its starting point.

Features can be displayed at different sizes

On a GIS map, you can zoom in to see features at a closer range. As you do, the scale of the map changes.

Scale, commonly expressed as a ratio, is the relationship between the size of the features on a map and the size of the corresponding places in the world. If the scale of a map is 1:1,000,000, it means that features on the map are 1/1,000,000th of, or 1 million times smaller than, their true size.

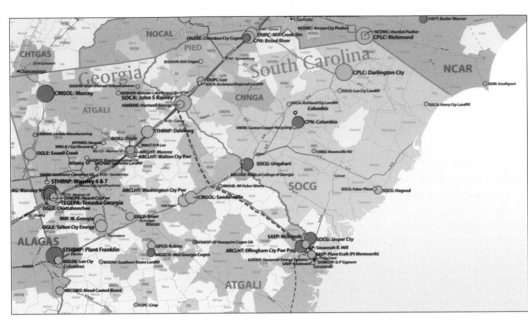

A utilities map at a scale of approximately 1:1,000,000. County, state, and shoreline boundaries are shown.

Zooming in gives you a closer view of features within a smaller area of the map. The amount of detail in the features does not change, however—it's a matter of whether you can see those details or not. A river has the same bends, and a coastline has the same crenulations, whether you are zoomed in and can discern them or are zoomed out and cannot.

A city transit map at a scale of approximately 1:12,000. Building footprints, roads, and transit lines are visible. Copyright RMSI & Universal Map Group LLC, 2009. Reproduced by permission.

How much detail features have depends on the layer you are using. Just as a paper map of the world generalizes the boundaries of Brazil more than a map of South America does, so do different GIS layers contain more feature detail or less.

Internet basemaps, such as those on the ArcGIS Online platform, cover a wide range of map scales—and a wide level of detail. A map can have several levels of detail that are correlated to the scale range. As you zoom in, layers at small scale with less detail will be turned off, and new layers at large scale with more detail will be turned on. Each zoom level can be perceived as being made up of seamless data— each level, however, is made up of layers composed of several datasets, all with varying amounts of detail, and all of which is contained within a spatial database.

Features are linked to information

There is more to a feature than its physical shape and location. There is also everything else that might be known about that feature. For a country, this might be its population, capital, system of government, leading imports and exports, average rainfall, mineral resources, and any number of other things. For a road, it might be the speed limit, the number of lanes, whether it is paved or unpaved, and whether it is one-way or two-way traffic. There is a great deal of information to be had about any feature, whether it be a humble length of sewer pipe or the vastness of an ocean.

Information about the features in a layer is stored in a table. The table has a record (row) of each feature in the layer and a field (column) for each category of information. These categories are called attributes, and the table is called an attribute table.

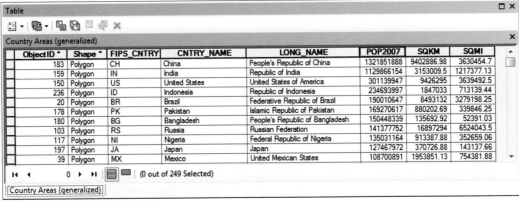

The attribute table of a layer of countries includes each feature's shape, ID number, name, population, and area, among other things. Data from Esri Data & Maps 2010, courtesy of ArcWorld, DeLorme, Esri, and ArcWorld Supplement; made with Natural Earth.

Features on a GIS map are linked to the information in the feature's attribute table. If you select Brazil on a map (see facing page), you can bring up all the information stored about Brazil in the attribute table of countries. If you select a record in the table, you will see the corresponding feature highlighted on the map.

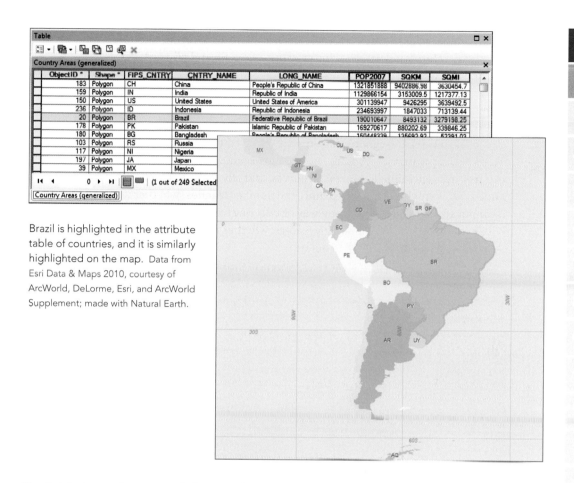

Brazil is highlighted in the attribute table of countries, and it is similarly highlighted on the map. Data from Esri Data & Maps 2010, courtesy of ArcWorld, DeLorme, Esri, and ArcWorld Supplement; made with Natural Earth.

The link between features and their attributes makes it possible to ask questions about the information in an attribute table and be able to display the answers on the map. It can help locate and identify very specific features.

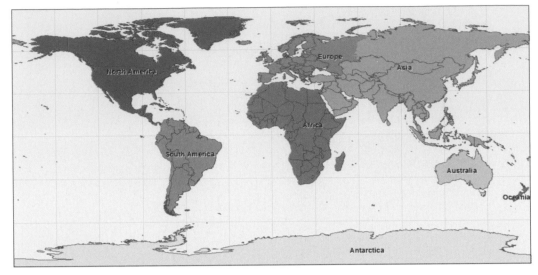

Which countries belong to the continent of Asia? Data from Esri Data & Maps 2010, courtesy of ArcWorld, DeLorme, Esri, and ArcWorld Supplement; made with Natural Earth.

Which countries have a population of 1 billion or greater? Data from Esri Data & Maps 2010, courtesy of ArcWorld, DeLorme, Esri, and ArcWorld Supplement; made with Natural Earth.

Which cities in Asia are national capitals and which are not? Data from Esri Data & Maps 2010, courtesy of ArcWorld, DeLorme, Esri, and ArcWorld Supplement; made with Natural Earth.

Which cities in Asia are capitals and have populations of 5 million or greater? Data from Esri Data & Maps 2010 courtesy of ArcWorld, DeLorme, Esri, and ArcWorld Supplement; made with Natural Earth.

Similarly, you can use attributes to create thematic maps—maps that use colors or other symbols to indicate the nature of feature attributes.

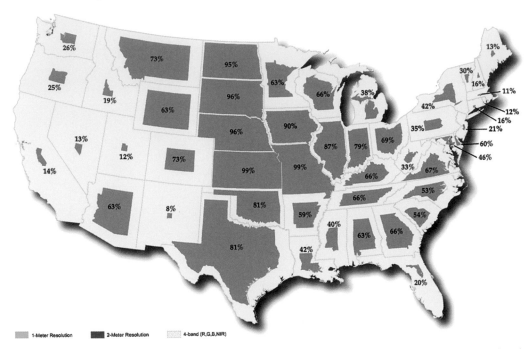

This map shows the percentage of a state covered by the common land unit, the smallest unit of agricultural land that can be enrolled in a farm program administered by the US Department of Agriculture. Map courtesy of US Department of Agriculture Farm Service Agency Aerial Photography Field Office, 2009.

Features have spatial relationships

Besides asking questions about the feature information stored in attribute tables, you can also ask questions about the spatial relationships between features—for example, which ones are nearest others, which ones cross others, and which ones are contained by others. The coordinates of features are used in a GIS to compare their locations. In the map shown on the facing page, the relationships between parcel locations, water lines, and valves are established in the GIS.

This utility map shows parcel locations, water lines, and valves for Cobb County Water System in Marietta, GA. Map courtesy of Cobb County Water System

It is critical to the utility's distribution operation to ensure the isolation of the high-pressure system from the normal-pressure system. The map provides detail for each of the essential valves located in the high-pressure water system. Which water valve belongs to which parcel? Which water line crosses a main road? Which valve affects the water pressure of which water line? By taking a close look at the map, engineers can quickly make informed decisions on pressure problems reported in the area and determine whether extending the high-pressure zone is a potential solution.

New features can be created from areas of overlap

Queries about attributes and spatial relationships identify existing features based on whether they have or do not have certain qualities. To solve some geographic problems, however, it is necessary to use a GIS to create new features from existing ones.

In the map shown on the next page, a GIS database containing layers such as soils, slopes, rivers and streams, wetlands, floodplains, parks and natural areas, significant habitat inventory, greenways, and natural hazard data was used to identify regionally significant areas.

Regionally significant landscape features are derived from other features in the area surrounding Portland, Oregon. Map courtesy of Matthew Hampton.

To find places that meet the specified conditions, the cooperative expertise of various federal, state, local, and private organizations was used in a GIS to find areas of overlap among features in the different layers. The map shows significant natural systems and land patterns that define the quality and character of the region and capture its sense of place. This allows for resource protection at a larger landscape and ecosystem scale and helps define where future growth should and should not occur. The new feature's boundary is the area of overlap, which is different from the boundaries of the features it was created from. The result of the analysis is a new layer that shows regionally significant landscape features.

You now have some idea of what a GIS is and what it can be used to do. In chapter 2, you'll learn a little about ArcGIS for Desktop, the latest GIS software from Esri.

Terms

vector

raster

graticule

Chapter 2
Introducing ArcGIS

Learning objectives

- *List the license levels of ArcGIS for Desktop*
- *Name the ArcGIS for Desktop applications used for general mapping and data management*
- *Briefly describe some ArcGIS extensions*
- *Know the components of the ArcGIS system*
- *Name some Esri data sources*

WHEN YOU SHOP FOR A CAR, A CORDLESS VACUUM, OR A CREDIT CARD, YOU must choose among different models in a product line. The models do essentially the same thing and usually look about the same—the difference is in the extras that come with the high-end models.

ArcGIS for Desktop is a GIS software product line. The entry-level model is the ArcReader application, which can be downloaded for free from the Esri website. ArcReader lets you view and print digital maps created by other users, navigate to different parts of the map, and zoom to various levels of detail. For professionals, there are three additional models, called ArcGIS for Desktop Basic, ArcGIS for Desktop Standard, and ArcGIS for Desktop Advanced. These license levels all share the same core applications, user interface, and development environment. Each license level provides additional GIS functionality as you move from Basic to Standard to Advanced. (The three professional licenses were once called ArcView, ArcEditor, and ArcInfo, respectively.)

In ArcGIS for Desktop Basic, you can query data; analyze spatial relationships, such as distance, intersection, and containment among map features; and overlay layers to discover how different types of data are interrelated at a particular location.

ArcGIS for Desktop Standard gives you the complete functionality of Basic and has additional methods for data creation, as well as additional editing tools and geoprocessing capabilities. Other features include versioning, which allows multiple users to edit data simultaneously, and disconnected editing, which allows data to be checked out of a database, edited in the field, and then checked back in.

At the top of the line, ArcGIS for Desktop Advanced gives you the complete functionality of Standard, plus a full set of spatial analysis tools to perform advanced GIS data analysis and modeling. It also provides high-end cartographic tools and advanced capabilities for data translation, creation, and conversion of a wide range of spatial file formats.

Applications and tools

ArcGIS for Desktop comprises several applications, in which you will do your work. The ArcMap and ArcCatalog applications are the most frequently used. The ArcScene and ArcGlobe applications are used for 3D visualization. Core ArcGIS for Desktop geoprocessing tools are available through the ArcToolbox window and the ModelBuilder application, which are accessible from both ArcMap and ArcCatalog.

ArcGIS for Desktop applications

GIS tasks can be broadly divided into two categories: (1) mapmaking, editing, and spatial analysis; and (2) database design and data management. This division of tasks is reflected in the functionalities of the ArcMap and ArcCatalog applications, the core components of ArcGIS for Desktop Basic, Standard, and Advanced.

ArcMap is the ArcGIS application for making maps and analyzing data.

ArcMap has a data view for creating, symbolizing, and analyzing maps. Map from the Florida Geographic Data Library, courtesy of the University of Florida GeoPlan Center.

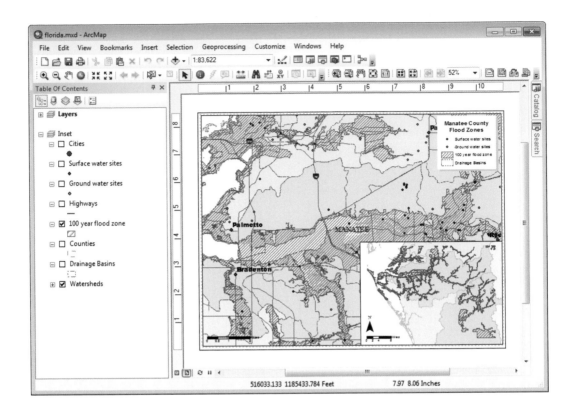

ArcMap has a layout view for composing maps to be printed. You can add a title, a scale bar, a legend, and other elements. Map from the Florida Geographic Data Library, courtesy of the University of Florida GeoPlan Center.

ArcCatalog is the ArcGIS application for data management.

In ArcCatalog, *left*, you can organize, browse through, document, and search for spatial data. ArcCatalog also has tools for building and managing GIS databases. Although there is a separate ArcCatalog application, the Catalog window in ArcMap, *right*, can meet most of your data organization needs. Data from the Florida Geographic Data Library, courtesy of the University of Florida GeoPlan Center.

ArcGIS for Desktop also offers two 3D visualization environments: ArcScene and ArcGlobe. ArcScene is a 3D visualization application for viewing your GIS data in three dimensions. ArcGlobe is part of the ArcGIS 3D Analyst extension (more about extensions later in this section) that is based on a global cube projection to create a seamless visualization of raster and vector features displayed at varying levels of detail. The data is organized into image *tiles* and can be *cached* (saved) to system memory for quick display and navigation.

ArcGIS for Desktop tools

Although ArcMap has many geoprocessing tools that are accessible through menus and various toolbars, the ArcToolbox window contains a collection of geoprocessing tools, models, and scripts organized in one interface. When a geoprocessing tool is run from ArcToolbox, a dialog box appears that contains parameters for the user to specify. Unlike ArcMap and ArcCatalog, which are stand-alone applications, ArcToolbox must be started from within one of these two applications.

ArcToolbox includes a collection of geoprocessing tools, models, and scripts. Here, the Extract toolset is highlighted.

ModelBuilder is a graphic user interface (GUI) for creating models to solve spatial analysis problems. You can use the ModelBuilder interface to organize the inputs, tools, and outputs needed to perform geospatial analysis. ModelBuilder can be used to string together geoprocessing tools from ArcToolbox to create a model to perform a complex workflow with a single click.

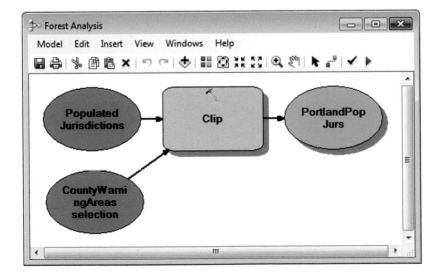

ModelBuilder can be used to create models to automate geoprocessing tasks.

Extending ArcGIS for Desktop functionality

Extension products that are fully integrated with ArcGIS for Desktop Basic, Standard, and Advanced add increased functionality and capabilities to ArcGIS for Desktop. These specialized tools allow you to perform tasks such as raster geoprocessing, three-dimensional (3D) analysis, and map publishing. The ArcGIS Spatial Analyst, ArcGIS 3D Analyst, and ArcGIS Network Analyst extensions, and related extension applications, are included in the 180-day software trial installation that comes with this book (available at http://www.esri.com/gtkArcGISforDesktop10-1). Of the three extensions, which are described as follows, only ArcGIS 10.1 Spatial Analyst is introduced in this book. You can explore the other extensions on your own by using the online Help and the tutorial that is included with the extension.

ArcGIS Spatial Analyst maps and analyzes measured continuous data such as elevation, rainfall, and chemical concentrations. By dividing geographic space into a matrix of square cells that store numeric values, ArcGIS Spatial Analyst allows you to represent, query, and statistically summarize this kind of data. You can also estimate values at unmeasured locations by mathematical interpolation of known sample values.

ArcGIS 3D Analyst gives you the ability to see spatial data in three dimensions. You can "fly through" terrain and examine it from any height or angle. You can model cities and neighborhoods by drawing buildings at their correct heights. Analysis tools let you solve visibility, volume, and downhill path problems. (Which areas can be seen from the mountaintop? Can tower 1 be seen from tower 2? How much earth is in the hill? Which course will water follow down a surface? These types of problems can also be solved with ArcGIS Spatial Analyst.)

ArcGIS Network Analyst provides network-based transportation spatial analysis tools to map fleet routing, travel directions, closest facility, service areas, and location allocation (determining optimal location sites). You can dynamically model realistic network conditions, including one-way streets, turn and height restrictions, speed limits, and variable travel speeds based on traffic data. You can easily build transportation networks from your existing GIS data and use it to define service areas based on travel time and distance.

The ArcGIS system

ArcGIS for Desktop is just one part of the ArcGIS system, which also includes ArcGIS for Mobile, ArcGIS for Server, and ArcGIS Online.

Mobile GIS

For people who record and use geographic data in the field—for example, surveyors, field technicians, delivery people, and public service and utility workers—the ArcPad extension for ArcGIS for Server is a mobile field mapping system. It includes advanced GIS and global positioning system (GPS) capabilities to capture, edit, and display spatial data. It integrates directly with ArcGIS for Desktop and ArcGIS for Server and is part of an enterprise GIS solution.

ArcGIS for Windows Mobile helps organizations deliver GIS capabilities and data from centralized servers to a range of mobile devices. It helps field staff who may have no GIS experience map, sketch, and edit using ArcGIS.

The ArcGIS apps for smartphones and tablets can be used to navigate maps, collect and report data, and perform GIS analysis. You can use them to browse through map galleries from ArcGIS Online, display maps and capture information, and develop your own custom applications.

Server GIS

ArcGIS for Server is used to build web services that deliver maps, apps, and information—such as driving directions or retail locations—over the Internet. ArcGIS for Server brings GIS to people who may have no GIS software themselves. Like ArcGIS for Desktop, ArcGIS for Server also comes in Basic, Standard, and Advanced license levels.

ArcGIS for Server is also designed for *cloud computing*, a rapidly emerging technology that allows you to use a cloud provider's computing infrastructure instead of your own. With cloud computing, you get the same capabilities for managing and delivering GIS services and applications as you do when you run ArcGIS for Server at an actual physical location.

Online GIS

ArcGIS Online is a family of cloud-based web products and resources that enhance and extend the ArcGIS user experience. With ArcGIS Online, you can create maps; access libraries of maps; store and manage maps, data, and other geospatial content; and share your content with others. ArcGIS Online contents and capabilities are built into ArcGIS for Desktop, ArcGIS Explorer Online, and ArcGIS.com.

Esri data

But when it comes down to it, what is software without high-quality data? Esri provides many kinds of geospatial data for organizations involved in GIS projects and delivers this data in a variety of ways.

ArcGIS Online Map Service and StreetMap Premium for ArcGIS deliver basemaps and enhanced street datasets that provide for geocoding (finding street addresses based on geographic location), routing, and high-quality cartographic display for the entire United States, Canada, and Europe.

Esri Tapestry Segmentation data classifies US neighborhoods into 65 segments, based on their socioeconomic and demographic composition, which helps companies and organizations analyze markets, profile customers, evaluate competitors, and more. The most recent version provides US demographic data from US Census 2010.

Esri Business Data provides business locations by industry, bank locations, major retail shopping centers, and crime statistics.

Data and Maps for ArcGIS is a set of annual map data that is included at no additional cost with ArcGIS software and is preconfigured to work with ArcGIS products.

In this book, you will learn the basics of ArcGIS for Desktop. To find out more about ArcGIS for Desktop extensions or about ArcGIS for Mobile, ArcGIS for Server, and ArcGIS Online, visit the Esri website at http://www.esri.com.

Terms

tiles

cache

cloud computing

Getting started

with maps and data

Chapter 3
Interacting with maps

Displaying map data
Navigating a map
Using basic tools
Looking at feature attributes

Learning objectives

- *Navigate the ArcMap interface*
- *Locate and open ArcMap windows, menus, and tools*
- *Turn map layers on and off*
- *Reorder layers in a map*
- *Change a layer name*
- *Examine layer properties*
- *View the source of map layers*
- *Explore a map document using zoom and pan tools*
- *Identify features and view related information*
- *Turn on MapTips*
- *Turn on feature labels and set a scale dependency*
- *Examine alternative views of the map area*
- *Find data on a map*
- *Create a spatial bookmark*
- *Measure distances on a map*
- *Open and explore layer attribute tables*
- *Examine attribute statistics*

ARCMAP IS THE MOST FREQUENTLY USED COMPONENT OF THE ARCGIS SYSTEM. It is an application for displaying maps and investigating them, for analyzing maps to answer geographic questions, and for producing maps that make analysis persuasive. The ArcMap application consists of a map display for viewing spatial data, a table of contents for listing the layers shown in the display, and a variety of toolbars for working with the data.

ArcMap application of ArcGIS for Desktop. Data from Esri Data & Maps 2010, courtesy of ArcWorld, ArcWorld Supplement, DeLorme, and Esri.

You can change the way ArcMap looks to suit your preferences and the kind of work you do. Toolbars can be hidden or shown. New commands can be added. Toolbars can be docked at different places in the application window or float independently of it.

To dock a floating toolbar, drag it to the interface. To undock it, use the dashed vertical gray bar at the toolbar's left edge to drag the toolbar away from the interface. To hide or show a toolbar, click the Customize menu, point to Toolbars, and click to select or clear the desired toolbar option.

Most embedded windows (Catalog, Search, ArcToolbox, Table Of Contents, and Python) can be floating, docked to the interface, or collapsed into a tab for easy access. (The exception is the ModelBuilder window, which is designed to be used only as a separate, floating window.) To dock a floating window, drag it to the interface. The window turns blue, and four arrows point to locations where the window can be docked. Drag the window to an arrow, and then release it to choose the docking location. For windows with the Auto Hide option, when the window is docked, click the Auto Hide button (it looks like a pushpin) to activate the tabbed behavior. To remove the tab, click the Auto Hide button and undock the window.

ArcToolbox window docked

Tools toolbar floating

Draw toolbar turned on

Catalog window (auto-hidden)

Search window (auto-hidden)

Table of contents floating

ArcMap toolbars and windows can be easily docked and undocked. Data from Esri Data & Maps 2010, courtesy of ArcWorld, ArcWorld Supplement, DeLorme, and Esri.

Most of the graphics in this book reflect the default interface. It is assumed that the Standard and Tools toolbars, which are used a great deal, are always turned on.

Top to bottom, the Standard and Tools toolbars are generally turned on.

Changes you make to the interface are applied to subsequent ArcMap sessions, so if you dock a toolbar or window in one session, it will be docked the next time you start ArcMap, and if you resize the application window, it will keep the new size in the next session. Changes such as this will not significantly affect the exercises but may give you slightly different results for operations that are influenced by the size and scale of the map display.

You can further customize the appearance of the ArcGIS for Desktop graphic user interface by removing unused tools, renaming tools, moving specialty extension tools to the Tools toolbar where they can be quickly accessed, creating custom toolbars that pertain to your organization's workflow, and more. For more information about tailoring the graphic user interface to your preferences, go to ArcGIS for Desktop Help. (From the Start menu, click All Programs > ArcGIS > ArcGIS for Desktop Help > ArcGIS 10.1 for Desktop Help.) On the Contents tab, navigate to Developing > Desktop interface > About configuring the user interface.

Data

The exercises in this chapter use the following datasets, stored in a geodatabase called World.gdb:

- Background—grid of polygons used to display a single-color background behind land features, such as the ocean *(Source: Esri)*
- Cities—point features representing major cities of the world with a population greater than 1 million *(Source: ArcWorld)*
- Countries—polygon features representing world countries, including demographic data and air pollution estimates *(Source: ArcWorld; US Census International Division, CIA Factbook; The World Bank)*
- Latlong—line features representing latitude and longitude *(Source: Esri)*

Exercise 3a
Displaying map data

You will be able to do the exercises only if you have installed the trial version of ArcGIS 10.1 for Desktop Advanced, available from http://www.esri.com/ gtkArcGISforDesktop10-1 with the authorization code printed on the inside back cover of the book, or if you have a licensed version of ArcGIS 10.1 for Desktop on your computer. Using a previous version of ArcGIS means that certain tools, functions, windows, or dialog boxes may or may not be present, and the map documents provided for the exercises may not open. You may encounter steps in the exercises where a workaround may be necessary.

In this exercise, you'll learn how to display data in ArcMap. You'll explore a map document that contains world data and learn the basics of how map documents work. The exercise instructions assume you have installed the data for *Getting to Know ArcGIS for Desktop* to the default directory (C:\EsriPress\ GTKArcGIS). If you have installed the data elsewhere, you'll need to substitute the correct paths.

1 **Start ArcMap by clicking the Start button on the taskbar, and then, on the Start menu, click All Programs > ArcGIS > ArcMap 10.1.** When ArcMap is open, you see the ArcMap - Getting Started dialog box on the main application window. This dialog box allows you to quickly start a new map or open an existing map or template.

2 On the Getting Started dialog box, under the Existing Maps section, click "Browse for more ...".

3 On the Open ArcMap Document dialog box, browse to your C:\EsriPress\GTKArcGIS\Chapter03 folder (or to the folder where you installed the GTKArcGIS data). Click ex03a.mxd, as shown in the graphic, and click Open.

When the map document is open, observe the display area (where you see the map — a map of the world, in this case), the toolbars, and the table of contents on the left.

You see a world map with grid lines and dark-blue circles. Taking a quick look at the table of contents and with some modest deciphering, you can surmise that the dark-blue circles represent cities with population greater than 1 million (with increasingly larger circles representing increasingly larger populations), and the grid lines represent lines of latitude and longitude. Countries are yellow, and oceans are light blue. Recall from chapter 1 that a *layer* is a spatial dataset represented by symbols on a GIS map. The visible layers on this map are named Cities, Latlong, Countries, and Ocean. You also see additional layers in the table of contents (World Population and Air Pollution by Country), although they aren't visible on the map at the moment. Next, you will take a closer look at each layer.

4 In the table of contents, select the check box next to Cities to turn
the layer off. The cities disappear on the map.

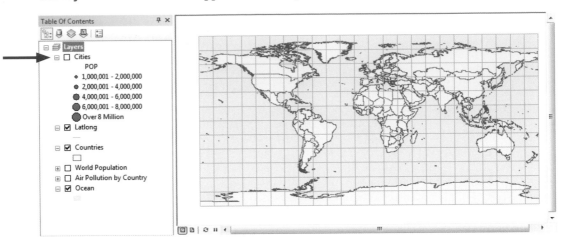

5 Turn Cities back on. In the same way, turn the following layers off
and then back on: Latlong, Countries, Ocean.

All map layers in ArcMap can be turned off and on with a click. Next, you will
observe the effects of changing the layer drawing order. Layers are drawn, one
on top of the other, like sheets of paper stacked together. Imagine that the
layers you are looking at are actually opaque symbols drawn on transparent
sheets of paper. If you placed cities, drawn as filled-in dots, below the countries,
which are filled-in blocks of color, you wouldn't see them. You can experiment
with this concept in ArcMap.

6 In the table of contents, drag Cities to the bottom of the list of layers, below Ocean. As you drag the layer, a horizontal black bar indicates its position in the list.

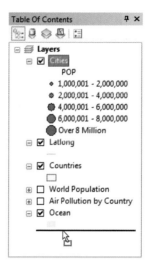

When you let go of the mouse button, Cities is at the bottom of the table of contents and is no longer visible on the map. The polygon layers using solid fill colors obscure the point layer of cities.

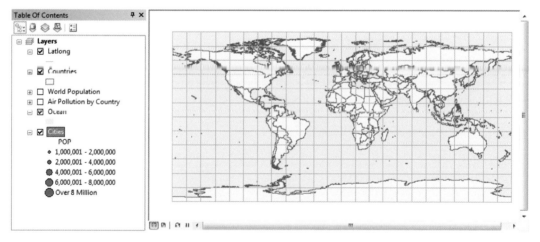

Lines and points should always be layered on top of polygons to avoid this problem. When you add data to ArcMap, the software automatically places points and lines above polygons.

7 Move Cities back to its original location at the top of the table of contents.

3a

On your own

Experiment by moving the Latlong layer to different locations in the table of contents. What happens when you drag it below the Countries layer? What happens when you drag it below the Ocean layer? When you're done, move the layer back to its original position, below Cities.

The layer name Cities isn't ideal, since it might imply that all cities of the world are included in the dataset. So next, you will give the layer a more descriptive name.

8 In the table of contents, right-click the Cities layer name. This opens a *shortcut (context) menu*. Many ArcMap operations are started from these shortcut menus. On the shortcut menu, click Properties to open the Layer Properties dialog box.

9 On the Layer Properties dialog box, click the General tab if it's not already active. (ArcMap remembers the last tab used and makes this the active tab; the very first time you open the properties, the General tab should be active.) The Layer Properties dialog box has several tabs for setting layer properties, most of which you'll use in this book.

10 **In the Layer Name text box, the name "Cities" is highlighted. Type** Cities > 1 Million **in its place. Make sure the dialog box matches the graphic, and then click OK.** The layer is renamed in the table of contents.

Layer Properties

| General | Source | Selection | Display | Symbology | Fields | Definition Query | Labels | Joins & Relates | Time | HTML Popup |

Layer Name: Cities > 1 Million   ☑ Visible

Description: World Cities represents the locations of the major cities of the world.

Credits:

Scale Range

You can specify the range of scales at which this layer will be shown:

● Show layer at all scales

○ Don't show layer when zoomed:

Out beyond: 1:80,000,000   ▾   (minimum scale)

In beyond: <None>   ▾   (maximum scale)

[OK]   [Cancel]   [Apply]

Here is another way to rename a layer: just click the layer name in the table of contents to highlight it, click again to activate the cursor, and then type the new name.

You should know that when you change a layer name in a map document, you are not changing the name of the underlying source dataset. A layer name does not have to match the name of the source dataset.

You will learn more about the connection between a layer and its source data shortly.

⊟ 🖾 **Layers**
  ⊟ ☑ Cities > 1 Milli
      POP
      • 1,000,001 - 2,000,000
      ● 2,000,001 - 4,000,000
      ● 4,000,001 - 6,000,000
      ● 6,000,001 - 8,000,000
      ● Over 8 Million

11 **Using the method taught in the preceding tip, in the table of contents, below Cities > 1 Million, change the legend name "POP" to** Population. This makes it clear that the size of a blue circle corresponds to the size of a city's population.

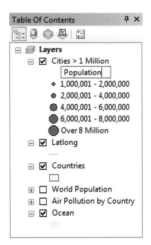

On your own

Change the Latlong layer name to **Graticule**, the technical name for lines of latitude and longitude on a map.

12 **In the table of contents, select the check box next to World Population to turn the layer on.** Nothing is changed in the display. Can you figure out why? If you think it may have something to do with the drawing order, which was discussed earlier in the exercise, you are correct. The World Population layer is composed of polygons that are the same shape as those in the Countries layer. Since Countries is the first polygon layer in the list of layers, it is obscuring the World Population layer below it.

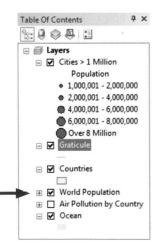

13 Turn off the Countries layer. The World Population layer is now visible.

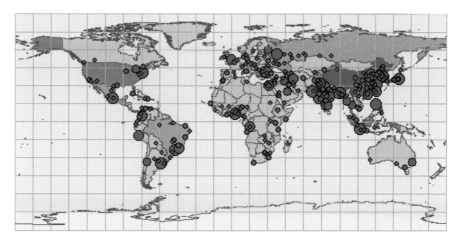

Now you can see something new on the map—in the World Population layer, countries appear in different colors. To understand what the colors mean, you will examine the layer's legend. But first, you will turn off the points and lines to get a clearer picture.

14 Turn off the layers Cities > 1 Million and Graticule.

15 Click the plus sign (+) next to World Population to expand the legend for this layer. Examine the colors on the map and compare them with the legend.

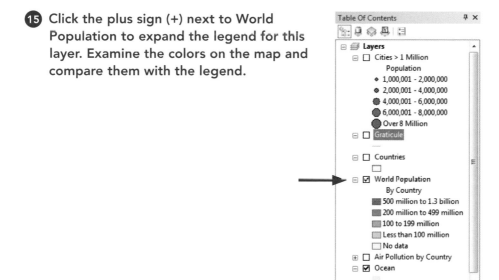

Two countries display populations greater than 499 million. Can you name them?

To make sure you answered the preceding question correctly, you will next use the Identify tool.

16 **On the Tools toolbar, pause the mouse pointer over the Identify tool (a blue circle with the letter *i*). Notice that the tool name and a short description appear.** In the rest of the book, tools generally will be shown as an individual button only, as follows: . When the Identify tool is used, the pointer is accompanied by an *i* symbol as it is moved around the map display. When you click a feature, it opens the Identify window.

17 **Click the Identify button and move the pointer over one of the two areas filled with the darkest shade of orange, and then click in that area.** The Identify window (see facing page) shows the name of the country you clicked, as well as other information, called *attributes*, which are stored in a table that is part of this dataset. (Depending on which country you clicked, you may have different results, of course.)

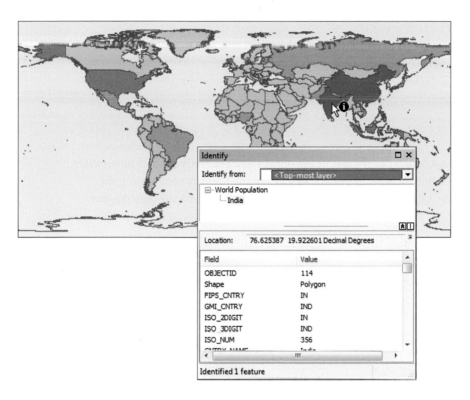

18 **Close the Identify window and turn off the World Population layer. Turn on the Air Pollution by Country layer and expand its legend.**

This layer shows concentrations of particulate matter (PM) in the air. Data was gathered for urban areas, from which countrywide estimates were calculated.

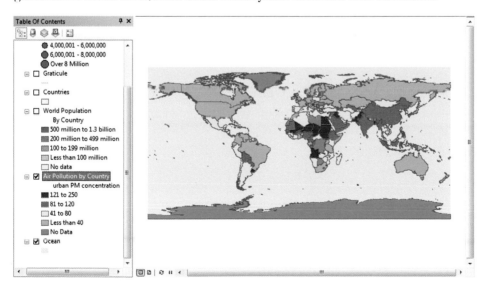

19 Turn the World Population on and off to compare it with Air Pollution.

Does high pollution always correlate with high population? What might be some reasons for this?

Now that you have seen some examples of various types of data being displayed on a map, you will next investigate the connection between *source data* and *map layers*.

20 In the table of contents, right-click Countries and click Properties. On the Layer Properties dialog box, click the Source tab and examine the Data Source information. (Notice that a layer does not have to be visible on the map to see its properties.)

Later in the book, you may simply be asked to "open the layer properties." You will have to remember how to do this.

What is the source data type?

A *feature class* is a collection of geographic features that have the same geometry type (such as point, line, or polygon), the same attributes, and the same spatial reference. Feature classes can be stored in a geodatabase or as shapefiles, coverages, or other data formats. (You will learn more about data formats in chapter 4.)

What is the name of the source feature class, and in which geodatabase is it located?

21 Using the same approach, look at the source dataset for World Population and Air Pollution by Country.

What are the names of the source feature classes for these two layers?

It is important to understand that a map may include multiple layers that point to the same underlying data. In the case of the current map document, the Countries, World Population, and Air Pollution layers are based on the same data source—a geodatabase feature class called Countries. (You will learn more about data types later in this book.) Layers and source datasets sometimes have the same name, but they don't have to, as demonstrated in step 10.

Despite sharing the same source dataset, the layers look different from each other. Why? Because they have different *symbology* schemes. Unlike source data, layers store symbology information in a map document. For instance, the Countries layer is symbolized using a single color for all country polygons. The World Population layer has graduated symbols based on population data, and the Air Pollution layer has graduated symbols based on particulate matter calculations. Population and particulate matter are all stored as attributes of the Countries feature class (there are many other attributes stored with the feature class as well). You will learn more about attribute data later in this chapter, and you will learn how to create symbology in chapters 7 and 8.

22 Close the layer properties. At the top of the table of contents, click the List By Source button 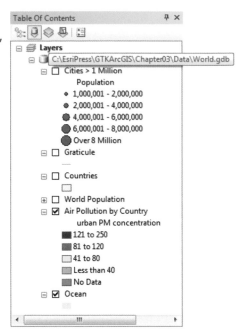 . Pause the mouse pointer over the data path to see the path in its entirety.

The List By Source option is especially helpful when you have data from different locations in the same map document.

23 Click the List By Drawing Order button to restore the table of contents to its default appearance.

24 If you want to save your work, on the menu bar, click File, and then click Save As. Browse to C:\EsriPress\GTKArcGIS\Chapter03\MyData. Give the file a new name and click Save. Saving your work is optional. When you begin a new exercise, however, it is important that you open the new map document specified in the instructions. The exercises are not always perfectly continuous, and if you work in the same map document throughout a chapter, you may get results that don't match the book.

25 If you are continuing to the next exercise, leave ArcMap open. Otherwise, click the File menu and click Exit. Click No if prompted to save your changes.

Exercise 3b
Navigating a map

Now that you understand some basic principles of how map documents work, it's time to do some exploring using ArcMap navigation tools. In this exercise, you will learn how to zoom and pan around a map, use MapTips, and identify features and examine their attributes.

1 Start ArcMap if necessary. If you see the Getting Started dialog box, click Cancel. On the menu bar, click File, and on the menu that appears, click Open. Browse to your C:\EsriPress\GTKArcGIS\ Chapter03 folder. Click ex03b.mxd, and then click Open.

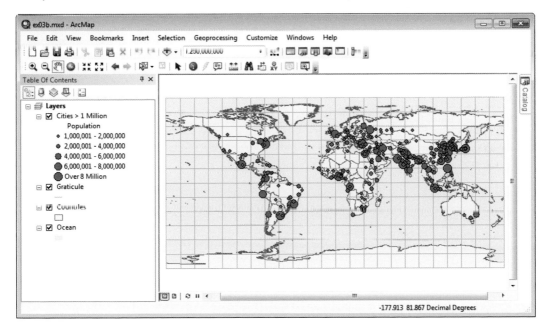

At the bottom of the Getting Started dialog box, there is a check box you may select if you would rather not see the dialog box when you open ArcMap.

3b

This map document looks similar to the one you worked with in exercise 3a but with fewer layers.

2 On the Tools toolbar, click the Zoom In tool ⊕ .

3 Move the mouse pointer over the map. The pointer becomes a magnifying glass. Press and drag the pointer to draw a box around the country of Australia, approximately as shown in the graphic. (If you make a mistake, click either the Full Extent button ⊙ or the Go Back To Previous Extent button ← and try again.)

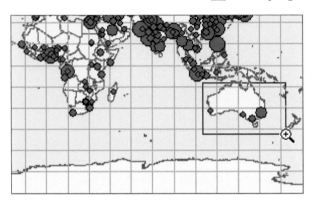

The display is zoomed in on Australia.

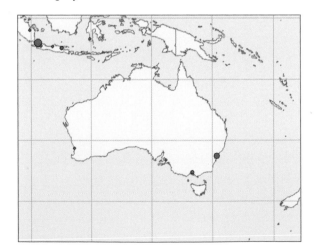

Zooming in or out changes the *display scale*, which is shown on the Standard toolbar. When the map showed the entire world, the scale was approximately 1:200,000,000. This means that map features are displayed at 1/200,000,000th of their actual size. The scale should now be somewhere in the vicinity of 1:40,000,000 or so. (Scale is also affected by the size of the ArcMap application window.)

You can easily zoom in or out by rotating the wheel of the computer mouse, if you have one. Try it to see how it works.

④ To avoid accidental zooming, it's best to use the default tool. On the Tools toolbar, click the Select Elements tool .

Remember, you can pause the mouse pointer over a tool to see its name.

⑤ Move the mouse pointer over the Australian continent. The country name is displayed as a MapTip. MapTips are a layer property. They can be turned on or off on the Display tab on the Layer Properties dialog box. You can see MapTips, regardless of which tool is active.

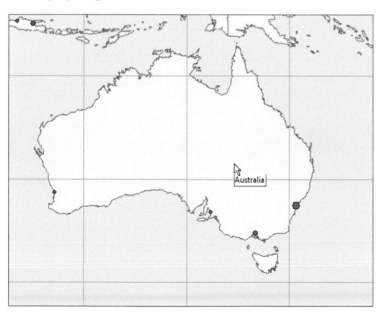

6 On the Tools toolbar, click the Pan tool 🖑 . When you move the mouse pointer over the display area, it turns into a hand symbol.

7 Pressing and holding the mouse button, drag the display to the southeast (so your travel direction is northwest) until you can see Iran (use the MapTips for help identifying the country). You will have to press and then release the mouse button a couple of times until you arrive at your intended destination.

8 On the Tools toolbar, click the Identify tool, and then click Iran on the map.

Notice that the feature flashes green when you click it with the Identify tool.

In the Identify window, you will find feature-specific information that is stored with the Countries dataset, including ISO (International Organization for Standardization) numbers and codes, population, area, and many other statistics.

What is the ISO number for Iran?

9 With the Identify window still open, click several other countries in the display area. Notice that the information in the Identify window is updated automatically as you click different features—you do not have to close the window and reopen it each time you want to identify a new feature.

10 Close the Identify window.

11 Click the Fixed Zoom In button ▓ once. The Fixed Zoom buttons maintain the center point of the display area as they increase or decrease the display scale.

12 Click the Fixed Zoom In button three more times and observe the scale bar. Now click the Fixed Zoom Out button 🔲 several times.

13 Click the Go Back To Previous Extent button several times and then the Go To Next Extent button ➡ several times.

On your own

Use the following tools and move around the map, until you are comfortable with ArcMap navigation: Zoom In, Zoom Out, Pan, Full Extent, Fixed Zoom In, Fixed Zoom Out, Previous Extent, and Next Extent.

---

14 When you are done navigating, zoom to the full extent of the map, and then click the Select Elements tool.

It's a good habit to activate the Select Elements tool whenever you finish using a tool. This will prevent unwanted zooming, measuring, or other tool activity once you are finished using the tool.

You didn't make any changes to this map, so there's no need to save your work.

15 If you are continuing to the next exercise, leave ArcMap open. Otherwise, on the menu bar, click File, and then click Exit. Click No when prompted to save your changes.

Exercise 3c
Using basic tools

In this exercise, you'll learn about many common ArcMap tools and functions
that facilitate data exploration, beyond navigation. You will learn how to turn
on MapTips, create feature labels that turn on only at a specified map scale,
create spatial bookmarks to save a specific view of a map, measure distances
between features, and use the viewer windows to see alternative views of the
data simultaneously.

1 **In ArcMap on the Standard toolbar, click the Open button ☜ .**
(Dismiss the Getting Started dialog box if necessary.) Browse to
your C:\EsriPress\GTKArcGIS\Chapter03 folder, click ex03c.mxd,
and click Open.

If you don't see the .mxd file extension, check your folder option settings in
Windows Explorer: Right-click the Start button and click Open Windows Explorer. Click
Organize > Folder and Search Options. Then click the View tab and clear the "Hide
extensions for known file types" check box. (Your system options may be different.)

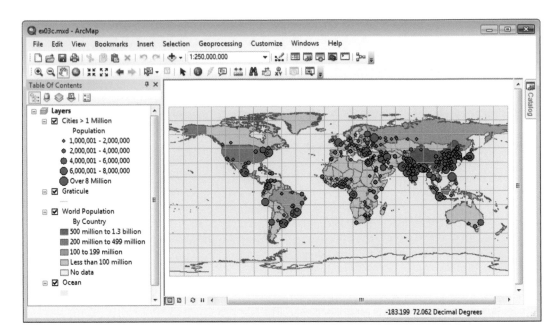

The map is similar to the one in exercise 3b, but instead of the Countries layer, it has a World Population layer, which you saw in exercise 3a. So that you can quickly obtain city names, you will next turn on MapTips.

2 Open the properties for Cities > 1 Million. Click the Display tab and select the "Show MapTips using the display expression" check box.

Notice that the Display Expression field is set to CITY_NAME. This is the attribute field ArcMap will use to pull the information when you use MapTips. You can change this field so that ArcMap will show different types of information when you pause the mouse pointer over features. By default, ArcMap selects the first string-type field in the layer's attribute table that includes the text "name" (case insensitive). If no field with that text in its name exists, ArcMap defaults to the first field of string type, then the first field of integer type, and then the first field of any type. (You will learn more about field types in chapter 11.) This is the case for both MapTips and labels.

3 **Click OK to close the properties. Then pause the mouse pointer over any city to see its name (zoom in first if necessary, and then zoom back to the full extent when you are done).**

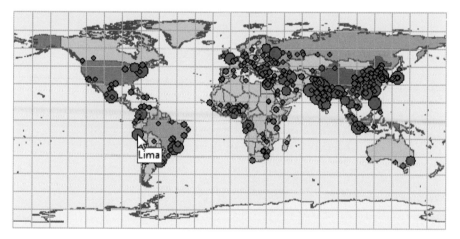

MapTips are helpful, but many times you'll want to see permanent labels on your map. You can turn labels on so they are always visible. However, at the full extent of the map, displaying all the city names would result in a cluttered map that is difficult to read. To avoid this problem, you will next set a visibility scale so the labels will appear only when the map is zoomed to a certain scale.

4 **Again open the properties of the Cities > 1 Million layer and click the Labels tab. Near the top of the dialog box, select the Label features in this layer check box. Notice that the label field is set to CITY_NAME.**

5 **Click the Scale Range button. On the Scale Range dialog box, click the second option, and next to Out beyond, type** 1:80,000,000. This means that labels will be visible only at scales larger than 1:80,000,000.

6 **Click OK to return to the Labels tab and OK again to close the Layer Properties dialog box.**

7 To see your labels, click the Fixed Zoom In button several times, until the city labels appear. Then look at the scale bar and notice that it is set to 1:80,000,000 or less (as the denominator gets smaller, the scale gets larger).

To quickly turn on labels for any layer without setting a scale range, simply right-click the layer in the table of contents and click Label Features.

8 Return to the map's full extent. On the Tools toolbar, click the Create Viewer Window button ⊞ . The mouse pointer turns into a crosshair symbol so that you can draw a box around an area you would like to observe in a separate window.

3c

4

5

9 **Press and drag the mouse pointer to draw a box around Europe, creating a Viewer window.** Depending on the size of your screen and the dimensions of the box you draw, your Viewer window may or may not display city labels.

The Viewer window allows you to examine different parts of the map without changing the main display. You can move the Viewer window so it doesn't obscure the main map, or you can enlarge it—this is helpful when you have a large computer monitor (or two monitors) and you want to see two views of the same map simultaneously.

10 **Use the basic navigation tools in the Viewer window—not the tools on the ArcMap Tools toolbar—to manipulate the display of the window.**

11 **When you're comfortable navigating in the Viewer window, click the right arrow in the upper-right corner of the window, and then click Magnifier.** Now the window acts like a magnifying glass.

12 Drag the Magnifier window title bar to move the window around the main map display. The area seen through the window is magnified four times (or 400 percent—a number you can modify).

On your own

On the menu bar, click Help > ArcGIS Desktop Help. Over the left column, click the Search tab. In the text box, type **viewer window** and click Ask. Choose the first returned topic, "Working with magnifier, viewer, and overview windows." Read about these windows to find out more about how they work.

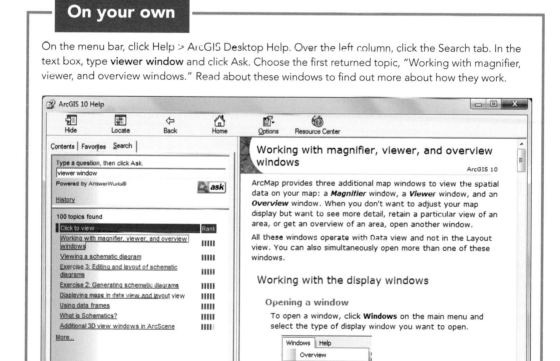

3c

4

5

Suppose, for this exercise, you are a pilot and intend to fly a small plane from Shanghai to New York. You want to measure the distance between the two cities and determine the shortest route, so you'll do that next.

13 Close any open windows. On the Tools toolbar, click the Find tool 🔍 .

14 On the Find dialog box, in the Find box, type Shanghai, and in the In box, click the Cities > 1 Million layer.

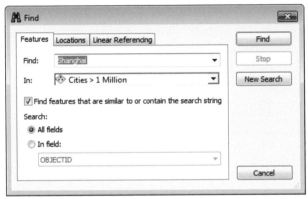

15 Click Find. The bottom of the dialog box expands to show the results. Right-click Shanghai and click Zoom To.

16 **Close the Find dialog box.** The map is centered on Shanghai.

17 **On the menu bar, click Bookmarks > Create.**

18 **Replace the existing Bookmark name with** Shanghai, **as shown in the graphic, and click OK.**

19 **Now use the Pan tool to go westward over Asia and the Atlantic Ocean until you reach the point representing New York City.**

20 On the Tools toolbar, click the Measure tool 🔢 . This opens the Measure window. Click the Choose Units arrow, point to Distance, and click Kilometers if it's not already selected. (ArcMap maintains the last setting selected.)

21 Move the mouse pointer over the display. The pointer becomes a ruler with a small crosshair. Place the pointer at the center of the point that represents New York City and click to begin a line. Without clicking anywhere else on the map, go to the Bookmarks menu and click Shanghai. (The Measure tool will remain active when you click the Bookmarks menu.)

If you accidentally click the wrong spot, double-click to end the measurement, and then start again.

22 When you see the point representing Shanghai, double-click to end the measure line. Shanghai is over 11,000 kilometers from New York. Due to differences in exact click locations, your results may vary slightly.

The Measure tool calculates the shortest distance between two points. Is the shortest distance between the cities found by traveling over the Atlantic Ocean or the Pacific Ocean?

On your own

Zoom to the map's full extent and again measure the distance between New York and Shanghai. Notice the behavior of the measurement line. Why do you think this is happening? You will have more opportunity to think about the mechanics of placing a spherical earth on a flat map surface in chapter 6.

23 If you want to save your work, on the menu bar, click File, and then click Save As. Browse to C:\EsriPress\GTKArcGIS\Chapter03\MyData. Give the file a new name and click Save.

24 If you are continuing to the next exercise, leave ArcMap open. Otherwise, click the File menu and click Exit. Click No if prompted to save your changes.

3c

4

5

Exercise 3d
Looking at feature attributes

In a GIS, a feature on a map may be associated with a great deal of information—more than can be displayed at any given time. This information is stored in an attribute table. A layer's attribute table contains a row (or record) for each feature in the layer and a column (or field) for each attribute or category of information. When you used the Identify tool in exercises 3a and 3b, you saw a feature's attributes in the Identify window.

In this exercise, you will examine the attribute table of the world's largest cities.

1 In ArcMap, open the ex03d map document from your C:\EsriPress\ GTKArcGIS\Chapter03 folder. This is a replica of the map you worked with in exercise 3c.

2 In the table of contents, right-click Cities > 1 Million and click Open Attribute Table.

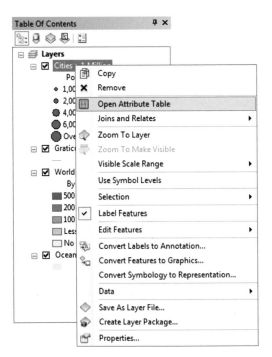

Later in this book, you may simply be asked to "open the attribute table." You'll have to remember how to do this.

The attribute table shows you all the data for the cities included with this map. It can be resized to your preference.

Table						

Cities > 1 Million

	OBJECTID *	Shape *	CNTRY_NAME	CITY_NAME	GMI_ADMIN	
▶	1	Point	Brazil	Brasilia	BRA-DFD	Distrito
	2	Point	Brazil	Goiania	BRA-GOI	Goias
	3	Point	Chile	Santiago	CHL-RMT	Region
	4	Point	Ecuador	Guayaquil	ECU-GUA	Guayas
	5	Point	Ecuador	Quito	ECU-PCH	Pichincl
	6	Point	Peru	Lima	PER-LIM	Lima
	7	Point	Brazil	Belem	BRA-PAR	Para
	8	Point	Brazil	Manaus	BRA-AMZ	Amazo
	9	Point	Bolivia	Santa Cruz de La Sierra	BOL-SCR	Santa C
	10	Point	Paraguay	Asuncion	PRY-CNT	Central
	11	Point	Argentina	Rosario	ARG-SFE	Santa F
	12	Point	Argentina	Buenos Aires	ARG-DFD	Distrito

I◀ ◀   1   ▶ ▶I   (0 out of 256 Selected)

Cities > 1 Million

③ **Scroll down through the table and examine the records (each row in the table is called a record). There are 256 records, one for each city with a population greater than 1 million. Now scroll across the table to look at the attributes, or fields (each column is called a field).** There are 11 attribute fields. The first is the ObjectID field—it contains a unique identification number for each record. The Shape field describes the feature geometry. Both of these fields are created and maintained by ArcGIS for Desktop. The rest of the fields are added and maintained by the user. Other fields include the name of each city, its state, country, population, population rank, and administrative abbreviations.

The intersection of a record and a field is a cell. A cell contains an attribute value. For example, the attribute value of the CITY_NAME field for the first record is "Brasilia." You can change the width of any field by dragging the divider between fields.

Next, you'll narrow the display width of the CITY_NAME field so you can see more attributes without having to scroll across the table.

4 Scroll back all the way to the left. Place the mouse pointer on the vertical black bar between the CITY_NAME and GMI_ADMIN field names. The pointer changes to a double-headed arrow.

5 Press and hold the mouse button and drag the pointer to the left. As you drag, the original field width is marked by a vertical red line and the new width by a vertical black line. Release the mouse button at some point before you start cutting off city names.

Some of the numbers of the POP_CLASS field are cut off, so next, you'll widen this field.

6 Scroll to the right side of the table. Place the mouse pointer on the vertical black bar to the right of the POP_CLASS field name. Drag the pointer slightly to the right to see the entire name of the field.

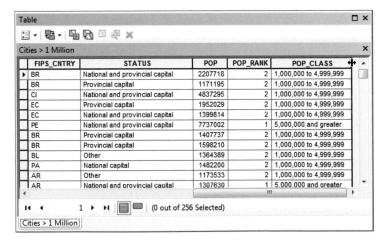

The FIPS_CNTRY field contains country abbreviations designated by Federal Information Processing Standards. If you don't want to see a field in an attribute table, but you don't want to delete it permanently either, you can simply hide it, which you'll do next.

7 Scroll to the left and place the mouse pointer on the vertical black bar between the FIPS_CNTRY and STATUS field names. Drag the pointer to the right edge of the ADMIN_NAME field and release the mouse button. The FIPS_CNTRY field is hidden. (Actually, you've simply made it so narrow that it can no longer be seen.) If you wanted to restore the hidden field, you would double-click on the border between the ADMIN_NAME and STATUS field names to see the field again.

3d

4

5

GMI_ADMIN	ADMIN_NAME	STATUS	POP	POP_RAI
BRA-DFD	Distrito Federal	National and provincial capital	2207718	
BRA-GOI	Goias	Provincial capital	1171195	
CHL-RMT	Region Metropolitana	National and provincial capital	4837295	
ECU-GUA	Guayas	Provincial capital	1952020	
ECU-PCH	Pichincha	National and provincial capital	1399814	
PER-LIM	Lima	National and provincial capital	7737002	
DRA-PAR	Para	Provincial capital	1407737	
BRA-AMZ	Amazonas	Provincial capital	1598210	
BOL-SCR	Santa Cruz	Other	1364389	
PRY_CNT	Central	National capital	1482200	
ARG-SFE	Santa Fe	Other	1173533	
ARG-DFD	Distrito Federal	National and provincial capital	1307630	

Cities > 1 Million

(0 out of 256 Selected)

Cities > 1 Million

You can also hide fields by right clicking the field name and clicking Turn Field Off. With this option, to restore hidden fields, you must click the Table Options arrow, and then click Turn All Fields On.

On your own

Hide the GMI_ADMIN field.

You can rearrange the order of fields as well. It would be more natural to have the city name appear before the country name, so next, you'll rearrange them.

8 **Click the column header of the CNTRY_NAME field.** The field is highlighted in cyan.

9 **Drag the column header to the right. The pointer becomes an arrow with a small rectangle, showing that a field is being moved. When the vertical red line is between the CITY_NAME and ADMIN_NAME fields, as shown in the graphic, release the mouse button.**

The field names are rearranged.

You can also freeze fields to make it easier to explore and understand feature attributes. To do so, right-click a field name and click Freeze/Unfreeze Column. This allows you to scroll across fields and maintain a view of the frozen column. It is especially useful for tables with many fields. Repeat the action to return the table to its default behavior.

Records, as well as fields, can be highlighted. When a record is highlighted in the table, the corresponding feature is highlighted on the map. A highlighted record or feature is said to be selected.

10 Click the gray tab to the left of the first record in the table. The record is selected, shown by the cyan highlighting.

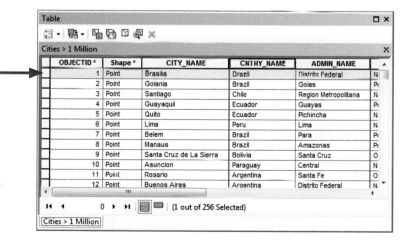

11 Move the attribute table so it doesn't cover the map display. Brasilia is highlighted on the map.

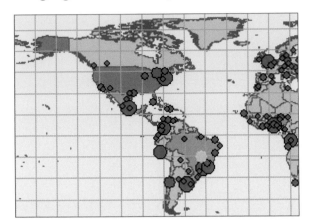

Next, you will clear the current selection, and then select the cities with a population of 10 million or greater.

12 At the top left of the attribute table, click the Table Options arrow and click Clear Selection.

13 In the attribute table, scroll to the POP field. Right-click POP and click Sort Descending.

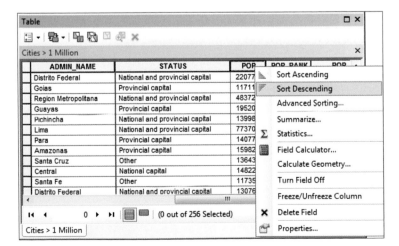

The records are ordered from the largest population to the smallest. Resize the attribute table if necessary to see more fields.

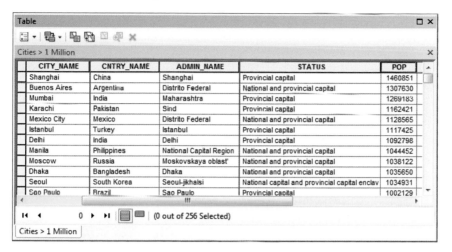

Next, you will select multiple records.

14 Press SHIFT, and then click the tabs to the left of the first 12 records in the table (all cities with a population of 10 million or greater).

Look at the bottom of the table. Notice that 12 of 256 records are selected.

15 Move the table out of the way to see the map. The 12 cities with populations of 10 million or greater are selected on the map.

On your own

At the bottom of the table, click the Show selected records button. This streamlines the table to show only the records of immediate interest.

(To restore the complete table, click the Show all records button to the left.)

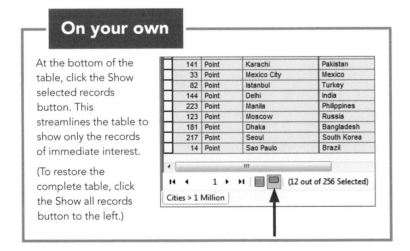

	141	Point	Karachi	Pakistan
	33	Point	Mexico City	Mexico
	82	Point	Istanbul	Turkey
	144	Point	Delhi	India
	223	Point	Manila	Philippines
	123	Point	Moscow	Russia
	181	Point	Dhaka	Bangladesh
	217	Point	Seoul	South Korea
	14	Point	Sao Paulo	Brazil

◄ ‖ ‖ ►

I◄ ◄ 1 ► ►I 🗐 🗐 (12 out of 256 Selected)

Cities > 1 Million

Sorting a field is useful for seeing high and low values, but you can get more detailed information from running statistics on a field. If records are selected in a table, the statistics will be calculated for those selected records only. Next, you'll run some statistics.

16 With the 12 records still selected in the table, right-click the POP field name and click Statistics. The statistics window displays the number of records for which statistics are being calculated (12) and the minimum, maximum, sum, mean, and standard deviation values.

What is the total number (sum) of people living in the 12 largest cities?

The Frequency Distribution chart represents the distribution of values graphically.

How many selected cities have values under 11,508,260?

(Hint: Compare your answer to the table to be sure you are reading the chart correctly.)

17 Close the statistics window and the attribute table. On the menu bar, click Selection > Clear Selected Features.

18 If you want to save your work, on the menu bar, click File > Save As. Browse to your C:\EsriPress\GTKArcGIS\Chapter03\MyData folder. Give the file a new name and click Save.

19 Click the File menu and click Exit. Click No if prompted to save your changes.

Summary

The intent of this chapter was to give you a broad introduction to map exploration using ArcMap, the mapping application of ArcGIS for Desktop. By now, you should feel comfortable navigating a map, opening shortcut menus to choose commands, and using basic tools that are commonly used to explore GIS data. You should now understand that GIS data includes feature geometry, as well as attribute data. You can obtain valuable information not only from the map display, but also from the feature attribute tables.

Go back to the beginning of the exercises and review the learning objectives. Do you feel that you are able to perform the tasks listed? If not, you may want to review some sections of this chapter as a reminder. You received a lot of introductory information in this chapter. If you feel a bit overwhelmed, don't worry—you will get more practice performing these basic tasks, and learning new ones, in the chapters ahead.

Terms

layer

shortcut (context) menu

attributes

feature class

symbology

display scale

Chapter 4
Interacting with data

Browsing through map data
Adding data to a map
Working with map layers

Learning objectives

- *Examine the contents of your GIS workspaces in ArcCatalog*
- *Differentiate data types according to their icons*
- *Preview geographic data*
- *Preview tabular (attribute) data*
- *Navigate and identify data in preview mode*
- *Create thumbnail graphics*
- *Read metadata*
- *Browse through data catalog from within ArcMap*
- *Add data using various methods*
- *Search for data*
- *Fix broken data paths in a map document*
- *Work with data frames in layout view*
- *Copy, paste, and remove layers*

ARCCATALOG IS AN ARCGIS FOR DESKTOP APPLICATION FOR MANAGING
geographic data. In ArcCatalog, you can copy, move, and delete data; search for
data; look at data before deciding whether to add it to a map; create new data;
and read detailed information about data. The ArcCatalog application window
includes the catalog display for looking at spatial data, the Catalog tree for
browsing through data, and several toolbars (see graphic on next page).

Menu bar

Standard toolbar

Geography toolbar

Location bar

Metadata toolbar

Catalog tree

Catalog display

(preview mode)

ArcCatalog interface. Data from Esri Data & Maps 2010, courtesy of ArcWorld, ArcWorld Supplement, DeLorme, and Esri.

You can also manage your data from within ArcMap by opening the embedded Catalog window (see facing page). This is the easiest way to manage data. From here, you can access the core functionality of ArcCatalog without having to open a separate application.

ArcMap interface. Data from Esri Data & Maps 2010, courtesy of ArcWorld, ArcWorld Supplement, DeLorme, and Esri.

Although the Catalog window in ArcMap can meet most of your data-browsing needs, the ArcCatalog application includes 3D navigation tools that aren't available in the ArcMap Catalog window. So if you need to preview 3D data, you'll want to use the ArcCatalog application. You will use both methods in this chapter, but in subsequent chapters where data browsing is required, the Catalog window method will be used.

Spatial data comes in many different formats, including geodatabase feature classes, coverages, shapefiles, CAD (computer-aided design) files, rasters, and TINs (triangulated irregular networks). Each format is identified by its own icon in ArcCatalog. The icon for a geodatabase, for example, is a cylinder. The feature classes within the geodatabase are represented by rectangles, with different patterns distinguishing point, line, and polygon features. (Geodatabases can also contain other special datasets such as annotation features, topologies, and relationship classes, which you'll learn more about in part 4 of this book.)

In this book, you'll use only the most common vector data types: feature classes stored in a data container called a *geodatabase*, as well as some *shapefiles*. You will occasionally work with raster surfaces, and you will also use layer files, which are not spatial datasets, but rather instructions for displaying spatial datasets by using certain colors, symbol markers, line widths, and so on. You'll learn more about layer files and spatial data types throughout the book.

Geodatabase icons. Data from Esri Data & Maps 2010, courtesy of ArcWorld, ArcWorld Supplement, DeLorme, and Esri.

Data

Chapter 4 uses the same datasets as those used in chapter 3, as well as one additional shapefile:

- US_Cities_PM.shp—point features representing US cities with a population of 100,000 or greater, including demographics and pollution data *(Source: ArcWorld; The World Bank)*

Exercise 4a
Browsing through map data

ArcCatalog gives you more information about spatial data than you can get from Windows Explorer or other file browsers. It can show you which folders contain spatial data and what kinds of spatial data they contain. It lets you preview features and attributes of data before you add the data to a map. It also lets you examine and edit *metadata*, which is information about your data, such as when and how it was created.

It is never a good idea to use the system file browser (such as Windows Explorer) to view or move GIS data. Spatial datasets appear differently in Explorer, with many files, whereas in ArcCatalog, you see only one. If you don't copy all the files, the dataset will be corrupted. ArcCatalog (or the Catalog window in ArcMap) is the best option for viewing, moving, and managing your GIS data.

In this exercise, you'll use ArcCatalog to look at GIS data and get information about it.

The exercise instructions assume you have installed the data for *Getting to Know ArcGIS for Desktop* to the default directory (C:\EsriPress\GTKArcGIS). If you have installed the data elsewhere, you'll need to substitute the correct paths.

1 To open ArcCatalog, on the taskbar, click the Start button, and then, on the Start menu, click All Programs > ArcGIS > ArcCatalog 10.1. This opens the ArcCatalog application window.

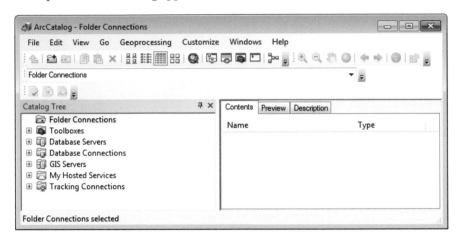

If this is the first time you've opened the ArcCatalog application, you may need to display the Catalog tree.

2 If necessary, click the Catalog Tree button 🗔 on the Standard toolbar. The Catalog tree lists the data and services that ArcCatalog is connected to. Your application may look different from the graphic in step 1, depending on the connections you may have already created.

You can connect to any local drive on your computer, as well as database servers, Internet servers, and other services, organized by folders in the Catalog tree. ArcMap also maintains a connection to ArcToolbox, which you will learn about later in this book.

Once you connect to a folder, you can access the GIS data it contains. In this exercise, it will be helpful to see the file extensions of different spatial datasets, so next, you will make sure ArcCatalog is set to display these extensions.

3 On the menu bar, click Customize > ArcCatalog Options (see facing page). This opens the Options dialog box. The Options dialog box lets you specify the types of data ArcCatalog displays and the information it shows about this data (file name, file size, date modified, and so on). You can distinguish folders that contain spatial data from those that don't and make many other customizations to the way data is displayed.

4 Make sure the General tab is selected. Clear the Hide file extensions check box if selected, and then click OK.

Next, you will create a connection to the GTKArcGIS exercise data so you can work with it in ArcCatalog.

5 On the Standard toolbar, click the Connect To Folder button [⊞]. This opens The Connect To Folder dialog box.

6 On the Connect To Folder dialog box, navigate to the local C drive to view its contents. (Systems are different, but you may find the C drive in a directory called My Computer or Computer.) Expand the EsriPress folder, click the GTKArcGIS folder, and click OK.

You have made a connection to C:\EsriPress\GTKArcGIS.

7 In the Catalog tree, click the plus sign (+) next to C:\EsriPress\
GTKArcGIS. Then click the plus sign next to the Chapter04 folder. It
contains two subfolders, Data and MyData, and a map document
called ex04c.mxd. Expand the Data folder by clicking the plus sign
next to it. The Data folder includes a geodatabase called World.gdb and a
shapefile called US_Cities_PM.shp.

8 **Click the plus sign next to World.gdb.**
The geodatabase contains four feature classes. A feature class is a group of points, lines, or polygons that represent geographic objects of the same kind. The Cities feature class contains point features, the Latlong feature class contains line features, and the Background and Countries feature classes contain polygon features, with each type of feature class symbolized by a specific icon.

The US_Cities_PM shapefile is a different data type. It contains points that represent cities and has more attributes than the Cities feature class stored in the geodatabase. It is not unusual to work with several different types of data, although at some point in your workflow, you may choose to convert these data types to the same format.

9 **On the Standard toolbar, make sure the Details button** [icon] **is active.**

10 **In the Catalog tree, click World.gdb. In the catalog display, make sure the Contents tab is active.** Four adjacent buttons on the Standard toolbar change how files look on the Contents tab. The Large Icons button displays large icons horizontally. The List button displays small icons vertically. The Details button is like the List button, except that it also shows the file type—in this case, file geodatabase feature classes. The Thumbnails button allows you to view small graphics of spatial datasets. Whenever a folder or a geodatabase is highlighted in the Catalog tree, these four buttons are activated.

11 On the Standard toolbar, click each of the buttons that change the look of the Contents tab: Large Icons ⊞ , List ⊞ , Details ⊞ , and finally, Thumbnails ⊞ .

At any time, pause the mouse pointer over a button to see its ToolTip. This also works in ArcMap and other ArcGIS for Desktop applications.

The display changes to show thumbnail graphics of the content of each file. A thumbnail graphic of a dataset can help you decide quickly whether you want to use the data or not. Notice there is no thumbnail graphic for countries. When a thumbnail graphic hasn't been created, you will see only a larger version of the feature class icon. Next, you'll create a thumbnail graphic for the Countries feature class.

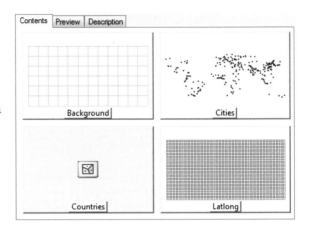

12 In the Catalog tree, click the Countries feature class in the World geodatabase. In the catalog display, click the Preview tab. The Countries polygons are displayed in pale yellow.

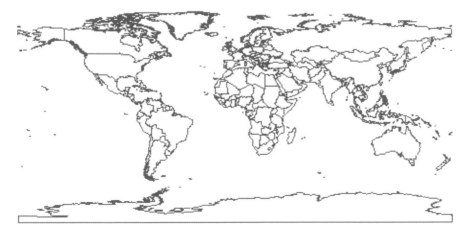

All polygons are displayed in pale yellow when previewed in ArcCatalog. Points are displayed in black and lines in navy blue.

13 On the Geography toolbar, click the Create Thumbnail button . An image of the data shown on the Preview tab is saved as a thumbnail graphic.

14 In the catalog display, click the Contents tab to see the new thumbnail graphic.

A thumbnail graphic gives you a quick visual of the data, but you may want to view the data in more detail. The Preview tab and the tools on the Geography toolbar let you investigate a dataset further before you decide whether you want to use it on a map. So, you'll try both of these options next.

15 Click the Preview tab to see the Countries dataset at full size.

16 On the Geography toolbar, click the Zoom In tool . Drag the tool to draw a box around the area north of Australia, as shown in the graphic.

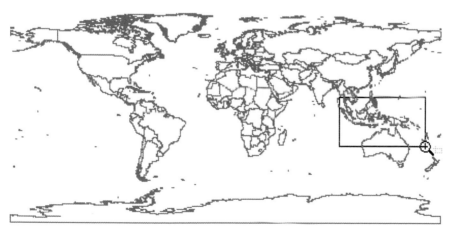

17 On the Geography toolbar, click the Identify tool 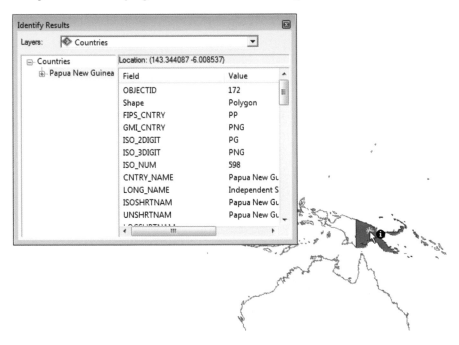 . Click the feature indicated in the graphic to identify it. If you don't see the feature flash green on the map, move the Identify Results window away from the display and click the feature again.

Is Papua New Guinea a UN member state?

(Hint: You may need to scroll down in the Identify Results window to view the Status attribute.)

18 Close the Identify Results window. Use the Pan tool from the Geography toolbar to move around the map. The Pan tool works here just as it does in ArcMap.

19 Click the Full Extent button to zoom to the full extent of the data.

You can preview the attributes of a dataset as well as its geography. So, you'll do that next.

20 **At the bottom of the catalog display, click the Preview arrow and click Table.** The display shows the attribute table of the Countries feature class.

OBJECTID *	Shape *	FIPS_CNTRY	GMI_CNTRY	ISO_2DIGIT	ISO
1	Polygon	GY	GUY	GY	GUY
2	Polygon	NS	SUR	SR	SUR
3	Polygon	TD	TTO	TT	TTO
4	Polygon	VE	VEN	VE	VEN
5	Polygon	AR	ARG	AR	ARG
6	Polygon	BL	BOL	BO	BOL
7	Polygon	BR	BRA	BR	BRA
8	Polygon	CI	CHL	CL	CHL
9	Polygon	EC	ECU	EC	ECU
10	Polygon	PA	PRY	PY	PRY
11	Polygon	PE	PER	PE	PER
12	Polygon	UY	URY	UY	URY
13	Polygon	CA	CAN	CA	CAN
14	Polygon	GT	GTM	GT	GTM
15	Polygon	MX	MEX	MX	MEX
16	Polygon	BB	BRB	BB	BRB
17	Polygon	DO	DMA	DM	DMA
18	Polygon	GJ	GRD	GD	GRD

Contents　Preview　Description

1 ► ►► (of 249)

Preview: Table

21 **Scroll through the attribute columns.** Sometimes attributes have cryptic names. Many organizations maintain a data dictionary, which decodes and describes attributes.

22 **When you are finished looking, click the Preview arrow again and click Geography.** The third tab on the catalog display is the Description tab, which displays metadata—essentially, data about data. ArcCatalog doesn't require you to maintain metadata, but it's considered best practice to do so, and many organizations require metadata to ensure data standardization and quality.

23 In the catalog display, click the Description tab. If needed, scroll down to see the thumbnail graphic you created, as well as keyword tags, summary, description, credits, and use limitations.

Notice the metadata shows a different name (World Countries instead of Countries). The original dataset, from Esri Data & Maps 2010, was called World Countries. When we the authors imported the dataset into our World geodatabase, we shortened the name to Countries. One technique for documenting data sources is to maintain the original name in the metadata, even if it doesn't match the new name provided upon import. Or you could change the source name using the Edit function described in step 24.

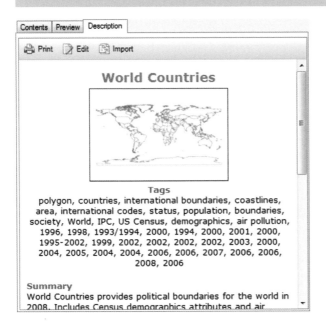

24 Click the Edit button on the uppermost bar inside the description to access metadata about the Countries feature class. Metadata documentation can be created and edited on the Description tab. Next, you'll edit some of the metadata.

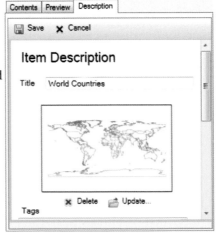

25 Scroll down to see what type of metadata can be entered. Set the Approximate Scale Range slider to Continent, and then click Save.

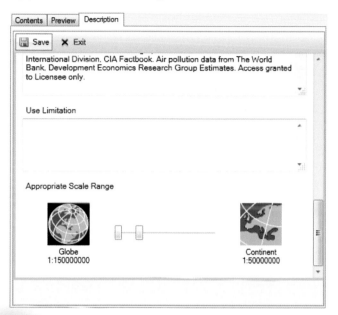

Many organizations require adherence to strict metadata standards. You can choose different styles of metadata to be in accordance with various national or international standards. For more information about metadata and the styles supported by ArcGIS, in ArcGIS for Desktop Help (from ArcCatalog, click the Help menu, and then ArcGIS for Desktop Help), on the Contents tab, navigate to Geodata > Data types > Metadata.

26 In the catalog display, click the Contents tab. In the Catalog tree, click World.gdb. On the Standard toolbar, click the Details button. This restores the display to its initial state.

27 If you are continuing to the next exercise, leave ArcCatalog open. Otherwise, exit the application.

Exercise 4b
Adding data to a map

You can easily add data to ArcMap from the ArcCatalog application or, easier still, from the Catalog window within ArcMap. Once feature classes are added to ArcMap, they become map layers.

ArcGIS applications include a Search window that allows you to search for geographic data on disk or across a network. You can search by name, file type, geographic location, creation date, or by other properties stored in the metadata.

Suppose you want to create a map that presents worldwide air pollution data. In this exercise, you will add layers to a new map document and search for more data that may be a useful addition to the world population map you studied in chapter 3.

1 Start ArcCatalog if not already open. If necessary, in the Catalog tree, browse to your C:\EsriPress\GTKArcGIS folder connection and expand Chapter04\Data\World.gdb.

2 On the Standard toolbar, click the ArcMap button ⚙ to start ArcMap.

3 If you see a Getting Started dialog box, under New Maps, click My Templates. In the My Templates panel, click the Blank Map symbol, and then click OK.

Next, you will add layers to your new, empty map document.

4 Position the ArcCatalog and ArcMap application windows so that you can see both the Catalog tree and the ArcMap display window.

5 From the Catalog tree, drag the Latlong feature class file anywhere over the ArcMap display window and release the mouse button. Bring ArcMap forward by clicking its title bar. The layer is added to ArcMap. ArcMap chooses random symbol colors when new datasets are added to a map, so your results probably won't match the graphics in the book.

6 Close ArcCatalog. You will add the rest of the map layers from within ArcMap.

7 In ArcMap, look for a Catalog tab on the right side of the display window. Click the tab to see the Catalog window. If you do not see a small Catalog tab, on the Standard toolbar, click the Catalog button to open the window. Dock the window to the right of the map display.

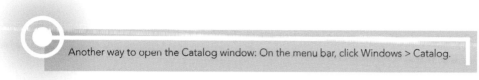

Another way to open the Catalog window: On the menu bar, click Windows > Catalog.

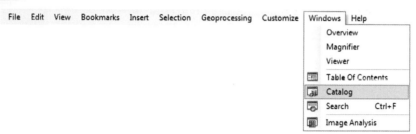

Most of the functionality of ArcCatalog is available in the ArcMap Catalog window. The window can be docked or undocked. If docked, you can use the Auto Hide function to collapse the window when you're not using it. The Catalog window shows your folder connections and also includes a Home folder, which points to the current map document's workspace, if you are working in a saved map document.

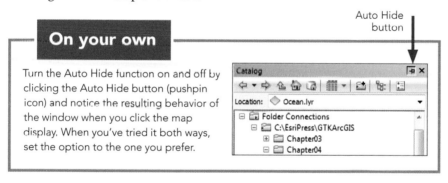

Auto Hide button

On your own

Turn the Auto Hide function on and off by clicking the Auto Hide button (pushpin icon) and notice the resulting behavior of the window when you click the map display. When you've tried it both ways, set the option to the one you prefer.

8 In the Catalog window, go to your C:\EsriPress\GTKArcGIS folder connection and expand Chapter04\Data\World.gdb. Click the Countries feature class to highlight it.

9 On the Catalog window toolbar, click the Toggle Contents Panel button to see both the file tree and the preview. (You may need to drag the divider between the panes to see the preview in its entirety.)

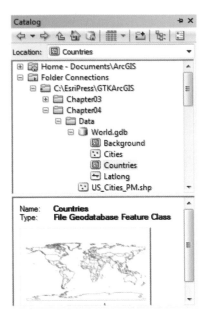

10 Drag the Countries feature class to the ArcMap display.

You can also add data using the Add Data button, which you'll do next.

11 On the ArcMap Standard toolbar, click the Add Data button ⊕ . (Click just the button, not the arrow next to it.)

12 On the Add Data dialog box, browse to \EsriPress\GTKArcGIS\ Chapter04\Data. Double-click World.gdb. Click the Background feature class, and then click Add.

13 In the ArcMap table of contents, drag Background to the bottom of the list so it does not obscure Countries.

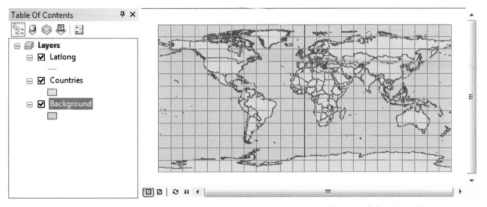

Your map clearly needs work—the default symbols need to be modified so the map looks nice and shows the attributes you are interested in. In particular, you want the Countries layer to show air pollution levels. However, you won't concern yourself with creating symbology in this chapter. That will come in chapter 7.

On your own

Open the attribute table of the Countries layer and verify that the dataset includes the air pollution field called pm10.

Perhaps you want to add more data that pertains to the worldwide pollution data you already have, but you don't know if there's anything available. So next, you will use the Search window, which is a tool to find maps, data, and tools that match your search criteria.

14 **On the Standard toolbar, click the Search button** 🔍 **.** This opens the Search window. By default, it is placed to the right of the display area. Like the Catalog window, the Search window can be hidden, docked, or left floating.

15 **Undock the Search window so it floats or arrange it as you prefer.**

You can limit a search to specific folders, which you'll do next. This spares ArcCatalog from looking for data in unlikely places.

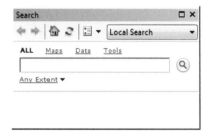

16 **In the Search window, click the Search Options button** 🔲▼ **. (Click just the button, not the arrow next to it.)** This opens the Search Options dialog box.

17 In the Register Folders and Server Connections frame, click the Add button, browse to your C:\EsriPress\GTKArcGIS\Chapter04 folder, and click Select. This limits the search to the Chapter04 folder. If you wanted to search other folders as well, you could do so by clicking the Add button again.

Before searching a specific data folder for the first time, you should create an index to facilitate the Search tool, so you'll do that next.

18 At the bottom of the Search Options dialog box, click the Index New Items button. The contents of the Chapter04 folder are indexed after a few moments.

Indexing Status	
Items Indexed:	7
Indexing Status:	Active
Last Indexing Start Time:	1/20/2012 11:29 AM
Last Indexing Duration:	31 seconds
Next Indexing Start Time:	1/20/2012 12:11 PM

Seven items are indexed, and the time and duration of the index process is recorded. Indexing duration depends on the performance of your system and the number of folder or server connections to be indexed.

19 Click OK to close the Search Options dialog box.

You will next concentrate your search for population data using keywords, or tags, stored in the metadata. If any matches are found, the dataset and its description will be returned in the search results.

20 In the Search window, make sure the drop-down menu is set to Local Search. In the text box, type pollution and press ENTER. When the results are returned, you may want to expand the window for easier viewing. The Search function returns several matches: a map document, two feature classes, and several tools.

Notice you can choose to search for data on your local drive, as well as throughout your enterprise (data on the server) or on ArcGIS Online (data on the web).

㉑ In the search results listing, click the description (navy-blue text) of the US_Cities_PM shapefile. This opens an Item Description window, which includes the metadata.

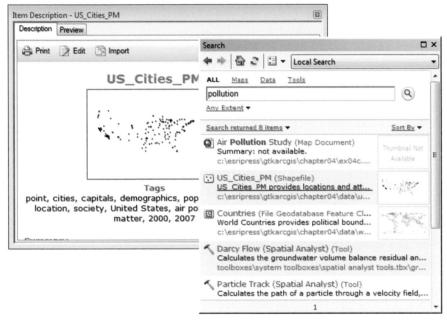

㉒ In the Item Description window, read the Summary statement. The summary informs you this feature class shows air pollution measurements for most major US cities (population of 100,000 or greater). If you hadn't used the Search function, you might not have realized this shapefile includes pollution information.

The Search function highlights the importance of good metadata. Without good metadata, including relevant tags, an item would only be returned if the keywords were found in the item's name.

23 In the Item Description window, click the Preview tab. Notice the tab
has geography navigation tools so that you can explore your data
more closely, just the same as in the ArcCatalog application.

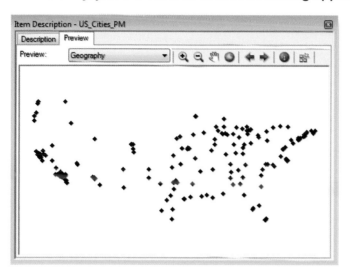

24 In the Preview list, click Table. Scroll to the right to see the last field,
pm10. This field stores particulate matter measurements in urban areas,
estimated by a research team at the World Bank.

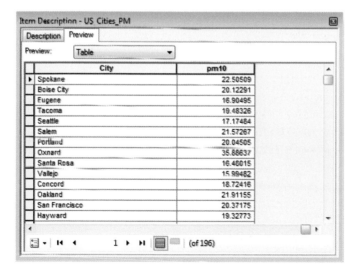

25 Close the Item Description window. From the Search window, drag US_Cities_PM (the light-blue link) to the map. Notice that the points go automatically to the top of the layers list in the table of contents.

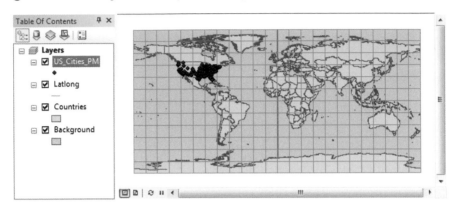

26 In the table of contents, right-click US_Cities_PM and click Zoom To Layer.

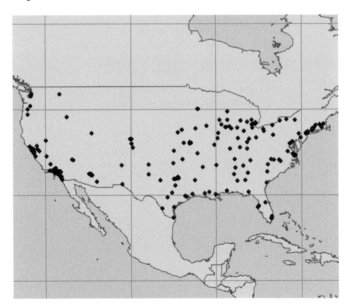

On your own

Identify a few point features and find the pm10 attribute value.

27 If you want to save your work, on the menu bar, click File > Save As. Give your file a new name and save to your MyData folder for chapter 4.

28 If you are continuing to the next exercise, leave ArcMap open. Otherwise, exit the application.

In exercise 4c, you will organize the layers to create a meaningful map layout.

1
2
3
4b
5
6
7
8
9
10
11
12
13
14
15
16
17
18
19
20

Exercise 4c
Working with map layers

When you add data to a map document, you can examine it as a navigable map display, as you did in chapter 3. This is called the *data view* of the map. You can also look at it as if it were on a page that you send to a printer. This is called the *layout view*. When you're in layout view, you can still interact with the map layers, but this view is best used for creating static presentation maps.

In layout view (see below), you see map layers organized in one or more rectangles on a larger background rectangle. The smaller rectangles are called *data frames* and the background is called the virtual page.

A map document can have one data frame or many. Data frames can contain different sets of data (usually related to a common subject), or they may show different views of the same data. If you wanted a layout to show more than one map view, you would use multiple data frames.

In data view (see below), you see just one data frame at a time.

The display window shows the contents of the active data frame. The name of the active data frame is in boldface in the table of contents. In the graphic, the active data frame is called Bioregions. The default name given to data frames is Layers.

In chapter 10, you'll learn in detail how to create maps for presentation. For now, you'll learn how to work with data frames in ArcMap to create a simple *layout*. Remember, your objective is to create a presentation that displays air pollution data.

1 **In ArcMap, open the ex04c map document from your C:\EsriPress\
GTKArcGIS\Chapter04 folder.** There are several things to note about this
map document, but of most pressing importance are the red exclamation
points in the table of contents. This is the sign of a broken data path, and
it often occurs if a map document is not saved with relative paths. If a user
creates a map document from data stored on the computer's desktop location,
and saves the map with absolute paths, ArcMap will look for that user's
desktop location in search of the source data, but to no avail.

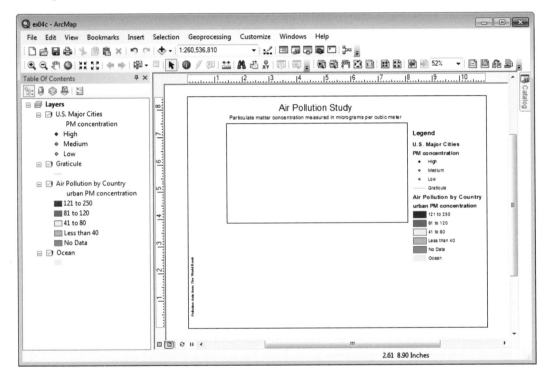

2 To fix the problem, in the table of contents, right-click U.S. Major Cities, point to Data, and click Repair Data Source.

3 On the Data Source dialog box, browse to C:\EsriPress\GTKArcGIS\ Chapter04\Data. Click US_Cities_PM.shp, and then click Add.

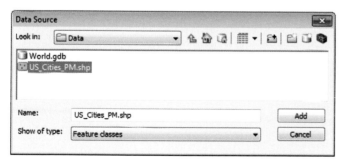

The broken path is repaired. In fact, ArcMap repaired the other broken links, too, by finding the correct source datasets in the folder you just specified.

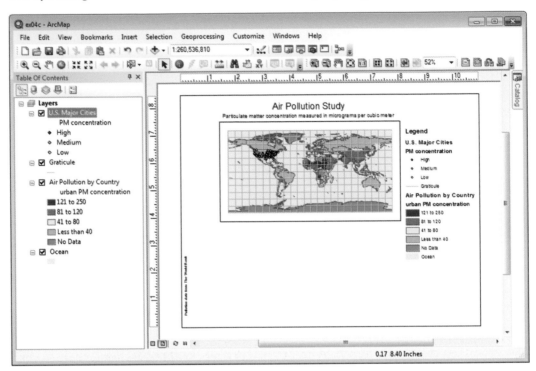

To avoid this problem in the future, you will next save the map document with relative paths.

4 On the menu bar, click File > Map Document Properties. In the lower portion of the dialog box (see facing page), select the "Store relative pathnames to data sources" check box, and then click OK.

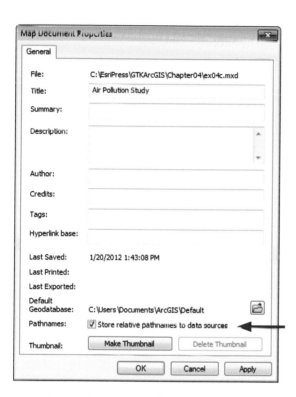

Now that you've fixed the data paths, you can take a look at the map display. The map is open in layout view. It includes the same data you worked with in exercise 4b, but the layers now have meaningful symbology, and they have been given more descriptive names. You see the map as if it were on a piece of paper. A new toolbar, the Layout toolbar, is open. It can be docked or left floating.

If you don't see the Layout toolbar, on the menu bar, click Customize > Toolbars, and then select the Layout check box to turn on the toolbar.

If you resize your ArcMap window in layout view, you can restore the appearance of the layout by clicking the Zoom Whole Page button ⊡ on the Layout toolbar.

The map document has a single data frame, occupying the upper half of the virtual page. Some other elements have been added to the layout for you. There is a title above the data frame, a subtitle, and a legend. The U.S. Major Cities points cannot be discerned on the map. You will zoom in to fix this, but you also would like to show the world data. To do this, you need two data frames.

5 In the table of contents, right-click U.S. Major Cities and click Zoom To Layer.

6 On the menu bar, click Insert > Data Frame.

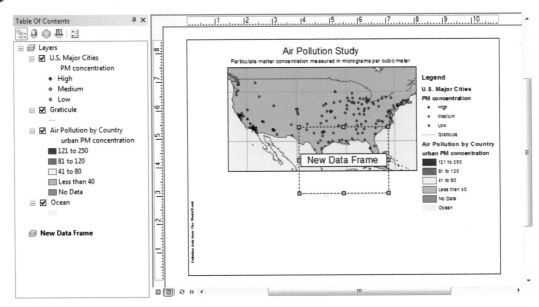

A data frame is added to the layout and its name, New Data Frame, appears in the table of contents. The name in the table of contents is in boldface, indicating that it is the active data frame—the one that new layers will be added to.

7 In the layout window, move the mouse pointer over the new data frame. The pointer changes to a four-headed arrow. Drag the new data frame below the first data frame.

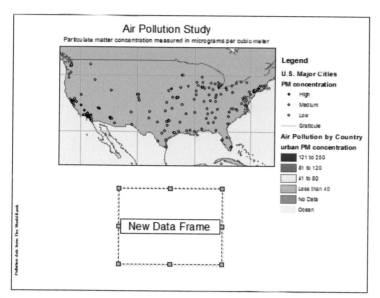

8 Move the mouse pointer over the lower-right anchor (the blue square). The pointer changes to a double-headed arrow. Drag the anchor to the lower-right corner of the layout page, enlarging the data frame window as shown in the graphic.

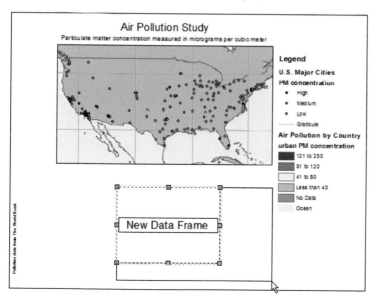

9 Now drag the upper-left anchor so the new data frame is approximately the same size as the one above it.

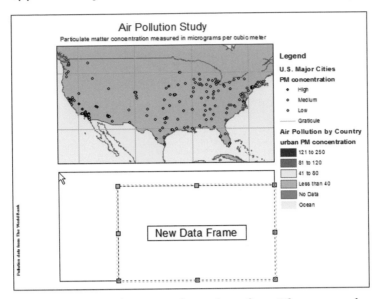

The name of a data frame is independent of any title you may choose to add to the layout. By default, the first data frame in a map document is called "Layers," the second is called "New Data Frame," the third is called "New Data Frame 2," and so on. It is often helpful to rename a data frame to something more descriptive, so you'll do that next.

10 In the table of contents, right-click the first data frame (Layers) and click Properties. On the Data Frame Properties dialog box, click the General tab. For name, enter United States, and then click OK.

11 Click the second data frame in the table of contents and click again to make it editable. Rename the second data frame World Countries.

Next, you'll switch back to data view to add layers to the new data frame. (You could do this in layout view as well.)

12 **Click the Data View button in the lower-left corner of the layout window. You use this button and the Layout View button next to it to switch views.**

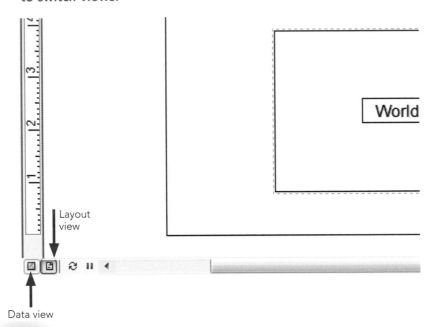

Layout view

Data view

You can also change views from the View menu. (On the menu bar, click View > Data View or View > Layout View.)

The display window is empty because the active data frame (World Countries) contains no layers.

To activate a data frame, right-click the frame name in the table of contents and click Activate. Or use the shortcut ALT+Click. In the layout display, a light-gray dashed line surrounds the active data frame. The navigation tools on the Layout toolbar work with whatever frame is activated.

13 **In the table of contents, right-click Air Pollution by Country and click Copy. Right-click the World Countries data frame and click Paste.** Now you have the same layer in two data frames—one layer is zoomed to the United States (the one in the United States data frame), and the other is zoomed to the full extent (the one in the World Countries data frame).

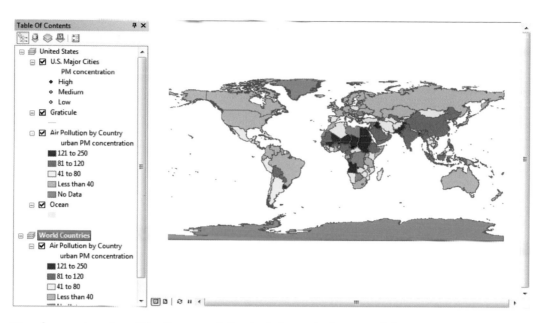

You also want copies of the Ocean and Graticule layers in the second data frame, so you'll add those on your own.

On your own

Copy the Ocean and Graticule layers to the World Countries data frame. Use the CTRL key to select both layers at once. Reorder the layers as needed for proper drawing in the World Countries data frame.

14 Switch again to the layout view if necessary (on the menu bar, click View > Layout View or click the Layout View button at the lower-left corner of the display window).

15 **Finally, click anywhere in the white area of the virtual page to clear the selection of the lower frame. (It will still have a dashed outline to show it is the active data frame.)**

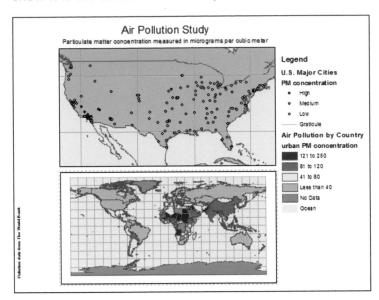

Your layout presents the air pollution data nicely. However, you decide the Graticule layer is unnecessary for the purpose of this map. You could just turn off the layer for both data frames in the table of contents, but next, you will remove it from the map document instead.

16 In the table of contents, under the United States data frame, right-click Graticule and click Remove. Also remove Graticule from the World Countries data frame.

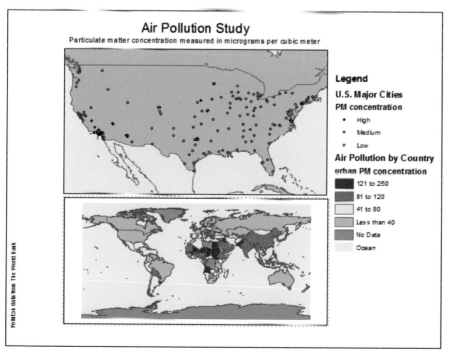

17 To close the Layout toolbar, right-click in an empty space near the ArcMap toolbars. From the list that appears, click Layout to clear the check box and turn off the layer.

18 To save your work, on the menu bar, click File > Save As. Give your file a new name and save to your MyData folder for chapter 4.

19 Exit the ArcMap application.

Summary

You now know how to bring data easily from ArcCatalog into ArcMap or—even easier—from the ArcMap Catalog window. Using the Search window, you learned how to find relevant data stored locally or on the web or available servers. Finally, you practiced working with multiple data frames, which is necessary to create a printed product with more than one map. The difference between data view and layout view should now be clear.

Terms

geodatabase

shapefile

metadata

data view

layout view

data frame

layout

Chapter 5
Exploring online resources

Creating a web map
Browsing through groups
Merging online and local layers

Learning objectives

- *Register for an Esri Global Account*
- *Browse the ArcGIS Online website*
- *Open and create an online map*
- *Explore shared online basemaps and thematic layers*
- *Examine map descriptions and properties*
- *Identify online map features*
- *Explore ArcGIS Online groups*
- *Search for shared online data*
- *Add ArcGIS Online layers to ArcMap*

ARCGIS ONLINE IS A WEB PORTAL FOR WORKING WITH MAPS AND OTHER TYPES of geographic information. It functions as an online repository of maps, data, applications, and tools and allows anyone to use a web browser to share GIS information with the online community. It also hosts developer resources for building web mapping applications. ArcGIS Online is an essential part of the ArcGIS system—it is integrated into ArcGIS applications, which makes it easy to search for and access geospatial information when you need it.

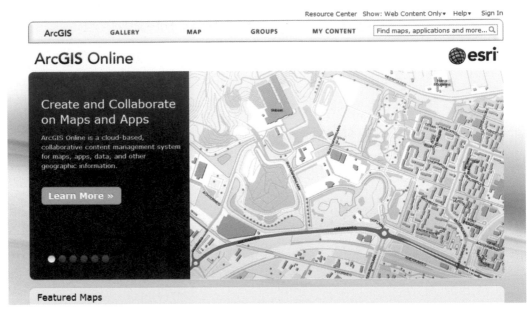

ArcGIS Online home page.

The ArcGIS Online Gallery has live, up-to-date maps created and shared by other users like yourself.

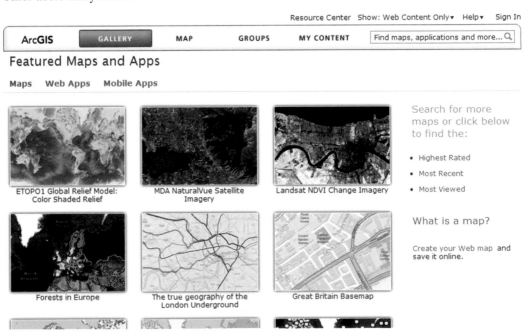

ArcGIS Online Gallery.

The web map viewer can be used to explore different parts of the world using the *basemaps* that are provided. You can add thematic layers, and save and share your map.

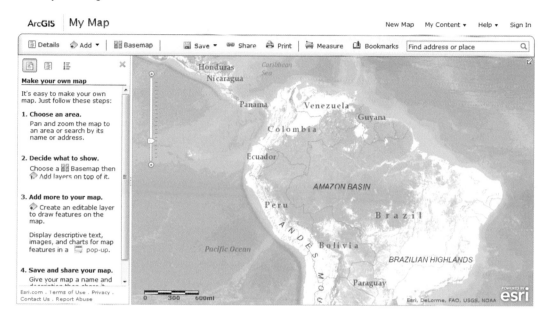

The Amazon Basin, as seen in the ArcGIS Online map viewer. Map courtesy of the Food and Agriculture Organization of the United Nations, National Geophysical Data Center, National Aeronautics and Space Administration, National Geospatial-Intelligence Agency, and Esri.

You can join different ArcGIS.com *groups* to collaborate, discuss, and share your work with other members.

Groups allow you to collaborate with other people.

Find groups of people that share your interests and join them. If you can't find the right group, create it!

Unsure what to look for? List all public groups.

Sign in to join and create groups.

First time visitor? Sign up now to enjoy these features:

- Create and share maps
- Contribute your content
- Join and create groups

Groups on ArcGIS Online.

Here are a few groups that you may find interesting.

National Maps for USA

This is a group formed by Esri for use in sharing information about data, maps and services of interest to the Federal Community. The maps may be used as basemaps and operational layers to support a variety of applications.

Details

created by Federal_User_Community on September 8, 2011

Esri Maps and Data

The Esri Maps and Data group provides access to online basemap and reference map services, map layers and documents, and tools to use in your GIS work.

Details

created by esri on January 16, 2012

ArcGIS for Local Government

Welcome to the ArcGIS for Local Government Group on ArcGIS.com. It provides useful maps, apps and best practices that will help you successfully implement GIS in your organization.

Details

created by ArcGISTeamLocalGov on July 20, 2012

You can manage your own storage space and content—such as web maps, ArcGIS Explorer documents, ArcGIS Server image services, ArcMap documents, and map layer packages—with your own account on ArcGIS Online.

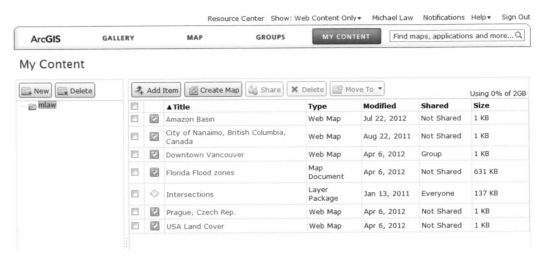

MY CONTENT space on ArcGIS Online.

To access the features of ArcGIS Online, you must have a working Internet connection, and you will need to become a member. To do this, you must first create an Esri Global Account.

There are several ways to access the maps and data you find in ArcGIS Online.

ArcGIS.com map viewer

You can use the free, lightweight map viewer provided in ArcGIS.com to create and interact with web maps composed of a basemap and additional layers that you find in ArcGIS Online, on the web, or through a GIS server. It's easy to create a map: choose an area of interest, decide which layers to show, and then save and share your map with others (within ArcGIS.com, and even through social media sites such as Twitter and Facebook).

ArcGIS Explorer Online

ArcGIS Explorer Online is a more robust web application than the ArcGIS. com map viewer, providing the ability to measure geography, add notes and bookmarks, and perform searches. With ArcGIS Explorer Online, you can create and save presentations to discuss specific issues related to your projects.

ArcGIS app for smartphones

The ArcGIS app can be used on smartphones, such as iOS (Apple), Android (Google), and Windows Mobile, to discover community-hosted web maps created in ArcGIS Online or to share your own maps. Maps are interactive, and you have the ability to collect data, access your own GIS data, and perform simple GIS functions, such as measuring distance and area.

ArcGIS Explorer Desktop

ArcGIS Explorer Desktop is a desktop client that has more features and is more powerful than ArcGIS Explorer Online. You can access web maps, basemaps, and layers shared by Esri and the ArcGIS community; view three-dimensional (3D) maps; and import KML, global positioning system (GPS) logs, text files, and other spatial data sources such as shapefiles and layer packages into your map.

ArcGIS for Desktop

ArcGIS for Desktop is the most powerful application for professional GIS work, and ArcGIS Online is fully integrated with it. You can use tools from ArcMap and ArcCatalog to create and edit maps, manage geographic databases, perform spatial analysis, create geographic models, and work with a wide range of data types.

This chapter concentrates on browsing ArcGIS Online to understand how it is used in conjunction with ArcGIS for Desktop. It will also provide information on how to register an Esri Global Account and access the *Getting to Know ArcGIS for Desktop* group. Due to the dynamic nature of websites, the graphics used in the book may not reflect the current appearance of ArcGIS.com.

Data

This chapter uses the following datasets:

- Topographic—basemap includes administrative boundaries, cities, water features, physiographic features, parks, landmarks, highways, roads, railways, and airports overlaid on land cover and shaded-relief imagery for added context. The map provides coverage of the world. *(Source: US Geological Survey (USGS), US Environmental Protection Agency (EPA), US National Park Service (NPS), Food and Agriculture Organization of the United Nations (FAO), Department of Natural Resources Canada (NRCAN), GeoBase, Agriculture and Agri-Food Canada, NAVTEQ, TomTom, Automotive Navigation Data (AND), DeLorme, and Esri)*
- World Imagery—basemap of low-resolution imagery of the world and high-resolution imagery of the United States and other areas around the world. *(Source: NASA Blue Marble, i-cubed, Getmapping, AeroGRID, GeoEye IKONOS, US Department of Agriculture (USDA), National Agriculture*

Imagery Program (NAIP), US Geological Survey (USGS), and Digital Ortho Quarter Quad (DOQQ))

- USA Tapestry Segmentation data—thematic map presenting tapestry segmentation in the United States in 2010. *(Source: Esri)*
- Intersections—point layer showing intersections in the City of Nanaimo, British Columbia. *(Source: City of Nanaimo)*
- Bike routes—line layer showing bike routes in the city of Nanaimo, British Columbia. *(Source: City of Nanaimo)*

Exercise 5a
Creating a web map

ArcGIS.com provides a robust map viewer that can be used to quickly create web maps containing a basemap and layers found throughout the site. In just three steps, you can create a web map, as follows: (1) choose an area to display, (2) add content, and (3) save your map. You can change the geography of interest (for example, you can zoom out from a country view to a continental view), explore the metadata behind the map, and add and remove layers.

Web maps always start with a basemap layer, and then you can add more data to the map with *operational layers*, which are the layers you can manipulate. This process is sometimes called a "*mashup*," because you are combining layers of information, often from different sources, to create a single map. These content layers can be any map, image, or feature service obtained from ArcGIS for Server and stored in ArcGIS Online or accessible from the web. Image services serve raster datasets that can be dynamically viewed. Feature services serve features and provide the symbology for displaying the features.

What's unique about these web maps is that you can view, create, and share them from within ArcGIS.com. In addition, other clients such as smartphones, ArcGIS Explorer Desktop, and ArcGIS for Desktop can access these maps without any additional programming or third-party applications or plug-ins. However, if you are looking to create web maps with additional functionality such as bookmarks, queries, and notes, you may want to consider using ArcGIS Explorer Online.

On your own

Exercises in this chapter require you to become a member of ArcGIS Online. To do so, you must first have an Esri Global Account. In a web browser, go to **http://www.arcgis.com** and click the Sign In link in the upper-right corner. If you have an Esri Global Account, simply click Register your Esri Global Account, or if you don't have a global account, click Create a Personal Account and follow the instructions provided. When you are finished, be sure to register your account on the ArcGIS.com sign-in page. Record your name and password somewhere safe. You will need to use them to sign in again.

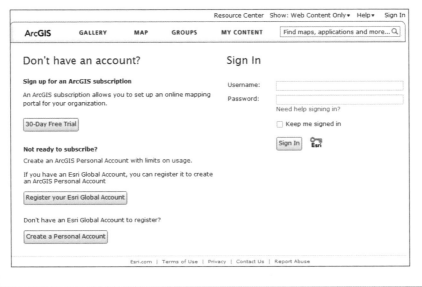

1. In a web browser, go to http://www.arcgis.com. In the upper-right corner, click Sign In. Enter your member user name and password and click the Sign In button. On the main navigation bar, click the MY CONTENT button.

The My Content page contains all the tools necessary to add, edit, share, and access all your saved content. Your map viewer may not show any content yet, but in the following steps, you'll learn how to create a map and save it to your My Content page.

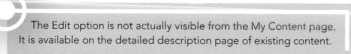

The Edit option is not actually visible from the My Content page. It is available on the detailed description page of existing content.

3
4
5a

② On the main navigation bar, click the MAP button. (If you have previously viewed or worked with maps in ArcGIS.com, a dialog box may appear to ask if you want to open a new map. Click Yes and open the map.)

The web map viewer displays a basemap showing a topographic map of the world. To navigate the map, which you are probably already familiar with, use the mouse pointer intuitively.

③ Drag the pointer to pan around the map.

④ Double-click a location on the map to center and zoom.

⑤ Rotate the scroll wheel of the mouse forward and backward to zoom in and out, respectively.

⑥ Pan and zoom to the Amazon Basin, which covers the majority of Brazil, and other countries such as Ecuador, Peru, Colombia, and Guyana (see next page).

The Amazon Basin, which covers over 8 million square kilometers, is drained by the Amazon River. It is home to the largest rain forest in the world, which is largely unexplored and contains many species yet to be discovered. To get a better sense of the land cover, you'll change the basemap next.

7 On the map viewer toolbar, click the Basemap button. The menu that appears shows previews of available basemaps. Click Imagery with Labels.

The basemap changes to show satellite imagery, country borders, and labels for countries and major cities. To find out more about this basemap, you'll next take a look at its description and metadata.

8 On the Details panel (if it is not visible, click the Details button), click the Show Contents of Map button ▣ .

ArcGIS | Amazon Basin

9 Pause the mouse pointer over the Imagery with Labels text. Notice that an arrow appears to the right of the label. Click the arrow, and then click Description.

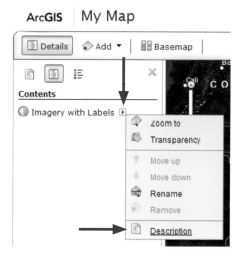

3
4
5a

A new page is loaded that displays a thumbnail graphic, description, and
sharing options for the Imagery with Labels basemap (which is composed of
World Transportation, World Imagery, and World Boundaries and Places).

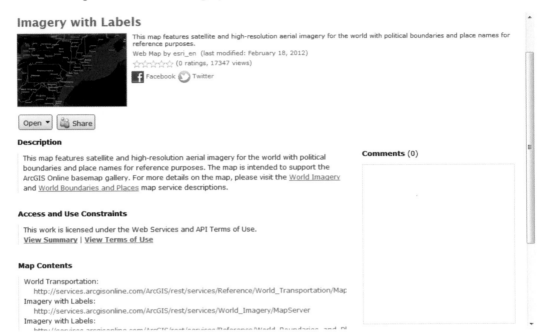

Imagery with Labels

This map features satellite and high-resolution aerial imagery for the world with political boundaries and place names for reference purposes.
Web Map by esri_en (last modified: February 18, 2012)
☆☆☆☆☆ (0 ratings, 17347 views)
Facebook   Twitter

Open ▾   Share

Description

This map features satellite and high-resolution aerial imagery for the world with political
boundaries and place names for reference purposes. The map is intended to support the
ArcGIS Online basemap gallery. For more details on the map, please visit the World Imagery
and World Boundaries and Places map service descriptions.

Comments (0)

Access and Use Constraints

This work is licensed under the Web Services and API Terms of Use.
View Summary | View Terms of Use

Map Contents

World Transportation:
    http://services.arcgisonline.com/ArcGIS/rest/services/Reference/World_Transportation/Map
Imagery with Labels:
    http://services.arcgisonline.com/ArcGIS/rest/services/World_Imagery/MapServer
Imagery with Labels:
    http://services.arcgisonline.com/ArcGIS/rest/services/Reference/World_Boundaries_and_Pl

10 Under the Map Contents heading, click the first Imagery with Labels
hyperlink (http://services.arcgisonline.com/ArcGIS/rest/services/
World_Imagery/MapServer).

Map Contents

World Transportation:
    http://services.arcgisonline.com/ArcGIS/rest/services/Reference/World_Transportation
    /MapServer
Imagery with Labels:
    http://services.arcgisonline.com/ArcGIS/rest/services/World_Imagery/MapServer ◄——————

ArcGIS Services Directory

Home > World_Imagery (MapServer)   Help | API Reference

World_Imagery (MapServer)

View In: ArcMap   ArcGIS Explorer   ArcGIS JavaScript   Google Earth   ArcGIS.com Map   Bing Maps (terms of use)   Google Maps (terms of use)

View Footprint In: Google Earth

Service Description: This map presents low-resolution imagery for the world and high-resolution imagery for the United States and other areas around the world. The map includes NASA Blue Marble: Next Generation 500m resolution imagery at small scales (above 1:1,000,000), i-cubed 15m eSAT imagery at medium-to-large scales (down to 1:70,000) for the world, and USGS 15m Landsat imagery for Antarctica. The map features i-cubed Nationwide Prime 1m or better resolution imagery for the contiguous United States, Getmapping 1m resolution imagery for Great Britain, AeroGRID 1m to 2m resolution imagery for several countries in Europe, IGN 1m resolution imagery for Spain, IGP 1m resolution imagery for Portugal, and GeoEye IKONOS 1m resolution imagery for Hawaii, parts of Alaska, and several hundred metropolitan areas around the world. Additionally, imagery contributed by the GIS User Community has been added in Alaska, New York and Virginia. i-cubed Nationwide Prime is a seamless, color mosaic of various commercial and government imagery sources, including Aerials Express 0.3 to 0.6m resolution imagery for metropolitan areas and the best available United States Department of Agriculture (USDA) National Agriculture Imagery Program (NAIP) imagery and enhanced versions of United States Geological Survey (USGS) Digital Ortho Quarter Quad (DOQQ) imagery for other areas. For more information on this map, visit us online.

Map Name: Layers

Legend

All Layers and Tables

Layers:

- World Imagery (0)
- Low Resolution 15m Imagery (1)
- High Resolution 60cm Imagery (2)
- High Resolution 30cm Imagery (3)

Tables:

This opens a new window (or tab). The service description lists information about the image sources, resolution, tiling detail, spatial extent, and ownership information. Notice the row of View In hyperlinks. You can click these links to open this basemap in other applications such as ArcMap, ArcGIS Explorer, and Google Earth. Or you can get more metadata about individual layers, which you'll do next.

11 Scroll down the page to view more metadata. Stop and review the Tile Info section.

How many levels of detail are there in this basemap?

Compare the resolution and scale values for two different levels. What does this tell you?

The basemaps in ArcGIS.com are in the form of individual image tiles. Multiple levels of detail provide the appropriate amount of detail at each map scale.

12 On the main navigation bar, click MAP. (If your description was opened on a new browser tab or in a new window, close it after you finish viewing the metadata. If you did not return to the exact same location, change the basemap to Imagery with Labels and zoom to the Amazon Basin again.)

| ArcGIS | GALLERY | MAP | GROUPS | MY CONTENT | Find maps, applications and more... 🔍 |

The map is now ready to be saved, so you'll do that next.

13 On the map viewer toolbar, click the Save button, and then click Save. On the Save Map dialog box, enter the title: Amazon Basin. Enter the tags: amazon, imagery, forest. Enter the summary: A basemap of the Amazon Basin and click the Save Map button. (The Save in folder list shows your own sign-in name.)

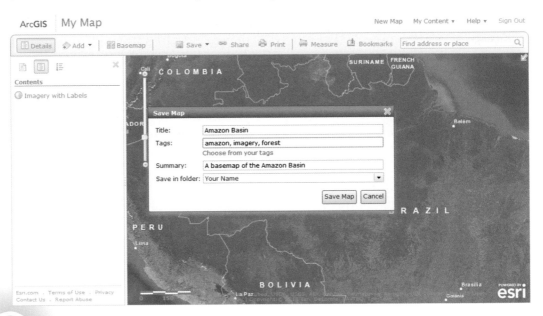

Tags (or keywords) are used to categorize content and make it searchable. Appropriately defined tags can help users find the content on ArcGIS.com or elsewhere on the web.

The map is saved, and the title of it is changed at the top of the map viewer.

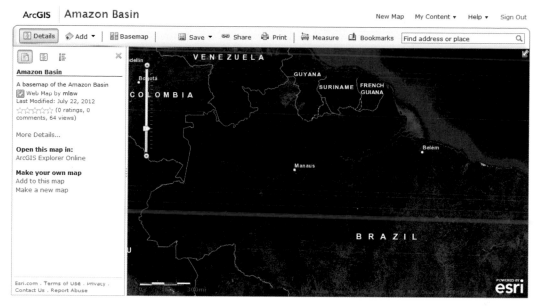

When you click the About this Map button, a short summary appears along with the type of document (Web Map) and the author (you). The rating system indicates which maps are popular. Additionally, there are options below the rating information to conveniently open your Amazon Basin map in ArcGIS Explorer Online.

14 On the ArcGIS.com uppermost menu (upper-right corner), click My Content, and in the drop-down list, click My Content.

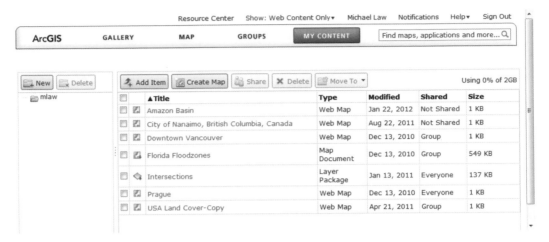

Notice that your Amazon Basin map is now listed, showing its document type, modified date, shared status, and file size. Next, you'll view more map details and options.

15 Click the Amazon Basin map title. This page is used to manage your saved map. It displays a thumbnail graphic of the map, its description, hyperlinks to the content used in the map (in this case, web services), shared status, tags, credits, file size, and spatial extent. In addition, this page also allows you, or anybody that you share your map with, to comment about the map. To further extend the reach of your map, you can share it with your social network through Twitter and Facebook.

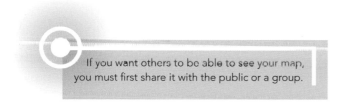

If you want others to be able to see your map, you must first share it with the public or a group.

Next, you'll view the edit page of this map to see its options.

16 **Above Description, click the Edit button.**

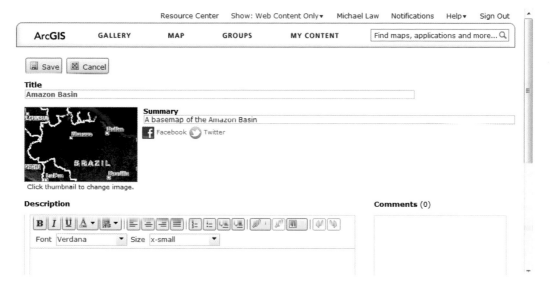

The edit page provides form fields where the thumbnail graphic, title, summary, description, access and use constraints, tags, credits, and spatial extent can be entered or modified. Next, you'll add some information to the Description box and save it.

17 **In the Description box, type** This map shows satellite imagery of the Amazon Basin.

18 At the top of the page, click the Save button. On the main navigation bar, click MY CONTENT to return to the content list. If you are continuing to exercise 5b, leave the browser window open; otherwise, you may close it now.

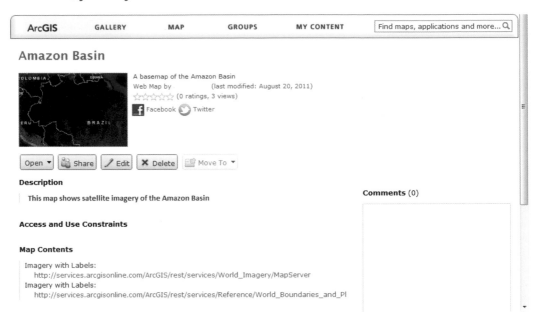

You have successfully created and edited an ArcGIS.com web map. You can edit the information in your map at any time, and the information will be updated in real time. This means that when it is shared with the public or a group, everybody will be able to see your edits immediately. This is very useful when working on a project with others in your community.

Now that you have learned how to create a map and edit its properties, it's time to learn about more features of ArcGIS.com, groups, and how to share maps, which you'll explore in exercise 5b.

Exercise 5b
Browsing through groups

ArcGIS.com groups allow you to collaborate with other members of the community. You can join groups to interact, discuss, and share content related to a project or communal activity. For example, you might be a humanitarian relief analyst working on a disaster map template of the affected area. To quickly find content, you might search for groups whose members are sharing geographic data or demographic data of residents in the disaster zone. You could join this group and share your map or collaborate as a group to create a standardized humanitarian relief map.

In this exercise, you'll learn how to browse through ArcGIS Online groups, add a basemap, add a data layer, and view its attributes.

1 Go to ArcGIS.com (if necessary, sign in with your Esri Global Account logon and password).

2 On the main navigation bar, click GROUPS.

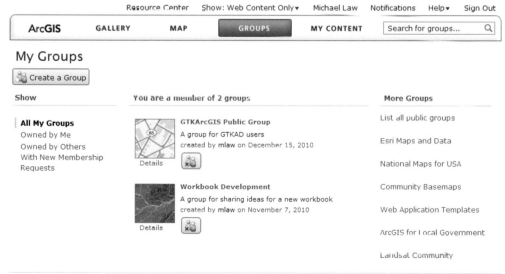

If you belong to any groups, they will be listed here. If you don't belong to any groups, you can browse through groups and explore the content they have to offer, which you'll do next.

3 On the right side of the page, click the List all public groups hyperlink.

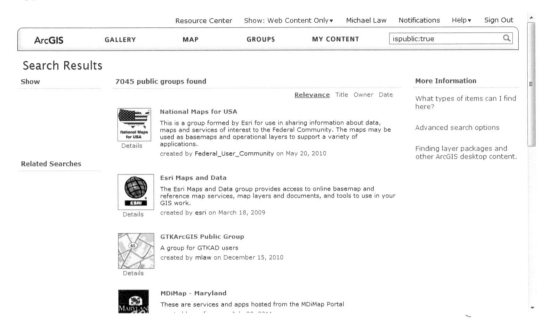

These are all the public groups currently on ArcGIS.com. The number of groups will continue to grow and change as more and more users join the community. For now, you'll browse through a group to see what kind of content it offers.

4 Locate the group called Esri Data & Maps and click the hyperlink (make sure it is the one with the ampersand in the title). If you do not see it, use the Search box and type the search term esri data maps. You may see different layers than what are shown in the graphic (on facing page).

This group is managed by Esri and contains lots of content available from Esri Data & Maps. Each map is accompanied by a thumbnail graphic and a brief description, as well as when it was last modified. The yellow stars represent the overall rating, as rated by the community. Underneath the thumbnail graphic are two links: Open and Details. Open allows you to view the map in the ArcGIS.com map viewer. Notice the drop-down arrow next to the Open link. It allows you to open the layer in the ArcGIS.com map viewer, ArcGIS Explorer Online, or ArcGIS 10 for Desktop and later versions. Details allows you to look at detailed information about the map. Next, you'll look at one of the maps in more detail.

5 At the top of the page, make sure Show: Web Content Only is selected. This filters maps that can be opened and viewed only on the web.

You can sort map lists by title, owner, rating, views, and date.

6 Find and click the map titled USA Topo Maps (it may be on the next page depending on how you sorted the list). Make sure you click the title or the Details hyperlink. (Clicking the thumbnail graphic will open the map.)

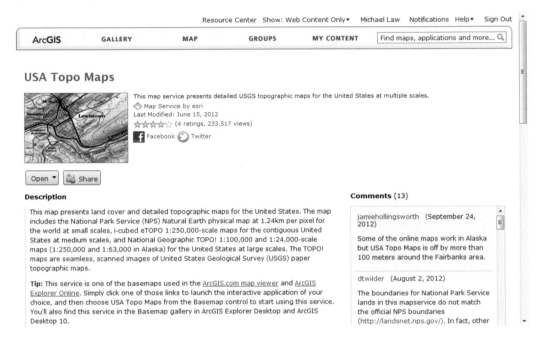

The map shows land-cover imagery for the world and detailed topographic maps for the United States. You'll open the map in the ArcGIS.com map viewer first and in ArcGIS for Desktop next.

7 Below the thumbnail graphic, click the Open button. On the drop-down menu, click Open in ArcGIS.com map viewer (alternatively, just click the thumbnail graphic). If you have an ArcGIS Online map already open, a warning will appear. Click Yes, open the map to continue.

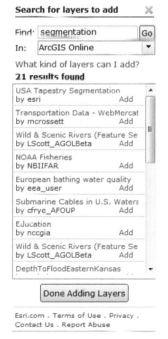

This opens the map viewer to a world map. To get a better sense of how groups can contribute to this map, you'll next add more layers to the map that are available in other groups.

8 **Click the Add button, and then click Search for Layers. In the Find box, type** segmentation **and click the Go button.** The panel is populated with maps shared by the community. Many of these maps may contain tags or keywords that match your search criteria.

9 In the list, locate the USA Tapestry Segmentation layer (published by Esri) and click the title. In the preview window that shows the summary, click Add to map. Then click the Done Adding Layers button.

The layer is added to the map but is hardly visible because of the current zoom scale. Next, you'll zoom in to properly view the layer.

10 Zoom in on the extent of the United States mainland (lower 48 states). Use the zoom control or double-click and pan the map so it is centered in the map viewer. You can also press and hold SHIFT and draw a rectangle around the United States mainland to zoom in or, alternatively, use the scroll wheel on your mouse.

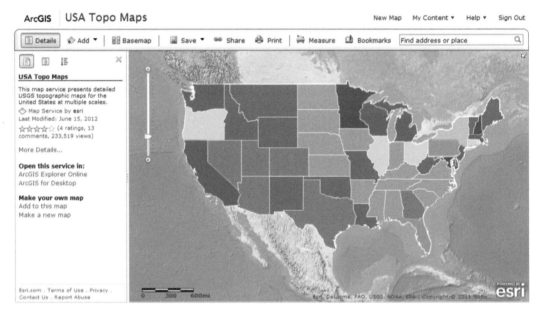

The USA Tapestry Segmentation layer is now clearly visible. Each state is represented by a tapestry category and a unique color. There are many layers to the tapestry data. Each layer is visible only within a scale range specified in the metadata. To see each level of detail, you'll need to zoom in closer, which you'll do next.

11 In the Find address or place box, type Washington state (it is the northwestern-most state, represented in blue on the map on the facing page) and click the Find Location button 🔍 or press ENTER. The map is redrawn and shows county boundaries this time. A callout bubble shows the state name, area type, and country (Washington State, United States).

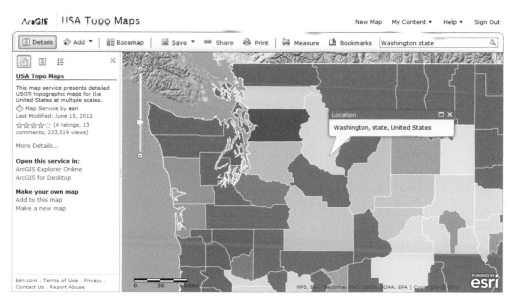

12 Locate the county colored purple and click it to identify it and see
a brief description. (You may have to close the callout bubble that
hides a portion of the county.) This window is populated with attribute
information that accompanies the layer. The county you identified is
Snohomish County, WA, and its tapestry is called Sophisticated Squires (6).
Tapestry segmentation attempts to describe socioeconomic and demographic
characteristics that characterize the area.

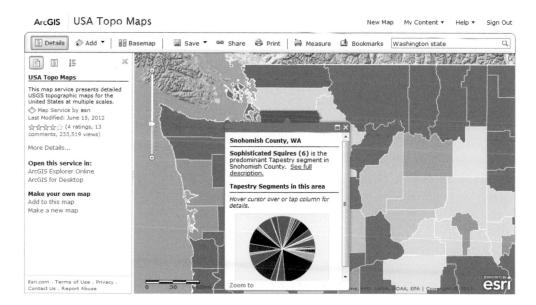

On your own

Learn more about Tapestry Segmentation by Esri at **http://www.esri.com/data/ esri_data/tapestry.html**. Find the predominant tapestry segment where you live.

13 Click the X to close the description window.

Next, you'll zoom in closer and explore the tapestry segmentation of Snohomish County by census tract.

14 Zoom in on Snohomish County until you see it labeled with the number 6.

15 At this point, zoom in again (see facing page) until census tracts are visible (they will be labeled with numbers as well). If you zoom in too close, you will see the detailed topographic layer.

16 In the details panel, click the Show Map Legend button. You may have to click the Details button first. The legend displays tapestry segmentation for the census tracts. The colors and numbers in the map correspond to those displayed in the legend.

On your own

Zoom to an even higher level of detail to show the census block groups. Click a block group first to show its brief description. Then click the See full description hyperlink to read more about that particular tapestry segmentation.

If you signed in, you will be able to save your map to your My Content page. You'll be able to access the map, share it with others, and add more layers to it. If you didn't sign in, you won't be able to save your map, although you can sign in now to save it.

You've learned how to browse through groups and to add data from groups to create a web map. In exercise 5c, you'll learn how to use ArcMap to retrieve and display data from ArcGIS Online.

Exercise 5c
Merging online and local layers

ArcMap has the ability to load basemaps directly from ArcGIS Online. This feature complements the online services available from Esri. The ArcGIS system now fully integrates all aspects of the web, allowing you to share information with others online. GIS professionals can compile and publish rich, interactive maps and geographic information straight from ArcMap. This exercise requires you to have an active connection to the Internet.

1 Open ArcMap. If you see a Getting Started dialog box, click My Templates and double-click Blank Map. Or if ArcMap is already open, click the New button and open a blank map.

2 On the menu bar, click File > Sign In.

3 On the ArcGIS Online Sign In dialog box, enter your user name and password credentials. If you don't have an account, click the Get new account hyperlink at the bottom of the dialog box. For this exercise, it is recommended that you sign in to your account, but it is not absolutely necessary.

You can still access maps and data if you don't sign in. You won't, however, be able to load your own maps and data, access groups, or share layers.

4 On the menu bar, click File > ArcGIS Online. This opens the ArcGIS Online window. Featured maps are listed in a tile view, each showing a thumbnail graphic, a short description, the publisher, and the date it was published. (Your screen will be different from the graphic shown on facing page.)

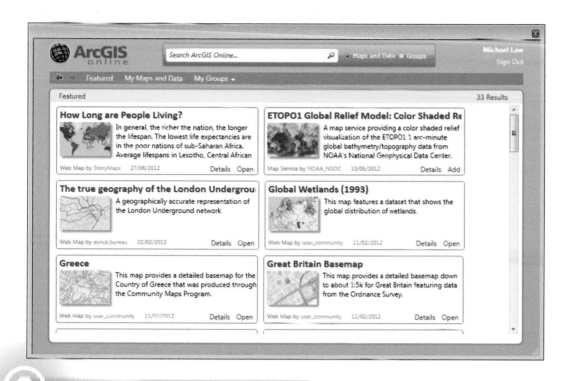

In ArcMap, you can also click the Add Data arrow, and then click Add Data from ArcGIS Online.

5️⃣ **In the Search box, type** Nanaimo **and press ENTER.**

6️⃣ **In the returned maps, locate the one titled Nanaimo, BC, Canada (a basemap authored by Esri Canada Limited) and click the Details link (see graphic on next page).** To the right is a description of the map that the publisher has provided. Notice that this map, and others like it, is interactive, allowing the community to rate it and leave comments. Through these shared user experiences, the community can see which maps are the most useful. This process also leads to improved maps and encourages publishers to update or publish new ones.

7 In the Description panel, click Open. If ArcMap asks you to save changes to your untitled map, click No. This loads the Nanaimo, British Columbia, Canada map into ArcMap. Depending on the size of your ArcMap window, your map may look different from the graphic shown.

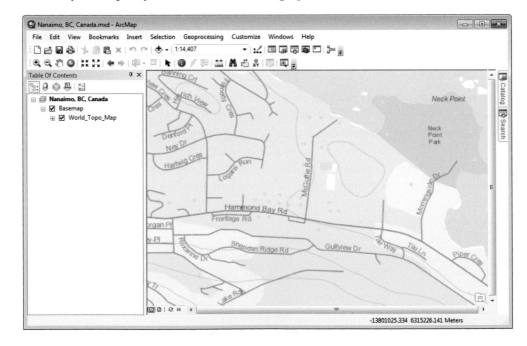

Only one basemap can be loaded at a time. You can, however, load as many data layers as you want.

8 **In the table of contents, expand the World_Topo_Map layer, and then expand Places Info.** Several layers from the country level to the city level are used to label features on the map.

On your own

View each layer's properties to see which map scale range is most appropriate for viewing.

Adding data to overlay an ArcGIS Online basemap is the same as adding data to any ArcMap map document. You'll next add several layers to the map to see how each one is displayed and then explore the different zoom levels of the map.

9 **On the Standard toolbar, click the Add Data button.**

10 **On the Add Data dialog box, browse to C:\EsriPress\GTKArcGIS\ Chapter05\Data. Click Bike Routes.lyr, and then click Add.** A line layer called Bike Routes is added to the map and displayed with preset symbols for each route. These routes indicate what kind of road the bike route shares. Arterials, commercial, and major roads are considered busy, and local roads, minor roads, and neighborhoods are considered not as busy.

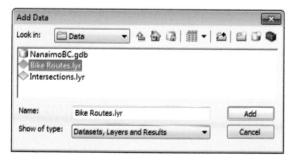

11 On the Tools toolbar, click the Fixed Zoom Out button until the map scale is approximately 1:200,000. If necessary, pan the map so the bike routes line layer is centered in the view. As you zoom out, notice how the different zoom levels affect how the map details look. This map and other maps in ArcGIS Online are loaded as tiles with multiple zoom levels at varying scales, also known as levels of detail. These tiles are cached by the system to load quickly when the level of detail is changed. The topographic layers that are drawn depend on the scale and effectively cover views from the city level to the world level. The basemap publisher defined the map scales at which certain details are visible—for example, streets and building footprints are visible at larger scales (greater than 1:200,000).

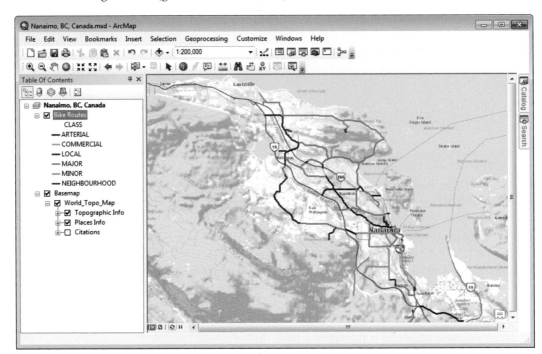

The bike routes are not affected by the zoom, because no rules for display based on map scale are set. Next, you'll add another layer to the map through ArcGIS Online.

12 On the menu bar, click File > ArcGIS Online. In the Search box, type intersections Nanaimo and press ENTER. In the search results, a layer package called Intersections for Nanaimo, BC, is available for download (it should be the only result listed). Click the Details link to see a short description of it.

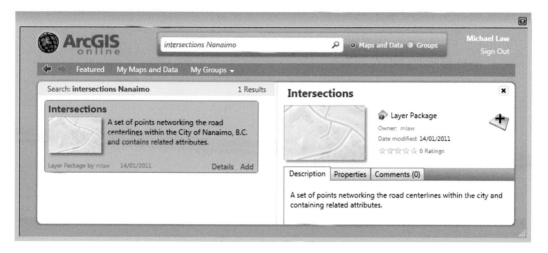

13 In the ArcGIS Online window, click the Add Data button.

What you added was a *layer package*. It encapsulates the data, symbology, and other properties of the layer as it's authored in ArcMap in one easy-to-share package. From the layer package, a point layer called Intersections is added to the map. It displays all the intersections that cross a bike route, including freeway, normal, and traffic light. This information could be useful for cyclists who want to know what type of intersections they might cross. Next, you'll zoom in closer to see the details.

14 Click the Zoom In tool and draw a rectangle around the downtown portion of Nanaimo, as indicated in the graphic.

Depending on how close you zoomed in, your map scale may be approximately 1:20,000. At this scale, it's easier to see how the basemap, intersections, and bike route layers come together.

15 Finally, zoom in until the map scale is at least 1:3,000. This zoom level has the greatest amount of detail included in the Nanaimo, British Columbia, ArcGIS Online map. The building footprints show slightly more detail, the lot parcel boundaries and numbers are visible, and contour lines can be distinguished.

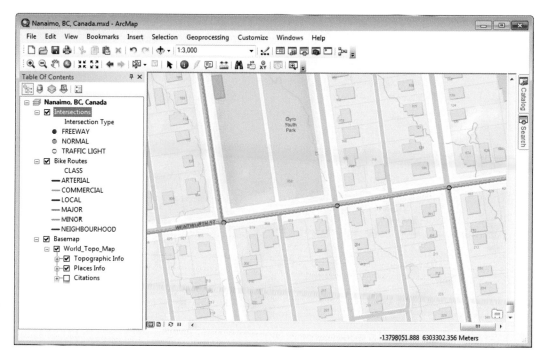

16 If you want to save your work, on the menu bar, click File > Save As. Name your map document my_ex05c.mxd and save to your MyData folder for chapter 5. As long as you have an Internet connection, the Nanaimo basemap will be recalled the next time you open the map, because it is stored in ArcGIS Online.

17 If you are continuing to the next chapter, leave ArcMap open. Otherwise, exit the application and click No if prompted to save your changes.

On your own

Create a group and become a group owner so you can decide whether you want it to be a public group, a nonmember public group, or a private group. These groups can be managed differently, depending on the task to be achieved. A public group can accept member requests and is open to all who want to participate. A nonmember public group shares content but doesn't let other users join or share. A private group is inaccessible to the public and you, as the owner, invite people to join. For example, you could work for a marketing firm and specialize in business analysis and create a private group for your department. The department is working with proprietary data to map ZIP Codes and demographic data in a new city and markets the information to local businesses but doesn't want public access. Group members can exchange their data layers with each other privately, but nobody outside the group can access it.

Do the following steps:

1. Explore ArcGIS Online and find out how to create a group. Then decide whether you want to share it publicly or only with members who join.

2. Learn how to invite your peers to the group to share content.

3. Learn about the different types of content that can be stored in ArcGIS Online and how this content can be managed, stored, and shared.

About sharing content

You are encouraged to share maps and applications if they are useful to others and viewable in a web browser, mobile device, or ArcGIS application. However, you should not share any geographic information that is confidential, copyrighted, violates laws, or would cause a security risk if somebody were to gain access to it. When working with an organization, you may find it appropriate to share some content, but other content may be proprietary or business-sensitive data and should not be shared with people outside the organization. In addition, map creation and data collection is often a work in progress and is not necessarily complete or run through final validation. However, once the content is finalized, feel free to share the data with the ArcGIS Online community and the public. For more information on best practices for sharing content, visit **http://help.arcgis.com/en/arcgisonline/ help/index.html**.

Summary

By now, you should have a good understanding of the fundamentals of using ArcGIS Online. You have learned how to register for ArcGIS Online, navigate the online mapping interface, and explore and open basemaps. If you explored ArcGIS Online groups, you were made aware of the potential of how maps can be shared with users both publicly and privately.

Review the learning objectives for this chapter and see if you are comfortable working with ArcGIS Online. If not, you may want to review some sections of this chapter as a reminder. Remember that ArcGIS is a large, integrated system of Esri products—including ArcGIS Online—and there is a great deal of information available online, including blogs, newsletters, and product demonstrations. If you're interested in learning more, go to http://www.esri. com to learn about the complementary aspects of the ArcGIS system.

Terms

basemap

group

operational layer

mashup

layer package

Displaying and presenting

data

Chapter 6
Working with coordinate systems and projections

Examining coordinate systems
Projecting data
Defining a map projection

Learning objectives

- *Become familiar with geographic coordinate systems and projected coordinate systems*
- *Explore the details of coordinate systems*
- *Define a coordinate system*
- *Identify projection parameters*
- *Project data on the fly*
- *Align datasets with different coordinate systems*

GIS SYSTEMS ARE HIGHLY DEPENDENT ON USING AND DEFINING PROPER **coordinate systems** and projections. Using the correct coordinate system ensures geographic accuracy, whereas using an improper coordinate system will result in unusable or error-prone mapping. Learning about coordinate systems and map projections can be challenging for beginners, but practice and experience will provide a solid foundation for practical use in ArcGIS applications.

A coordinate system is a reference framework that uses points, lines, and surfaces to determine positions in two- or three-dimensional space. In simpler terms, it helps you to find your way around a map. Defining a coordinate system is essential to every GIS mapping project, regardless of subject, projection, or scale. All coordinate systems are based on coordinate pairs that belong to a geographic grid, the location reference system for spatial features on the earth's surface. On a globe, locations on the earth's surface are defined in reference to lines of parallels and meridians, more commonly known as latitude and longitude, respectively. Latitude lines run parallel to the equator (at 0 degrees latitude) and measure how far north or south you are from the equator. Longitude lines run from pole to pole and measure how far east or west you are from the prime meridian (the meridian of 0 degrees longitude that passes through Greenwich, England).

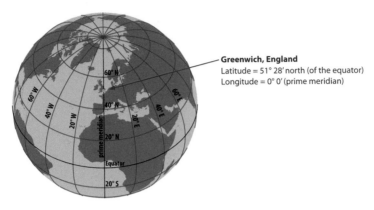

Lines of latitude and longitude. Data from Esri Data & Maps 2010, courtesy of ArcWorld Supplement and Esri.

Latitude and longitude are measurements of angles, not of distance. Latitude is the angle between the equator and the parallel on which the point you are locating lies, as measured from the center of the earth. Longitude is the angle between the prime meridian and the meridian on which the point you are locating lies, as measured from the center of the earth.

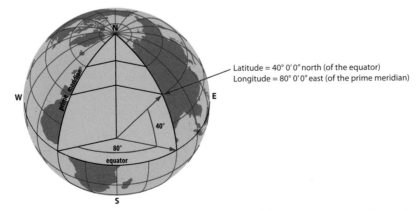

Angles are used instead of distance, because as meridians converge toward the poles, the distance between them decreases. A degree of longitude is about 69 miles near the equator; about 49 miles at the 45th parallel, and zero at the North Pole. In other words, distance varies between meridians, and only the angle between them remains constant. Data from Esri Data & Maps 2010, courtesy of ArcWorld Supplement and Esri.

The earth is divided into 360 degrees around the equator. One degree equals 60 minutes, and one minute equals 60 seconds. Starting from the prime meridian, longitude values go to +180° eastward and to –180° westward. (The +180° and –180° meridians are the same.) Starting from the equator, latitude values go to +90° at the North Pole and to –90° at the South Pole. Because latitude and longitude are angles, their values are commonly expressed in *degrees-minutes-seconds* (DMS) or sometimes as *decimal degrees* (DD). For example, in DMS, Tokyo is located at 35° 41' 6" N (latitude) and 139° 45' 5" E (longitude). In DD, those values would convert to 35.685° and 139.7513889°, respectively.

Latitude and longitude are the basis of a *geographic coordinate system* (GCS), a system that defines locations on the curved surface of the earth. Because different estimates have been made of the earth's shape and size, there are a number of different geographic coordinate systems in use. Although these systems are similar, the precise latitude-longitude coordinates assigned to locations vary from one system to the next.

Coordinate systems and projections are based on a defined spheroid and datum. A *spheroid* (also known as an ellipsoid) is a mathematical model that approximates the irregular shape of the earth using two axes. Because the earth bulges at the equator as a result of the earth's rotation, the model uses a major axis dissecting the equator and a minor axis dissecting the poles. The relationship between the shape of the earth and the shape of a spheroid is described using a *datum*, which essentially describes how a model fits into reality.

Left, some geographic coordinate systems are based on the assumption that the earth is a sphere. *Right,* most geographic coordinate systems are based on the assumption that the earth is a spheroid (exaggerated). This is a more accurate model because the earth bulges slightly at the equator and is flattened at the poles. Data from Esri Data & Maps 2010, courtesy of ArcWorld Supplement and Esri.

Hundreds of datums are in use, and each one uses different methods to measure the earth. This poses a challenge to map makers, because each datum measures and accounts for different features, such as flatness ratio, circumference, or topographical differences. The goal of the datum is to provide the optimal fit between the **geoid**, a measure of the earth's radius, and the chosen spheroid used for a map.

Some commonly used datums in North America include NAD27 (North American Datum of 1927), WGS84 (World Geodetic System of 1984), and NAD83. NAD27 is based on the Clarke 1866 datum, a ground-measured spheroid. Technological advancements and satellites are now used to determine more precise spheroids and measurements. Many GIS data providers and web map producers are moving toward WGS84 as a common standard. However, a considerable amount of existing data uses NAD27 and NAD83. So, it is important to be aware of the potential difference in feature positions between datums.

A **map projection** is used to represent the earth on a flat surface. In more technical terms, mathematical formulas are applied to a geographic coordinate system defined by a spheroid and datum to convert latitude (north–south orientation along the y-axis) and longitude (east–west orientation along the x-axis) values to a plane surface representing a geographic grid. The result is a two-dimensional (2D) surface that is composed of a grid of intersecting lines. One set of lines runs parallel to a horizontal x-axis, and the other set runs parallel to a vertical y-axis. The coordinates of any location or point are expressed as a distance value along the x-axis (longitude) and a distance value along the y-axis (latitude)—from the intersection of the axes. Therefore, longitude and latitude values are expressed as x and y, respectively.

 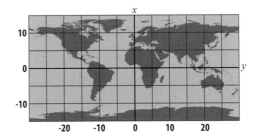

Left, a geographic grid with longitude values along the x-axis and latitude values along the y-axis. *Right*, on a flat map, coordinate values can represent any kind of linear units. Data from Esri Data & Maps 2010, courtesy of ArcWorld Supplement and Esri.

A **projected coordinate system** (PCS) is a system that defines locations on a flat map based on x,y coordinates. The x,y coordinates of any given point depend on the map projection, the units of measure (meters, feet, or whatever), and on which point the map is centered. If Cairo is made the center of the map, for instance, its x,y coordinates would be 0,0. Thus, the number of possible projected coordinate systems is infinite.

Keep in mind that not all data will be or must be projected, but it must have a geographic coordinate system. To understand how certain map projections work, you'll also need to know how geometric objects are used to construct a flat map. Imagine a piece of paper being shaped to fit against a lighted globe (see graphic) in three configurations: a cylinder, a cone, and a flat plane. A cylindrical projection is possible when a cylinder is used, a conic projection when a cone is used, and an azimuthal projection when a flat plane is used.

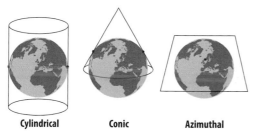

Cylindrical   Conic   Azimuthal

• Blue dot represents point of tangent

Common geometric shapes used to construct a map projection. Data from Esri Data & Maps 2010, courtesy of ArcWorld Supplement and Esri.

The projections are possible when the geometric object makes contact with the globe at a line of tangency or at two lines of tangency, also called a secant. These lines of tangency are called *standard parallels* if they follow a parallel or *standard meridians* if they follow a meridian. Depending on the case of tangency and orientation of the geometric object (see graphic), different projections can be produced, including secant conic, polar azimuthal, transverse cylindrical, and many more. The distortion of each projection varies, but as a general rule, a projection is the least distorted along the tangent and grows more distorted leading away from it. Secant projections allow for greater accuracy in measurements, particularly on large-scale maps where both standard parallels run within the constraints of the map. Secant projections, however, are not as suitable for small-scale maps.

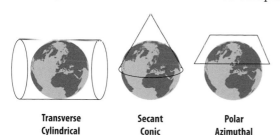

**Transverse   Secant   Polar
Cylindrical   Conic   Azimuthal**

Variations of common geometric shapes create different map projections. Data from Esri Data & Maps 2010, courtesy of ArcWorld Supplement and Esri.

The transformation of the earth's surface to a flat surface is inherently not perfect and causes distortion to shape, area, distance, or direction, depending on the projection. To counter distortion, map projections can be grouped into four types of preserved properties: conformal, equal area, equidistant, and azimuthal. Conformal projections preserve shape. Equal-area projections display an area at its proper relative size. Equidistant projections preserve scale, and thus distance. Azimuthal, or true-direction projections, maintain accurate directions.

Choosing the correct map projection is an important consideration. It allows you to control the type of distortion in a map. In many cases, mapping at a large scale requires a different map projection than mapping at a small scale. For an area the size of a small country such as Switzerland, for example, the use of an appropriate projection will result in little to no distortion. For an area the size of a large continent, or even the world, however, the use of a compromise or specialized map projection can minimize distortion. In the end, distortion is inevitable, but using good judgment and an appropriate map projection will minimize distortion of your map.

The figures shown here exhibit several examples of popular map projections.

The Mercator projection is a cylindrical projection that preserves shape but distorts area. Any straight line drawn at this projection represents an actual compass bearing. Data from Esri Data & Maps 2010, courtesy of ArcWorld Supplement and Esri.

The World Winkel Tripel projection is used for world maps as a compromise that seeks to minimize shape and area distortion. It has been used by the National Geographic Society since 1998 for general and thematic world maps. Data from Esri Data & Maps 2010, courtesy of ArcWorld Supplement and Esri.

The Goode Homolosine projection is a combination of the Mollweide and Sinusoidal projections. There is no distortion along the central meridians of the interrupted lobes where areas are represented accurately. Data from Esri Data & Maps 2010, courtesy of ArcWorld Supplement and Esri.

World Bonne is an equal-area projection that has true scale along the central meridian and all parallels. Shapes are distorted as you move away from these lines. Data from Esri Data & Maps 2010, courtesy of ArcWorld Supplement and Esri.

For more information about coordinate systems and map projections, in ArcGIS for Desktop Help, on the Contents tab, navigate to Guide books > Map projections > What are map projections?

Projecting data in ArcMap

ArcMap is able to handle an infinite number of map projections—both predefined and customized—to give you the most flexibility in working with your datasets. Your data will be either unprojected or projected. Unprojected data means that it is simply stored in a defined geographic coordinate system, such as WGS84, with latitude and longitude coordinate values shown in decimal degrees, or *map units*. You'll find out how to look up the coordinate system of a dataset in the exercises in this chapter.

Projected data means that a map projection has been applied to your dataset. The definition of a projected coordinate system includes the projection it's based on, the measurement units, and other details. Each projected coordinate system is also associated with a particular geographic coordinate system. Typically, the map units are converted from latitude-longitude to x,y coordinates in meters or feet. Additional information that can be found with a defined projection includes the central meridian, *false easting* and *false northing*, standard parallel, and central parallel.

Common projections used in ArcMap are the state plane and universal transverse Mercator (UTM) systems. The state plane coordinate system (SPCS) divides each state into manageable regions for large-scale mapping and bases it on three projections: Lambert Conformal Conic, Transverse Mercator, and Oblique Mercator. The UTM system divides the world into 60 zones—each one is 6 degrees and designated by a number and a letter N (northern hemisphere) or a letter S (southern hemisphere). The UTM system works best for maps that include small areas that are entirely encompassed by a zone. It attempts to minimize distortion, is useful for mapping areas that extend in a north–south direction, and provides a standardized method for projecting data. Depending on the geographic location of your data, a suitable map projection is most likely available in ArcMap.

You can locate the coordinate system of a dataset in ArcCatalog, on the Description tab (to view its metadata). You can also find it on the feature Properties dialog box (right-click a feature class and click Properties), on the XY Coordinate System tab. In ArcMap, you can find it on the Layer Properties dialog box, on the Source tab. Alternatively, find coordinate system information for a data frame in ArcMap by clicking the Coordinate System tab on the Data Frame Properties dialog box.

This data frame has an Albers Equal Area Conic projected coordinate system, which is based on the NAD83 datum.

The first layer added to a data frame in ArcMap sets a precedent for the coordinate system used in that data frame, meaning that additional layers will have to be projected to it. When the first layer is added, it is displayed according to the coordinate values of its features—geographic or projected. When a second layer is added, ArcMap reads its coordinate system and determines whether it matches the coordinate system of the data frame, based on the first layer. If it matches, there's no problem. If it doesn't, ArcMap attempts to automatically resolve the conflict with a map projection. Using a mathematical operation that recalculates geographic coordinates to projected coordinates, or vice versa, ArcMap determines which equation (forward or backward) is suitable for the projection. So if the second layer you add has a different coordinate system from the first, ArcMap recalculates the layer's coordinates to match the data frame. It is important to note that only the display position of the data is being changed, and not the data itself. This process is called *on-the-fly projection*.

On-the-fly projection doesn't permanently alter the feature coordinates of the dataset on disk or the dataset's coordinate system information. On-the-fly projection affects only a single data frame. It works best when layers are based on the same geographic coordinate system. If any layer added to the map has a different geographic coordinate system from the data frame, you'll get a message like the example in the figure.

ArcGIS warning for conflicting geographic coordinate systems.

The message tells you that ArcMap can't convert one geographic coordinate system to another without additional parameters that you must set. It can display the layer, but the spatial alignment may be slightly off. This difference is usually too small to notice at the scale of a world map, or even a continental map. But if you are making a map that demands highly accurate feature positioning, you'll need to perform a geographic transformation (or a *datum transformation*).

Projecting data on the fly is very useful and is an essential part of ArcMap. However, it may slow the redrawing of the map on screen because of the complex projection equations that are required. It is also not permanent, which may satisfy certain uses, but it can standardize and simplify workflows to have data in the appropriate projection from the start. ArcToolbox has all the tools required to project data from one coordinate system to another, and then save on disk.

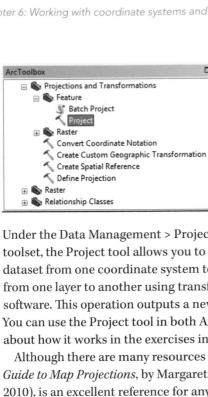

Projections and Transformations tools in ArcToolbox.

Under the Data Management > Projections and Transformations > Feature toolset, the Project tool allows you to transform the map projection of a dataset from one coordinate system to another. It converts the x,y coordinates from one layer to another using transformation instructions defined in the software. This operation outputs a new layer, retaining the original dataset. You can use the Project tool in both ArcCatalog and ArcMap. You'll learn more about how it works in the exercises in this chapter.

Although there are many resources available, *Lining Up Data in ArcGIS: A Guide to Map Projections*, by Margaret M. Maher (Redlands, CA: Esri Press 2010), is an excellent reference for any GIS user looking to learn techniques to solve data misalignment, dive deeper into data projections, and create custom projections to align data.

The Define Projection tool overwrites the coordinate system information (map projection and datum) stored within a dataset. It should be used only on datasets that have an unknown or incorrect coordinate system defined. Also, you must know the correct coordinate system of the dataset to successfully define a projection. This tool should not be confused with the Project tool.

Data

This chapter uses the following datasets:

- Canada_Mexico.shp—polygons of areas that make up Canada and Mexico *(Source: ArcWorld Supplement)*
- Cities.shp—points of cities across the United States *(Source: US Census)*
- ColoradoCounties.shp—polygons of areas that consist of counties in the state of Colorado *(Source: Esri, derived from Tele Atlas, US Census)*
- Lakes.shp—polygons of lakes in North America *(Source: ArcWorld)*
- Latlong.shp—lines of latitude and longitude *(Source: Esri)*
- USstates.shp—polygons of areas that consist of states in the United States of America *(Source: ArcUSA, US Census, Esri)*

Exercise 6a
Examining coordinate systems

The most recent US Census Bureau statistics have been released, and for this exercise, you will be creating a map of the United States that shows the change in population per square mile between 2000 and 2010. First, you will use ArcCatalog to explore where the coordinate information is found and to gain some familiarity with the properties of map projections.

1 Start ArcCatalog. In the Catalog tree, double-click Folder Connections and browse to the contents of your C:\EsriPress \GTKArcGIS\Chapter06\Data folder.

2 **Click USstates.lyr and click the Preview tab.** The US states layer is symbolized using a color ramp that shows the change in population (per square mile) in the last decade—light green for smaller increases and dark green for larger increases. You may notice that the appearance of the United States is distorted; its shape and area may look strange compared to other maps you may have seen in an atlas or in a published map. This is because the data is being displayed according to its geographic coordinate system and not a common US map projection such as Lambert Conformal Conic. Notice that the northern boundary of the United States, also known as the 49th parallel, is displayed as a straight line. Alaska has an east–west stretch that is not proportional, and Texas looks much smaller than its actual size.

If you don't see the file extension, on the menu bar, click Customize > ArcCatalog Options and clear the Hide file extensions check box.

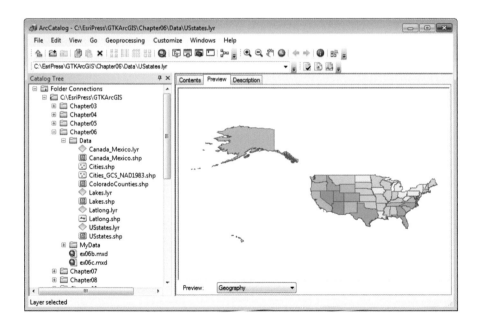

3 Click the Preview arrow at the lower-left corner of the window and click Table. Scroll to the right. The table has 2000 and 2010 population figures for each state, including population per square mile (POP00_SQMI and POP10_SQMI) and the change in population per square mile (in percentages) represented by the POP_CHANGE attribute. Every state experienced population growth during the past decade. The layer is classified and symbolized based on the POP_CHANGE attribute.

| Contents | Preview | Description |

POP2000	POP2010	POP00_SQMI	POP10_SQMI	SQMI	POP_CHANGE	
1211537	1309580	110.8	119.8	10931	8.1	
5894121	6756150	82.7	94.8	71297	14.6	
902195	983932	6.1	6.7	147038	9.1	
1274923	1338645	32.2	33.8	39555	5	
642200	662194	9.1	9.4	70698	3.1	
754844	827263	9.8	10.7	77115	9.6	
493782	548154	5	5.6	97813	11	
5363675	5741617	89.7	96	59822	7	
1293953	1581697	15.5	18.9	83570	22.2	
608827	626078	63.3	65.1	9614	2.8	
4919479	5334772	56.7	61.4	86835	8.4	
3421399	3865839	34.8	39.3	98381	13	
1235786	1329915	132.2	142.2	9350	7.6	
2926324	3057995	52	54.3	56272	4.5	
6349097	6555736	622.1	642.3	10200	3.3	
1711263	1822473	22.1	23.6	77352	6.5	

|◄ ◄   1 ► ►| (of 51)

Preview   Table

How was the POP_CHANGE attribute calculated?

(Hint: It involves two attributes and a simple math formula.)

Next, you'll confirm that the US states shapefile has a geographic coordinate system.

4 **In the Catalog tree, right-click USstates.shp and click Properties. On the Shapefile Properties dialog box, click the XY Coordinate System tab.** The Shapefile Properties dialog box shows that the current coordinate system is called GCS_North_American_1983. This is a common geographic coordinate system for spatial data covering the United States and is used by many data providers. The datum it uses is called D_North_American_1983 (NAD 1983), and the prime meridian is defined at 0 degrees longitude. It is from this dialog box that the parameters of a coordinate system can be modified. However, you will not be modifying this particular layer.

What are the angular units of this coordinate system?

What is listed as the prime meridian?

There are four other shapefiles related to this exercise in the Data folder: Cities, Latlong, Canada_Mexico, and Lakes. The latter three shapefiles also have a corresponding layer file that contains a symbolization scheme and a path to the source data. You will view the properties of the first three layers (if you want, you could explore the fourth, but only the first three are necessary to show the differences between them).

5 Click Cancel to close the USstates Shapefile Properties dialog box. In the Catalog tree, right-click Cities.shp and click Properties. The Shapefile Properties dialog box shows that this layer is a geographic coordinate system called GCS_WGS_1984 (WGS 1984). It is another popular geographic coordinate system that is in wide use. Since it is different from the coordinate system of the US states layer, it may not align exactly. ArcMap, however, is designed to make it work.

What spheroid does this layer use?

Next, you will see how ArcMap reacts to the coordinate system inconsistencies that you will find in exercise 6b, where you will also use ArcToolbox tools to solve the problem.

6 Click Cancel to close the Cities Shapefile Properties dialog box. In the Catalog tree, right-click Latlong.shp and click Properties.

The Shapefile Properties dialog box shows that this layer is in the GCS_North_American_1983 coordinate system, like the US states layer. The data should align properly with the US states layer without any coordinate system transformation or projection on the fly.

7 Click Cancel to close the Latlong Shapefile Properties dialog box. In the Catalog tree, click Canada_Mexico.shp and preview its geography. Then right-click Canada_Mexico.shp and click Properties.

Although a preview of Canada and Mexico can be seen, the Shapefile Properties dialog box shows that this layer is not in any geographic coordinate system and simply states that it is unknown (see facing page). How is this possible? ArcCatalog still shows a preview of the actual data and reads the known feature coordinate values, even though a geographic coordinate system is not defined. If this file is added to ArcMap, it will be drawn, but it will not be known which geographic coordinate system it is, and thus it cannot be projected. You will fix this situation in ArcCatalog.

6a

In ArcCatalog, the coordinate system information can be viewed in the data file properties, but not in the related layer file (lyr), if there is one.

In the Preview display, notice that the southern border of Canada has a long straight line similar to the northern border of the US states layer. Although it may not be the case in every situation, you've now been provided with a hint that the Canada and Mexico shapefile shares the same geographic coordinate system as the US states shapefile.

In many cases, trying to estimate the geographic coordinate system of a dataset that has none is based on trial and error, using best estimates. If the data is in the United States and shows a geographic extent in which the coordinates to the left of the decimal point are six, seven, or eight digits, the data is probably projected to the state plane or UTM coordinate system. You can view the feature extents in ArcCatalog (Shapefile Properties > Feature Extent tab) or in ArcMap (Layer Properties > Source tab). It is also possible to import the coordinate system of another feature layer that contains a defined state plane or UTM boundary and apply it to your shapefile or feature class with an unknown coordinate system to confirm whether it aligns properly.

8 On the Canada and Mexico Shapefile Properties dialog box, on the XY Coordinate System tab, click the Add Coordinate System button and click Import. Browse to \Chapter06\Data, click USstates.shp, and click Add.

The Canada_Mexico shapefile now assumes the coordinate system information of the US states layer—GCS_North_American_1983. This means that when the shapefile is added to ArcMap, it will be recognized and will have no issues of alignment or need further geographic projection.

9 Click OK to accept the change.

10 To close ArcCatalog, on the menu bar, click File > Exit.

On your own

It is common to find data in all sorts of coordinate systems and projections, or even data that is missing coordinate system information altogether. Learn more about how a shapefile or geodatabase stores coordinate information. There are many good resources online, including the ArcGIS Resource Center at **http://resources.arcgis.com**.

Now that you have a better understanding of the data and coordinate systems involved in this chapter, you will be able to make proper decisions on how to handle them in exercise 6b.

Exercise 6b
Projecting data

The map you will be creating in this exercise will consist of a shapefile of the United States displayed in three data frames, with a different projected coordinate system for each data frame (one for the mainland states, one for Alaska, and one for Hawaii). The data frames will also include other supporting layers: one that shows latitude and longitude lines and another that shows Canada and Mexico to complete the look of the map.

How is it possible to use one dataset in three data frames? Once data layers are added to ArcMap, they can be moved and copied between data frames. As you learned in chapter 3, multiple layers (even in different data frames) can be created from a single source dataset. This is more efficient than adding separate data layers with separate projected coordinate systems to the same map and is more practical than showing all 50 states in the same data frame. If that were the case, there would be many files to manage, and you would have to zoom to a map scale that would make it difficult to see the detail of all the states together.

In the exercise, you will be using three versions of the Albers equal-area conic projection—one for each data frame. This is a popular projection of the United States because it maintains area and keeps distortion of the shapes of the states to a minimum.

1 **Start ArcMap and open the ex06b map document from your \GTKArcGIS\Chapter06 folder.** ArcMap opens a layout that has three empty data frames: Alaska, Mainland States, and Hawaii. Each data frame is identified by a label at its center. An active data frame is indicated in the table of contents by a data frame name in boldface and on the layout page by a dashed outline. A small legend that shows the percentage increase of population change is visible in the lower-right corner of the page.

You'll add the US states layer to each data frame and learn how map projections apply to each one. You will also learn how to add other supporting layers.

2 Add USstates.lyr to the Mainland States data frame. (Use the Add Data button on the Standard toolbar or drag the layer file from the Catalog window.)

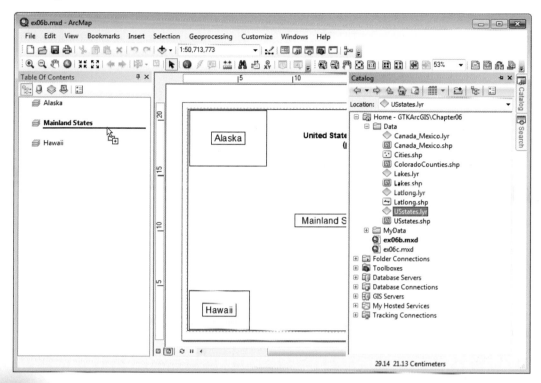

When using the Catalog window, you may need to refresh the map display after dragging over data. To do so, click the Refresh button ⟳ at the lower-left corner of the display area.

The US states layer is added to the Mainland States data frame. It is in the GCS_North_American_1983 geographic coordinate system that you previewed in ArcCatalog. By default, ArcMap includes the entire spatial extent of the layer. Next, you'll add a few more layers before projecting the data and adjusting the map scale.

3 Add the Canada_Mexico.lyr to the Mainland States data frame and position it below the US states layer (see next page). In exercise 6a, you imported a coordinate system for Canada_Mexico.shp. Because it is the same coordinate system as the US states layer, there are no alignment or accuracy issues.

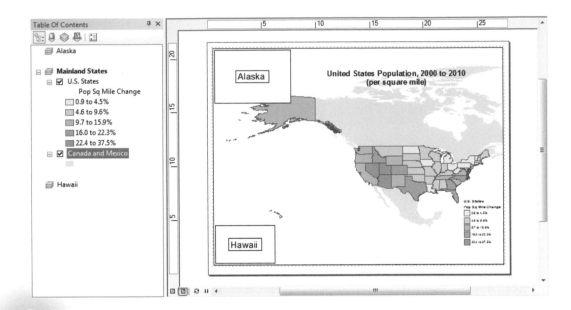

If you didn't import a coordinate system for Canada_Mexico.shp, an Unknown Spatial Reference dialog box will appear because there is no coordinate system information. (If this is the case, return to step 8 of exercise 6a.)

The Canada and Mexico layer is styled in a light-gray fill color to act as a subtle background to the thematic US states layer. Next, you'll add two more layers to the data frame.

4 **Add Lakes.lyr to the Mainland States data frame. It should be placed above all the layers in the table of contents.** (The Lakes layer is symbolized using a white color on purpose and is used to add a visual aesthetic to the map.)

5 **Add Latlong.lyr and drag it below all the other layers in the Mainland States data frame in the table of contents.** Because these two layers are also in the same coordinate system as the US states layer, everything lines up as it should. The Latlong layer provides reference for the geographic coordinates displayed on screen. It represents the x,y coordinate system from −180 to +180 degrees longitude and −90 to +90 degrees latitude, and the lines are perpendicular to each other at 90 degrees. However, viewing the world in this type of coordinate system shows significant distortion, especially in shape and area.

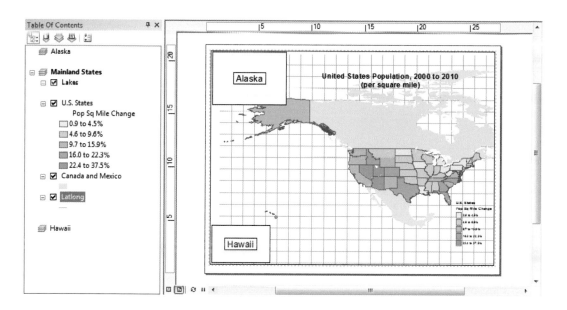

Use the Measure tool to measure the distance between two lines of latitude or two lines of longitude. What is the difference in degrees?

6 On the Standard toolbar, click the Add Data button and add Cities. shp to the Mainland States data frame. A warning appears when the layer is added. The message tells you that this layer has a different geographic coordinate system from the data frame.

Geographic Coordinate Systems Warning

The following data sources use a geographic coordinate system that is different from the one used by the data frame you are adding the data into:

Data Source	Geographic Coordinate System
Cities	GCS_WGS_1984

Alignment and accuracy problems may arise unless there is a correct transformation between geographic coordinate systems.

You can use this button to specify or modify the transformation(s) used by this data frame: [Transformations...]

The Transformations dialog can also be accessed from the Data Frame Properties dialog's Coordinate Systems tab after you have added the data.

☐ Don't warn me again in this session
☐ Don't warn me again ever   [Close]

If you don't see this warning, it's likely that you selected the Don't warn me again ever check box.

You previewed this layer in ArcCatalog and know that it is in the GCS_
WGS_1984 geographic coordinate system. When you added the US states
layer to the Mainland States data frame, the data frame was automatically set
to GCS_North_American_1983. Even though the Cities layer has a different
coordinate system, ArcMap can still display it, but the spatial alignment may
be slightly off. At this map scale, it may not even be noticeable. However, it may
cause further problems when it is projected.

7 **Read the messages in the warning dialog box, and then click Close.**

8 **In the table of contents, double-click the Cities layer name. On the
Layer Properties dialog box, click the Source tab and look at the
Data Source panel.**

This is a quick verification that the Cities layer is indeed in the GCS_WGS_1984
geographic coordinate system. You'll use an ArcToolbox tool to transform it
from its current coordinate system to the same coordinate system as the US
states layer, GCS_North_American_1983.

9 **Click Cancel to close the Layer Properties dialog box.**

10 **On the Standard toolbar, click the ArcToolbox button [icon] .**

11 In the ArcToolbox window, expand the Data Management Tools > Projections and Transformations > Feature toolset.

12 Double-click the Project tool to open its dialog box.

13 Click the Input Dataset or Feature Class arrow and click Cities. Click the Browse button next to the Output Dataset or Feature Class box and browse to your \GTKArcGIS\Chapter06\MyData folder. In the Name box, type Cities_GCS_NAD1983.shp, and then click Save.

The Input Dataset or Feature Class list contains layers that can be projected. Alternatively, the Browse button allows you to browse to the data source.

6b

7

8

9

10

14 Click the Properties button next to the Output Coordinate System box.

On the Spatial Reference Properties dialog box, you can choose from coordinate systems defined by ArcGIS, import a coordinate system from another dataset, or create an entirely new one (the latter two options are found using the Add Coordinate System button). You need to project the Cities layer to GCS_North_American_1983, so next you'll browse for an existing definition.

15 In the browse panel, click the plus sign (+) symbol to expand the Geographic Coordinate Systems folder, then expand the North America folder, and then scroll down and click NAD 1983. Click OK. The Current coordinate system panel on the Spatial Reference Properties dialog box is updated with the new coordinate system details.

You can also choose the coordinate system from your Favorites list.

16 Click OK to close the Spatial Reference Properties dialog box.
ArcMap automatically populates the Project dialog box with an appropriate
transformation using the coordinate system you selected. However, you might
need to use your knowledge of the data (for example, are you focusing on the
48 contiguous states?) to choose an appropriate transformation code. The
geographic transformation option, in this case, provides ArcMap with the
correct transformation to convert WGS84 to NAD83.

The Geographic Transformation box is optional. The Project tool can project data
from PCS to PCS (for example, state plane to UTM), transform GCS to GCS (NAD83
to NAD27), or PCS to GCS, and vice versa. In the event you are projecting data from
PCS to PCS, and they share the same datum, you are not transforming the data and
therefore do not need a transformation code.

17 Click OK. The Project tool runs, and when it is finished, a pop-up notification
briefly appears in the notification area, at the far right of the taskbar, indicating
the process was successful.

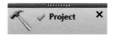

ArcMap adds the transformed Cities layer to the map. It may not look much
different, but it is in a new coordinate system. It is important to remember
that the new Cities layer now shares the same coordinate system as the other

layers, and changing the coordinate system of the data frame will not yield the
warning messages you saw before.

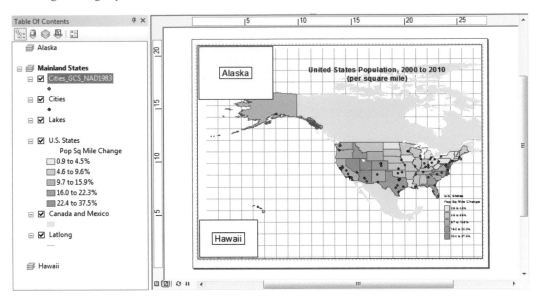

18 Close ArcToolbox or move it out of the way. In the table of contents,
right-click Cities (not Cities_GCS_NAD1983) and click Remove.

This layer is not needed anymore because you now have a cities layer with
the appropriate coordinate system to use. Next, you'll change the coordinate
system of the data frame to project layers on the fly.

19 In the table of contents, double-click the Mainland States data frame.
On the Coordinate System tab, expand the Projected Coordinate
Systems folder, then expand Continental, and then expand North
America. A variety of North American map projections are listed.

20 Scroll down and click USA Contiguous Albers Equal Area Conic,
and then click OK (see facing page). The coordinate system of the data
frame is now changed, and all the layers are projected on the fly to the USA
Contiguous Albers Equal Area Conic projection. The lines of latitude and
longitude exhibit a curvature similar to the shape of a cone. The overall shape
of the US states layer has changed and may look more familiar to you.

Which lines of latitude are defined as standard parallels for this projection?

(Hint: View the data frame properties.)

㉑ On the menu bar, click Bookmarks > Mainland States.

The map scale is changed to 1:20,000,000, and the layers are centered in the data frame. From here, you will begin the process of copying the data layers to the Hawaii and Alaska data frames.

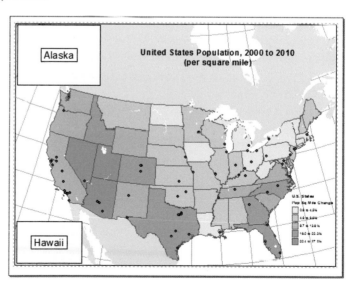

22 In the table of contents, hold down CTRL and click the Cities_GCS_NAD1983 layer, U.S. States layer, and the Canada and Mexico layer to select them. Right-click any selected layer and click Copy. Still in the table of contents, right-click the Alaska data frame and click Paste Layer(s). In the Alaska data frame layer list, hold down CTRL and click Canada and Mexico to clear the selection, then right-click any selected layer, and click Copy. Finally, right-click the Hawaii data frame and click Paste Layer(s).

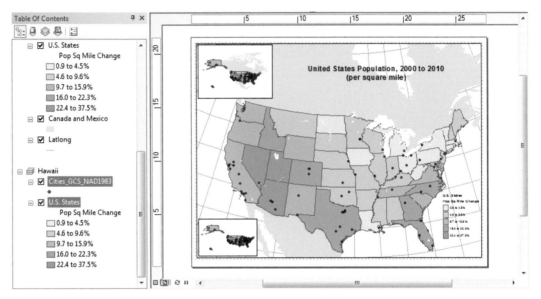

All three data frames now have content in them. The Alaska and Hawaii data frames have the data displayed in the GCS_North_American_1983 geographic coordinate system. Next, you'll give these data frames map projections that are specifically designed for these areas of the United States.

23 In the table of contents, double-click the Hawaii data frame. On the Coordinate System tab, expand Projected Coordinate Systems > Continental > North America and click Hawaii Albers Equal Area Conic. Click OK when your dialog box matches the graphic on facing page. The layers are projected on the fly. The mainland states look as if they are rotated. This projection minimizes distortion, specifically for Hawaii. The Canada and Mexico layer was not included in this data frame because you wouldn't be able to see it once the map scale and position of the data frame are adjusted.

24 Activate the Hawaii data frame (right-click it in the table of contents, and then click Activate). On the menu bar, click Bookmarks > Hawaii.

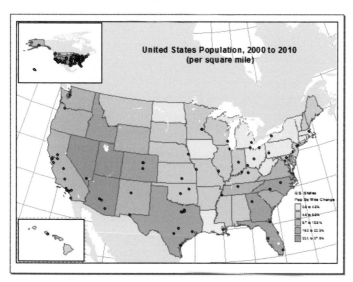

25 Repeat the same steps to project the Alaska data frame to Alaska Albers Equal Area Conic. Activate the Alaska data frame and zoom to Alaska using the bookmark. Each of the three data frames has a variation of the Albers Equal Area projected coordinate system. Data frame properties can be adjusted at any time. It is also possible to work on individual data frames in Data View mode. To change the map display, simply activate a data frame.

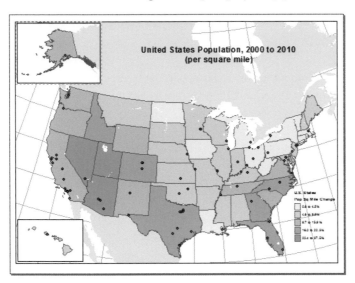

On your own

Examine other projected coordinate systems that could be applied to the Mainland States data frame. Change the map projection and see how it affects the four map properties: shape, area, distance, and direction.

26 If you want to save your work, rename your map document and save to your MyData folder for chapter 6.

27 If you are continuing to exercise 6c, leave ArcMap open. Otherwise, exit the application and click No if prompted to save your changes.

Exercise 6c
Defining a map projection

Gathering data from various sources can turn into a large task very quickly. It is even more challenging when you realize some of the data does not include the proper coordinate system information. For shapefiles, this information is stored in a separate projection file with the extension .prj. For geodatabases, coordinate system information is stored as a spatial reference, which is set as a property of a stand-alone feature class or of a feature dataset.

ArcMap reads feature coordinate values when a layer is added. When the coordinate system information is missing, it attempts to display the layer based on these x,y coordinate values. If the x-values have a range of –180 to +180 and the y-values have a range of –90 to +90, it can be assumed that the layer is in a geographic coordinate system.

Unfortunately, ArcMap won't know which one, so you will have to assign one. If you can't determine the correct coordinate system by researching the metadata or questioning the data source, you will have to make an educated guess, and then compare the results with another dataset to see if it aligns properly. Alternatively, if the x,y coordinates are six-, seven-, or eight-digit numbers, the layer could be in a projected coordinate system, but again, the software can't determine which one. This leads ArcMap to display a warning message like the one in the figure.

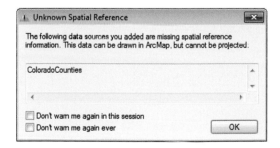

ArcMap can display the layer, but it cannot project its coordinate system. This means the layer will be displayed incorrectly, and its units will be unknown. As is usually the case when you use trial and error, the results are unpredictable. If a layer is added to a data frame set to a projected coordinate system, the layer may be displayed, but it may be seriously misaligned with the data that is in the correct location on the map. Other times, misalignment may be only minimal, a result that might be acceptable depending on the accuracy required.

In ArcToolbox, the tool used to assign coordinate system information to a dataset is called Define Projection. This tool overwrites any existing coordinate system information (map projection and datum) that is stored and should be used only if the coordinate system information is unknown or incorrect.

The most important point is, you must know the correct coordinate system of the dataset before using this tool. Often, to find the missing coordinate system information, you may have to contact the data provider and look through the metadata and supporting documentation.

1 **Start ArcMap. Open the ex06c map document from your C:\EsriPress\GTKArcGIS\Chapter06 folder.** The map is open in layout view. There are two data frames, each in a projected coordinate system: state plane and universal transverse Mercator. The State Plane data frame is active. Both frames include a US states layer, zoomed to the state of Colorado. Notice the slight difference in curvature of the state boundaries. The left data frame is projected to the Colorado North State Plane Coordinate System, and the right data frame is projected to the UTM Zone 13 Coordinate System.

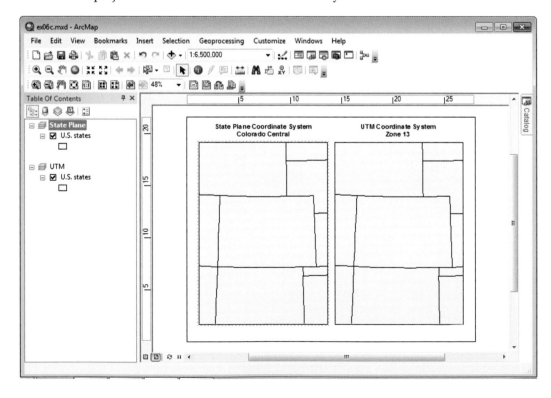

The SPCS was developed in the 1930s by the US Coast and Geodetic Survey of the National Oceanic and Atmospheric Administration to provide a conformal mapping system for the country. It is commonly used for county and statewide surveying and data applications. It is not recommended, however, for use across state lines. To maintain accuracy and minimize distortion, larger states have multiple zones.

The UTM divides the earth into 60 zones, each 6 degrees of longitude wide. There is almost no distortion at the center of each zone and gradually increasing distortion outward from the center. UTM grid coordinates are expressed as a distance in meters to the east, known as "easting," and a distance in meters to the north, known as "northing."

On your own

Explore more about the differences and uses of the state plane and UTM coordinate systems.

2 **Add ColoradoCounties.shp from your \GTKArcGIS\Chapter06\Data folder to the State Plane data frame.** ArcMap displays a warning that the layer is missing spatial reference information.

If you don't see the warning, you may have dismissed these types of warnings in a prior ArcMap session. To reset the default behavior of coordinate system warnings, close ArcMap and browse to C:\Program Files\ArcGIS\Desktop10.1\Utilities. Run the AdvancedArcMapSettings.exe file. On the ArcMap Advanced Settings Utility dialog box, click the TOC/Data tab and make sure the "Skip datum check" check box is cleared.

3 **Click OK to close the warning message.** You may notice quickly that something is wrong: ArcMap added the layer (you can see it in the table of contents), but it is not visible in the State Plane data frame. Because the Colorado counties layer did not have defined coordinate system information, ArcMap does not know where to properly place it.

4 **Right-click the Colorado counties layer and click Zoom To Layer. On the Standard toolbar, click the Fixed Zoom Out button several times until you can see surrounding states.**

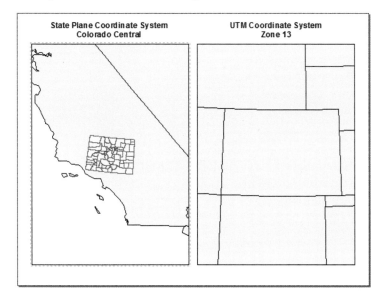

What are the spatial extent values of the Colorado counties layer?

(Hint: View the Source tab on the layer properties dialog box.)

Since ArcMap had no information to use, it drew the Colorado counties layer using a projection with latitude and longitude coordinates as if they were x,y coordinates in the state plane coordinate system. As a result, the layer is being placed in California, which is obviously incorrect. In a situation like this, it is up to you to find the appropriate coordinate system information. For this exercise, you're given the information that the coordinate system of Colorado counties is Albers Equal Area Conic. Next, you'll define the correct coordinate system information for the Colorado counties layer using ArcToolbox.

5 On the Standard toolbar, click the ArcToolbox button.

6 In the ArcToolbox window, expand Data Management Tools and then the Projections and Transformations toolset.

7 Double-click the Define Projection tool. Click the Input Dataset or Feature Class arrow and click Colorado counties.

> The Define Projection tool differs from the (Feature) Project tool used in exercise 6b. The Project tool projects spatial data from one coordinate system to another. The Define Projection tool overwrites the coordinate system information (map projection and datum) stored with a dataset and should be used only for datasets that have an unknown or incorrectly defined coordinate system. If this tool is used on data that already has a defined coordinate system, you will be warned that running the tool will change the coordinate system.

The coordinate system is listed as unknown. So next, you'll browse for the coordinate system information to apply to the layer.

8 Click the Properties button next to the Coordinate System box.

9 In the Spatial Reference Properties dialog box, expand Projected Coordinate Systems > Continental > North America. Scroll to and click USA Contiguous Albers Equal Area Conic. The Current coordinate system panel is updated. Notice that the geographic coordinate system used is GCS_North_American_1983 (you may have to scroll down to see it). This is the same system as in the State Plane data frame.

10 Click OK. This coordinate system information will be saved to a projection file (.prj) associated with the Colorado counties shapefile.

11 Click OK to run the Define Projection tool. While the Define Projection tool is running, ArcMap locks the layer so it cannot be accidentally modified or moved. A pop-up notification indicates the process was successful. ArcMap projects the Colorado counties layer to the Colorado North State Plane Coordinate System projection. The State Plane data frame is updated (see facing page), and the Colorado counties layer disappears from the map display.

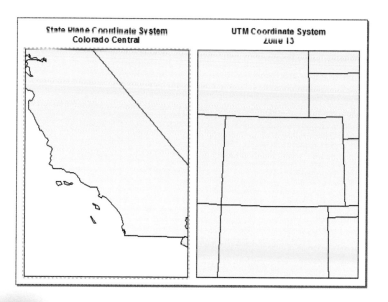

If you still see Colorado counties being drawn in California, click the small Refresh button at the bottom of the display area.

12 **On the menu bar, click Bookmarks > Colorado State.** The counties are now aligned with the US states layer.

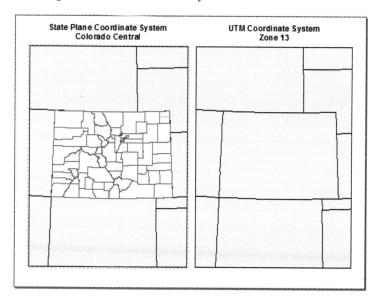

13 **In the table of contents, drag the Colorado counties layer to the UTM data frame, above the US states layer.** The Colorado counties layer is copied to the UTM data frame and is projected to UTM Zone 13. This is possible because the projected UTM coordinate system is also based on the GCS_North_American_1983 geographic coordinate system.

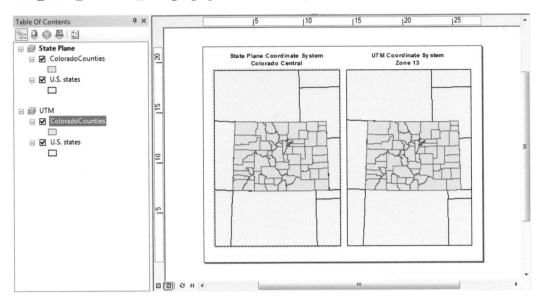

How does mapping at different map scales using SPCS and UTM coordinate systems affect the overall distortion and accuracy of your map?

14 **If you want to save your work, give it a new name and save to your MyData folder for chapter 6.**

15 **If you are continuing to the next chapter, keep ArcMap open. Otherwise, exit the application and click No if prompted to save your changes.**

Summary

In this chapter, you learned some of the fundamentals of projections and coordinate systems. You should have a good understanding of how latitude and longitude form the basis of a geographic coordinate system and how this affects locations on the curved surface of the earth. Similarly, you should have learned that map projections represent the earth on a two-dimensional, flat surface. Every map projection introduces some level of distortion to shape, area, direction, or distance.

It is important to know that a spheroid is a mathematical model that approximates the irregular shape of the earth. A datum is used to provide an optimal fit between the geoid and the chosen spheroid used for a map. Projected coordinate systems such as state plane and UTM are widely used in the United States. Each system provides projections that are used for local and state-level mapping.

You have learned how to use ArcMap and ArcCatalog to explore and manage the coordinate system information about data. ArcMap can hold multiple data frames, and data can be moved between them. The data is projected on the fly: the dataset's coordinates are recalculated, and the display position of the data is changed to match the data frame's coordinate system.

Geoprocessing tools can be used to project data permanently, as well as to define a projection for GIS data that may have unknown coordinate system information.

Review the learning objectives of this chapter and see if you are comfortable working with projections and coordinate systems in ArcGIS. If not, you may want to review some sections of this chapter as a reminder. When working on your own in the beginning, it may be a matter of trial and error as you work with projections that are not familiar to you. That's OK—it's often the best way to learn. Be sure to make a copy of any data to preserve the original, if you are experimenting and learning.

Terms

coordinate system

latitude and longitude

degrees-minutes-seconds

decimal degrees

geographic coordinate system

spheroid

datum

geoid

map projection

projected coordinate system

standard parallel

standard meridian

map unit

false easting

false northing

on-the-fly projection

datum transformation

Chapter 7
Symbolizing features

Creating custom symbology
Symbolizing features by categorical attributes
Using styles and creating layer files
Symbolizing rasters

Learning objectives

- *Change the default ArcMap symbology for points, lines, and polygons*
- *Reorder layers to influence effects of drawing order*
- *Symbolize features using the same symbol for each feature*
- *Change the full-extent specifications of a map*
- *Symbolize features based on descriptive attributes*
- *Create a definition query to limit the features being displayed*
- *Quickly label features*
- *Select symbols from a predefined style*
- *Create layer files*
- *Choose a new color scheme for a raster layer*
- *Use layer transparency effectively*

SYMBOLIZING FEATURES MEANS ASSIGNING COLORS, MARKERS, SIZES, WIDTHS, angles, patterns, transparency, and other properties that will make them recognizable on a map. Symbols often look like the objects they represent—for example, a lake polygon that is blue and fire hydrants that are marked by icons that look like fire hydrants. Sometimes the relationship is less straightforward. On a street map, varying line thicknesses may be used to show whether a street is a local road, an arterial, or a highway without implying that the widths on the map are proportional to the widths of the actual streets. Symbology may also be purely conventional. Cities, for example, are commonly symbolized as circles, even though city boundaries are seldom round.

By varying symbol properties, you convey information about features.

Left, fill patterns and colors differentiate perennial lakes, intermittent lakes, and salt pans; *center*, line thickness, color, and solidity show a hiking trail and main, secondary, and utility roads; *right*, unique icons mark major and minor cities.

Symbology is also influenced by scale. A city may be a circle on one map and a polygon on another.

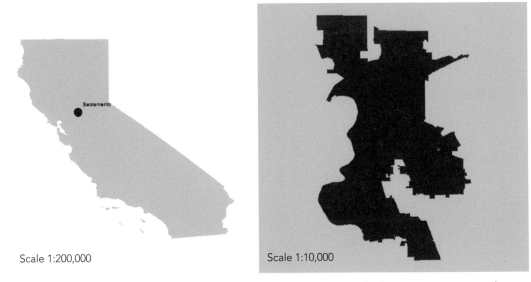

Scale 1:200,000   Scale 1:10,000

Left, Sacramento, California, represented as a circle (indicates location); *right*, Sacramento represented as a polygon (indicates boundary). Data from Esri Data & Maps 2010, courtesy of ArcWorld, ArcUSA, US Census, Esri (Pop2010 field), and the City of Sacramento, California.

ArcGIS has thousands of symbols for common map features. Symbols are organized in the Symbol Selector by style (Environmental, Transportation, Weather, and so on) to make it easy to find the ones you need. You can also create your own symbols.

When spatial data is added to ArcMap as a layer, all features in the layer have the same symbology. Points are displayed with small circles, for instance, and polygons are blocks of color in the shape of the objects they represent. Line features are displayed using a thin, simple line. The colors for points, lines, and polygons are randomly chosen.

To change the default symbology, you use information from a chosen field in the layer attribute table.

Terrestrial ecoregions in Madagascar. *Left,* by default, all features (polygons, in this case) have the same symbology. *Right,* each feature has a different fill color, indicating its ECO_NAME attribute value. Map from World Wildlife Fund/David M. Olson et al., "Terrestrial Ecoregions of the World: A New Map of Life on Earth," *Bioscience* 51 (2001): 933–38.

You can save a layer's symbology by making a layer file (.lyr). Then, when you add that layer file to a map, the features are already symbolized the way you want them.

Symbolizing raster layers is similar to symbolizing feature layers, but with fewer options. The only property that can be controlled is the color of the cells that make up the raster.

In this chapter, you will create the symbols for a Florida tourism map, and in chapter 9, you will create the labels. The grahic below shows you what the completed map will look like.

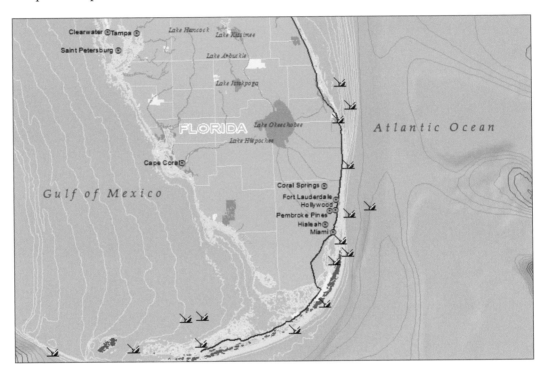

Data

The data used in the following exercises includes 10 feature classes (stored in the Florida geodatabase) and three raster datasets, as follows:

- Bathymetry—line features that represent changes in ocean depth *(Source: Florida Fish and Wildlife Conservation Commission's Fish and Wildlife Research Institute and National Oceanic and Atmospheric Administration)*
- Cities—major cities of Florida *(Source: National Atlas of the United States)*
- Coral—polygons representing coral populations off the coast of Florida *(Source: Florida Fish and Wildlife Conservation Commission's Fish and Wildlife Research Institute)*
- Counties—polygons representing Florida county boundaries *(Source: US Census Bureau)*
- Hwy1—a single line feature depicting a major state highway that traverses the Atlantic coastline and Florida Keys *(Source: Florida Department of Transportation)*
- Lakes—polygons representing Florida's major lakes *(Source: Florida Department of Environmental Protection)*
- Rivers—line features representing Florida's major rivers *(Source: Florida Department of Environmental Protection)*.   Shipwrecks—a selection of historical shipwrecks along the southern Atlantic coast of Florida and the Florida Keys *(Source: National Oceanic and Atmospheric Administration Office of Coast Survey's Automated Wreck and Obstruction Information System)*
- Southeast—a polygon layer that includes demographics *(Source: ArcUSA, US Census, Esri)*
- StParks—polygons representing Florida state parks *(Source: Florida Department of Environmental Protection)*
- florida_coast—an image of seafloor topography *(Source: Natural Earth)*
- shdrl—a hillshade raster showing continuous elevation values *(Source: University of Florida GeoPlan Center)*
- usgsdem—a digital elevation model raster for the southern portion of Florida *(Source: US Geological Survey)*

Exercise 7a
Creating custom symbology

For this exercise, you are a graphic designer creating a marketing product for the Florida Commission on Tourism. Your map, highlighting the southern Atlantic coast of Florida as well as the Florida Keys, will be printed in a brochure for prospective coastal tourists and scuba divers.

To symbolize, it's best to have your ArcMap window maximized if possible.

1 **Start ArcMap and open the ex07a map document from your C:\EsriPress\GTKArcGIS\Chapter07 folder.** At this point, the map includes only a raster dataset representing the ocean floor terrain along the Florida coast.

2 On the Standard toolbar, click the Add Data button.

3 Browse to \EsriPress\GTKArcGIS\Chapter07\Data.

4 Double-click Florida.gdb. Click Bathymetry, hold down SHIFT, and click StParks. This selects all the feature classes stored in the geodatabase. Next, click Add.

Your map will look a little different from the graphic below, because ArcMap randomly assigns symbol colors when data is added to a map.

You've got some work to do to make your map look presentable.

5 In the ArcMap table of contents, click anywhere in the white area to clear the selected layers.

Next, you will change some layer names for clarity.

6 Recall what you've learned in earlier chapters and rename the following layers:
- **Shipwrecks to** Historical Shipwrecks
- **Hwy 1 to** U.S. Route 1
- **Rivers to** Major Rivers
- **Lakes to** Major Lakes
- **StParks to** State Parks
- **florida_coast.tif to** Ocean Background

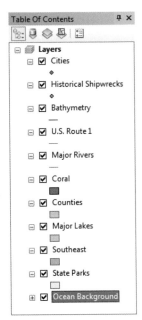

Remember, you are not renaming the source data files, but only the map layer names as they exist in the current map document.

7 Clear the Cities and Historical Shipwrecks check boxes to turn the layers off. You'll come back to these layers later.

8 In the table of contents, click the line symbol under Bathymetry to open the Symbol Selector. Bathymetry lines indicate changes in ocean depth, like the elevation lines on a topographical map.

9 In the Symbol Selector, click the Color arrow, and in the color palette, click Sugilite Sky, as shown in the graphic, and then click OK.

Pausing your mouse pointer over the color chips will reveal the color names.

The new color is fine, but you may notice that the bathymetry lines somewhat obscure the coast boundaries. To remedy this, you will next change the default drawing order.

1
2
3
4
5
6
7a
8
9
10

10 In the table of contents, drag Bathymetry below State Parks.

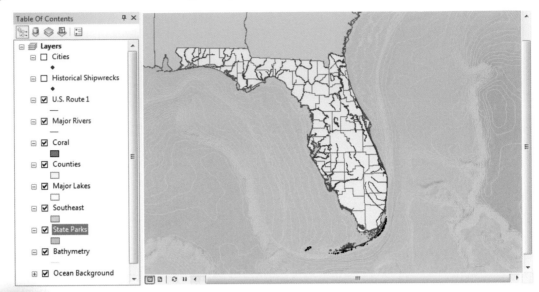

You can reorder layers only when the table of contents is in the default List By Drawing Order view.

Although lines are typically placed on top of polygons, why does it make sense to put bathymetry lines below polygons?

For simple color changes, a shortcut follows on the facing page.

11 Right-click the symbol for the Major Rivers layer to open the color palette. In the palette, click Oxide Blue.

You want the same color for lakes, but because lakes are polygon features, there are two color options—one for the outline and one for the fill. For more complex symbology such as this, you can't use the shortcut method you just used for Major Rivers.

12 Click the symbol below Major Lakes to open the Symbol Selector. For Fill Color, click the color chip, and then click Oxide Blue. For Outline Color, click No Color. When your dialog box matches the graphic, click OK.

You can't see lakes because the lake features are obscured by another polygon layer. As you work on your map, a more logical layering order becomes obvious. You want lakes to be the same color as rivers, and you also want them close together in the table of contents.

13 **Drag Major Lakes just below Major Rivers. Also drag State Parks above Counties.**

Next, you will modify the width of a line layer.

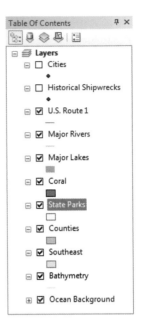

14 **Open the Symbol Selector for U.S. Route 1. For Color, click Mars Red. For Width, highlight 1.00 and replace it with** 1.5. **Click OK to apply the changes.**

On your own

On your own, symbolize the remaining layers as follows:

- Coral:
  Fill Color: Medium Coral
  Outline Width: 1.5
  Outline Color: Medium Coral

- State Parks:
  Fill Color: Macaw Green
  No outline

- Counties:
  No fill
  Outline Width: 0.40 (default)
  Outline Color: Gray 10%

- Southeast:
  Fill Color: Light Apple
  Outline Width: 0.40 (default)
  Outline Color: Gray 20%

15 Turn on Historical Shipwrecks. The wrecks are located on the southern Atlantic coast and throughout the Florida Keys. This will be the focus of your map.

16 In the table of contents, right-click Historical Shipwrecks and click Zoom To Layer.

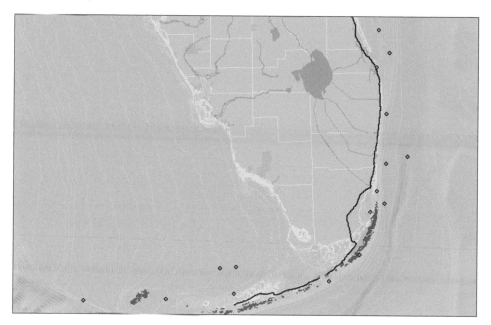

You will save this extent as the full extent of your map. That means whenever you click the Full Extent button, you will be brought back to this view.

> ### On your own
>
> Fine-tune the display: Change the scale to exactly 1:3,000,000, and then use the Pan tool to drag the landmass to the center of the display. Be sure to maintain the visibility of the coral reefs in the Keys.

17 Right-click Layers (the data frame name), and then click Properties. Click the Data Frame tab. In the Extent Used By Full Extent Command frame, click Other (see dialog box on next page).

6

7a

8

9

10

18 Click Specify Extent. On the Full Extent dialog box, click the Current Visible Extent option, and then click OK, and OK again to complete the task.

An alternative for opening the data frame properties is to go to the menu bar and click View > Data Frame Properties. Or double-click the data frame name.

19 Finally, in the table of contents, click the data frame (Layers) twice slowly to make it editable and change its name to South Florida Recreation.

> Alternatively, open the Data Frame Properties dialog box and change the name on the General tab.

Notice that the data frame title is in boldface in the table of contents to show that it's active.

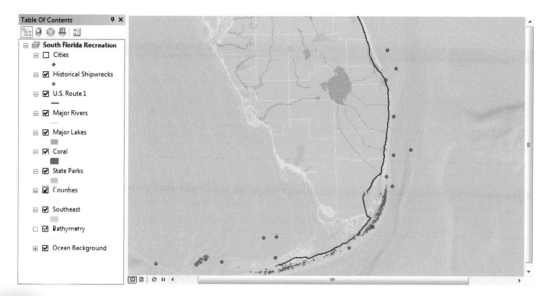

> *Cartography* is the field of study devoted to the art and science of mapmaking, a vast subject that focuses on design philosophy, color theory, and effective graphical presentation of spatial information according to the map's objectives and audience. Such concepts are beyond the scope of this book, but if you are interested in cartographic principles, you can find many books, websites, and courses devoted to the subject.

20 To save your work, rename your map document and save to your MyData folder for chapter 7.

21 If you are continuing to the next exercise, keep ArcMap open. Otherwise, exit the application and click No if prompted to save your changes.

Exercise 7b
Symbolizing features by categorical attributes

In exercise 7a, all rivers had the same marker symbol, and all state parks were drawn in the same color. You can make the symbology of a feature class more informative by assigning a different symbol to each unique value in an attribute field or to ranges of values. For example, you might symbolize a polygon layer of countries according to their names. Each feature would have a unique color, because every country has a different name. You might also symbolize countries by their political systems. Features with the same system would have the same color: republics might be blue, constitutional monarchies purple, and communist states red.

Attributes that are names or descriptions are called categorical attributes (or qualitative or descriptive attributes). They are usually text but may be numbers if the numbers are codes standing for descriptions or if they are ranges of numbers that compose a category, such as population ranges (rather than a specific population count). Attributes that are measurements or counts of features are called quantitative attributes. Using the previous example, a country's population is a count, and its area in square kilometers is a measurement.

In this exercise, you'll symbolize features by categorical attributes. You'll symbolize by quantitative attributes in chapter 8.

Sometimes you'll want to limit the number of features displayed on the map, but you won't want to change the source data by editing it. Instead, you would create a definition query, which you'll do in this exercise.

1 In ArcMap, open the ex07b map document from your Chapter07 folder. This is a map like the one you created in exercise 7a.

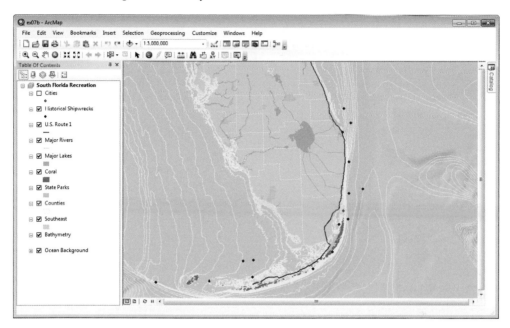

In this exercise, you want to make the symbology for State Parks more helpful. It would be nice if map readers could quickly determine whether a fee is required to visit the park.

2 In the table of contents, right-click State Parks and click Open Attribute Table. Scroll to the right and examine the PUBLIC_ACC (public access) field.

There are two public access categories: Open – Fee Required and Open – No Fee Required. You will choose a different symbol for each one so the categories can be easily discerned on the map.

3 Close the table and open the layer properties of State Parks. Click the Symbology tab.

4 Under Show, click Categories. Unique values is selected by default.

5 In the Value Field list, click PUBLIC_ACC.

6 Click Add All Values. Clear the <all other values> check box.

Adding values

You may want to focus on certain features in a layer and deemphasize or ignore others. The Add Values button (next to Add All Values) lets you select specific attribute values and assign symbology to features that have these values. Other features are assigned a single symbol—the one designated for <all other values>. If the <all other values> check box is cleared, these features are not displayed at all.

7 Double-click the symbol to the left of OPEN-FEE REQUIRED. In the Symbol Selector, you see default Esri symbol styles. Click Sun in the first row of color chips, set the outline color to No Color, and click OK to close the Symbol Selector.

8 Double-click the symbol next to OPEN-NO FEE REQUIRED. Click the Green color chip from the Esri style choices, set the outline color to No Color, and click OK to close the Symbol Selector. When your Layer Properties dialog box matches the graphic, click OK.

9 In the table of contents, click PUBLIC_ACC twice slowly to make it editable. Rename the label Public Access.

Now you know how to give features different symbols based on categorical attribute values. Next, you will limit the number of symbols used in the Cities layer.

On your own

Explore the other polygon symbols in the default Esri symbol styles: for example, give Major Lakes the symbol for Swamp or Glacier. Use either the Symbology tab on the Layer Properties dialog box or just click the symbol in the table of contents to open the Symbol Selector. When you are done experimenting, change the symbol back to Oxide Blue fill, no outline.

10 **Turn on Cities.**

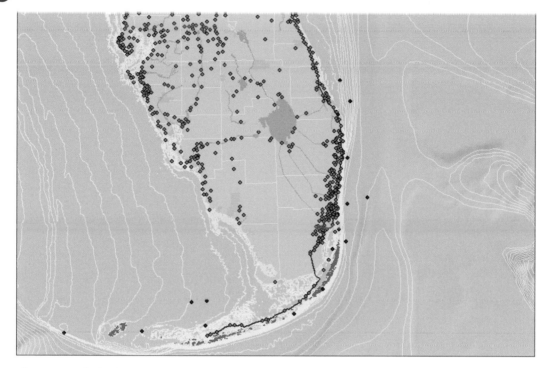

The city symbols clutter the map. Since this map is a marketing tool for tourism, you want to display only the larger cities that can accommodate visitors and recreation enthusiasts. You will create a **definition query** to limit the display of cities to those with a population of 100,000 or greater. This affects only the features shown on this map; it will not permanently delete features from the dataset. With a definition query, you can limit the display based on any attribute, whether categorical (such as population range) or quantitative (such as population).

11 Open the properties of the Cities layer and click the Definition Query tab.

12 Click Query Builder.

13 In the upper portion of the Query Builder dialog box, double-click "POP_RANGE" to add it to the query expression box.

14 Click the equals sign (=), click Get Unique Values, and double-click '100,000 - 249,999'. Then click the Or button.

"POP_RANGE" = '100,000 - 249,999' OR

15 Again double-click "POP_RANGE", click the equals sign, and double-click '250,000 - 499,999'. Click the Or button. Then double-click "POP_RANGE", click the equals sign, and double-click '500,000 - 999,999'. (There are no cities over 1 million.) The expression should read:

"POP_RANGE" = '100,000 - 249,999' OR "POP_RANGE" = '250,000 - 499,999' OR "POP_RANGE" = '500,000 - 999,999'.

16 Click OK and OK again to close the Query Builder and Layer Properties dialog boxes.

If any part of the expression is missing, you will see an error message. If this happens, go back to the Definition Query tab to ensure the expression is correct.

The number of cities has been dramatically reduced.

Alternatively, you could have created this simpler query:

"POP_2000" ≥ 100,000

You were given a more complex approach to show how to base your display on categories, as opposed to quantitative attributes.

17 Change the symbol for Cities. Use the Esri style Circle 3 and change the size to 10.

18 In the table of contents, right-click Cities and click Label Features. This is a fast way to label features, but you need to be sure the labeling field is set to the correct attribute on the Layer Properties dialog box, on the Labels tab. ArcMap will choose a default field, but it may not always be the one you want. In this case, the default choice was correct (see next page). In chapter 9, you will modify label properties to maximize readability.

1
2
3
4
5
6
7b
8
9
10
11
12
13
14
15
16
17
18
19
20

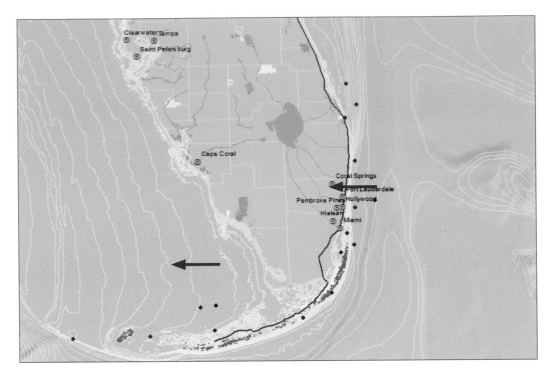

19 Finally, rename the **Cities layer** Cities Pop 100,000+.

⊟ ⬗ **South Florida Recreation**
  ⊟ ☑ Cities Pop 100,000+
      ⊚
  ⊟ ☑ Historical Shipwrecks
      ◆
  ⊟ ☑ U.S. Route 1
      —

20 To save your work, rename your map document and save to your MyData folder for chapter 7.

21 If you are continuing to the next exercise, leave ArcMap open. Otherwise, exit the application and click No if asked to save your changes.

In exercise 7c, you will learn how to use other symbology styles and save your symbology to layer files.

Exercise 7c
Using styles and creating layer files

In exercises 7a and 7b, you modified colors, sizes, and outlines of point, line, and polygon symbols. You also picked predefined symbols and colors from the default options provided in the Symbol Selector —these symbols and colors belong to the Esri style library. A *style* is a collection of predefined symbols, colors, and other map elements such as labels, north arrows, scale bars, and borders. ArcGIS has more than 20 style libraries, and you can make new ones that contain often used symbols from other styles or custom symbols you create yourself.

Left to right: Some symbols used in the Esri Conservation, Public, and Hazmat style libraries.

For information on creating and modifying styles, in ArcGIS for Desktop Help, on the Contents tab, navigate to Mapping > Symbols and styles > A quick tour of symbols and styles.

Having invested time in symbolizing a layer, you may want to save it as a *layer file* (a file with a .lyr extension). Layer files store symbology information for a specific spatial dataset—they do not contain the actual data.

In this exercise, you'll use a shipwreck symbol from the Esri Civic style library to represent historical shipwrecks on the Atlantic coast of Florida and the Florida Keys. Then you'll save the symbology to a layer file.

The ships on this map all went down from 1942 to 1944—the start of World War II. They represent only a sample of the numerous wrecks in these waters. Mapping the location of wrecks is important to navigation, historical studies, and marine recreation. Shipwreck diving is part of Florida's tourism economy.

1 **In ArcMap, open the ex07c map document from your Chapter07 folder.** You recognize the map you've been working on in exercises 7a and 7b.

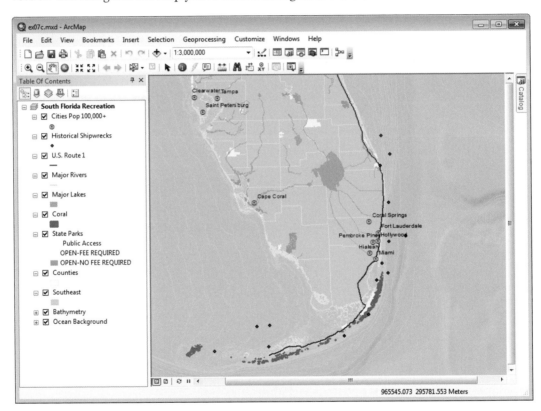

2 **In the table of contents, click the point symbol below Historical Shipwrecks to open the Symbol Selector. If you scroll through the list of symbols, you'll find there's nothing that looks like a shipwreck.**

3 At the lower right of the dialog box, click Style References to display the list of styles. When a style is loaded, it has a check mark next to it. By default, only your personal style library and the Esri style library are loaded. (Your personal style library is empty unless you've added symbols to it.)

There are many styles collected in style libraries that are intended for specific types of maps. If you know you're looking for a specific symbol but aren't sure where to find it, you can do a search, which you'll do next.

4 Click Cancel to close the Style References dialog box. In the Symbol Selector, in the search box, type wreck and press ENTER.

To see all the symbols in a style library, select the style on the Style References dialog box to load the complete set to the Symbol Selector.

5 Scroll down and click the Wreck symbol in the Civic style library, and then click OK to implement the change (see graphic on facing page).

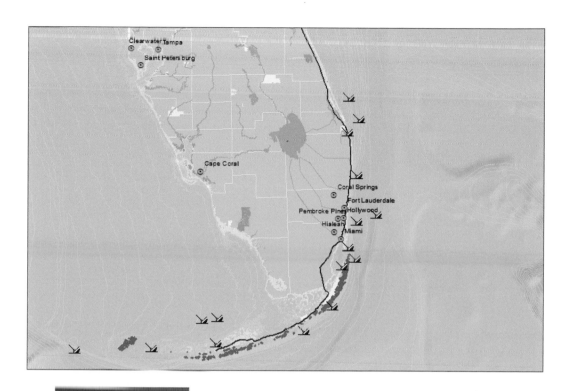

On your own

Use the Identify tool and click some of the shipwrecks to see their attributes. Keep in mind, when using graphic symbols, the center point is not always easy to find. With this symbol, you must click right at the bottom of the V formed by the mast and the ship body, as shown in the graphic.

If you wanted to reuse the symbolized layer in other map documents—or send it to a colleague—you would create a layer file. Layer files allow you to store symbology information—the colors, shapes, and sizes you choose for features—so you never have to re-create it. Just as a person can wear many different outfits, a spatial dataset can have many different layer files.

6 In the table of contents, right-click Historical Shipwrecks and click Save As Layer File.

7 On the Save Layer dialog box, browse to your \GTKArcGIS\ Chapter07\MyData folder. Accept the default file name of Historical Shipwrecks.lyr and click Save. Layer files can be given unique names—they don't have to share the same name as their source data.

8 Repeat step 7 to save the State Parks symbology to a layer file. Use the default name and save to your MyData folder for chapter 7. The Historical Shipwrecks and State Parks layer files can now be added to any map. Because they reference their source feature classes (Shipwrecks and StParks), the feature classes must also be accessible on disk or across a network. If you send a layer file to someone else, you must also send the data source that it references.

On your own

Can you figure out how to symbolize the bathymetry layer using graduated colors based on depth measurements in meters? Shallower waters should be light blue, and deeper waters, dark blue. Your results will resemble the graphic. If you succeed, you're ahead of the game—symbolizing features based on quantitative attributes is the focus of chapter 8. (*Hint:* Use the Layer Properties dialog box, not the Symbol Selector. You will have to change the sample size to 25,000 (Classify > Sampling).)

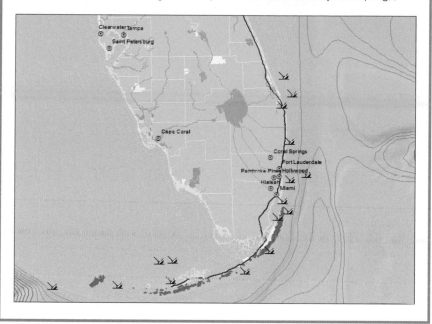

On your own

Create layer files for the other datasets in your map.
Save the layer files to your MyData folder for chapter 7.

6

7c

8

9

10

9 If you want to save your work, rename your map document and save to your MyData folder for chapter 7.

10 If you are continuing to the next exercise, leave ArcMap open. Otherwise, exit the application and click **No** if prompted to save your changes.

Cartographic representations

Using ArcGIS for Desktop Standard or ArcGIS for Desktop Advanced, you can leverage the power of the geodatabase to store cartographic representations, often referred to simply as representations.

You can convert any geodatabase feature class to a representation class, which stores symbology rules for features. The Representation toolbar provides special tools that allow mapmakers to improve how features look. You can also override the appearance or shape of features stored as representations, or remove them from the map display, without changing the actual source data.

For example, suppose that you are responsible for creating trail maps for all the state parks in Florida. The maps should share the same symbology scheme. You have a trails feature class that you symbolize with gray dashed lines, and then convert to a representation class. Each park map uses a definition query to display only the trails within its boundaries. A few state parks have unique needs, however. Some have equestrian trails, which should be differentiated by a unique color on the map. Sometimes parks have to delete a trail from a map or change its course because of maintenance. You don't want to permanently edit the trails database, so instead you will select the trail that requires modification and change its properties or shape or remove it from the map display. Later, if you need to, you can go back to the map's original appearance—because the source data remains the same.

For more about representations, in ArcGIS for Desktop Help, on the Contents tab, navigate to Mapping > Working with layers > Displaying layers > Cartographic representations > What are representations?

Exercise 7d
Symbolizing rasters

In chapter 1, you learned that a raster is a matrix of identically sized square cells. Each cell in a raster stores a value, usually a quantity of something such as elevation, rainfall, or temperature that has been measured at the location represented by the cell.

When you symbolize a raster, you assign colors to cell values or ranges of cell values. Raster values usually lie on a continuous scale, such as the scale of integers or real numbers, and are symbolized by color ramps. Rasters can have a single band or multiple bands of data. Most satellite imagery has multiple bands, whereas raster surface layers typically have one band of data.

The shades of red in this color ramp become darker as the values increase from 0 to 100.

Rasters can represent many kinds of data and be symbolized using different color ramps.

Left to right: Rasters of elevation, hillshade, and temperature for the same area. Raster images based on data provided by the US Department of Defense Mojave Desert Ecosystems program.

A *hillshade* raster shows an elevation surface in relief. It can be mathematically derived from an elevation raster by setting a certain angle and altitude for a light source.

Satellite images and aerial photographs are rasters in which the cell values are measurements of reflected light. Scanned maps are rasters in which the cell values are measurements made by the scanning device.

Left, an air photo and, *right*, a scanned map of the area around San Diego's Mission Bay. If you zoom in close enough on any raster, its cell structure will be revealed, as shown by the inset in the red box on the air photo. Satellite image and map courtesy of the US Geological Survey.

You can display raster data in ArcGIS for Desktop, but to create it, you need one of the ArcGIS extensions for working with raster data—ArcGIS Spatial Analyst, ArcGIS 3D Analyst, or ArcGIS Geostatistical Analyst. Raster data is available in many formats, on the Internet and from commercial vendors.

Multiple rasters can be organized in a raster catalog. If you wanted to view multiple adjacent raster images at a time, you would store them in a mosaic dataset. Both raster catalogs and mosaic datasets can be created in a geodatabase, if desired. Mosaic datasets can be very large, and require an ArcGIS for Desktop Standard or ArcGIS for Desktop Advanced license to create.

This chapter provides a first look at some types of raster data. In chapter 20, you will create raster surfaces using ArcGIS Spatial Analyst. For more information about raster data, in ArcGIS for Desktop Help, on the Contents tab, navigate to Geodata > Data types > Rasters and images.

In this exercise, you will symbolize an elevation raster and a hillshade raster to show the *topography* of southern Florida.

1 In ArcMap, open the ex07d map document from your Chapter07 folder. This is like the Florida map you've been working on.

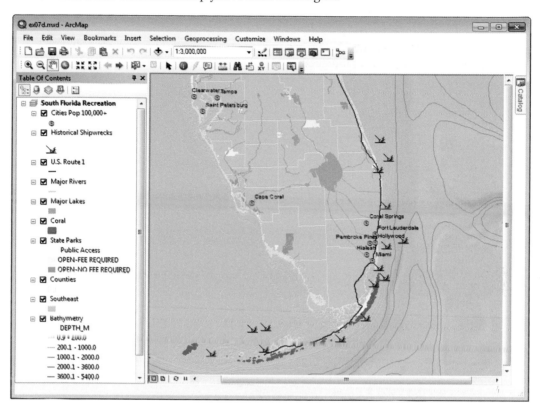

2 In the table of contents, expand the Ocean Background legend. You can see that the Ocean Background image is a three-band raster, symbolized by a Red-Green-Blue (RGB) value for each band.

Next, you'll add elevation and hillshade raster layers to the map.

3 Click the Catalog tab to expand the Catalog window. (If the tab is not visible, click the Catalog button on the Standard toolbar.)

4 Under the Home location, expand the Data folder and click the shdrl raster.

5 Drag shdrl from the Catalog window to the ArcMap display area. If prompted to build pyramids, click Yes.

6 Now drag usgsdem from the Catalog window to the ArcMap display area. If prompted to build pyramids, click Yes. A *pyramid* is a reduced-resolution raster layer that copies the original data in decreasing levels of resolution to enhance performance. The coarsest level of resolution is used to draw the entire dataset quickly. As the display zooms in, layers with finer resolution are drawn; drawing speed is maintained because fewer pixels are needed to represent the successively smaller areas. You need to build pyramids only once for a raster dataset. They are stored with the data as a file with the extension .rrd.

The raster layers are added near the bottom of the table of contents, above the Ocean Background layer. Next, you'll rename the raster layers.

7 In the table of contents, rename the usgsdem layer Elevation.

8   **Now rename shdrl** Hillshade.

To see the rasters, you'll need to change their position in the table of contents, which you'll do next.

9   In the table of contents, click the Elevation layer. Hold down CTRL and click Hillshade. Both rasters should be highlighted. Drag the layers below the Counties layer, but above the Southeast layer. The raster layers are repositioned in the table of contents and displayed on the map.

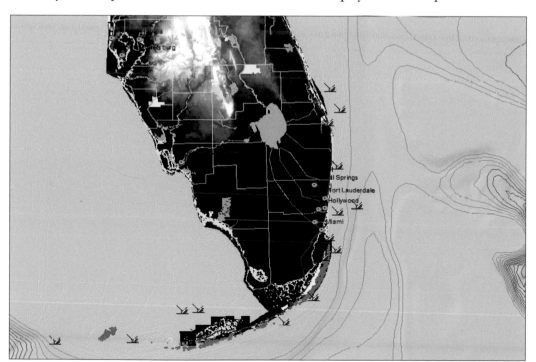

By default, rasters are drawn in shades of gray. Next, you'll choose a different color ramp for the Elevation layer.

10 In the table of contents, double-click the Elevation layer. On the Layer Properties dialog box, click the Symbology tab. There are four raster display methods in the Show panel. In the Stretched method (the default), ArcMap makes subtle transitions along the selected color ramp but doesn't give you precise information about which data values are associated with which shades of color. In the Classified method, color transitions are less subtle, but you can see exactly which value ranges correspond to which shades. In the Unique Values method, a different color is assigned to each data value in the raster. In the Discrete Color method, each unique value is symbolized randomly from a set of colors. This method is useful when a very large number of unique values are present.

11 Maintain the Stretched method. In the Color Ramp list, right-click the colors (not the drop-down arrow) and click Graphic View to turn it off.

12 Click the Color Ramp arrow, and then click the Yellow-Green Light to Dark color ramp. Select the Invert check box. Make sure your dialog box matches the graphic, and then click OK to close the Layer Properties dialog box.

The color ramp is applied to the layer. The colors range from dark olive green at lower elevations through lighter shades of olive green at higher elevations.

How would you describe the topography of southern Florida?

13 Click the Identify tool, and then click any land area of the map. In the Identify window, in the upper portion, change the Identify from value from <Top-most layer> to Elevation. Then click the map again, in several locations, to see the elevation value.

14 Close the Identify window. Zoom in closely (approximately 1: 60,000) on an area with varied elevation. If you look closely, you will notice that the raster is made up of an array of square cells, or pixels.

To see surface relief on your map, you will next make the Elevation layer partially transparent so that the Hillshade layer shows through.

7d

15 Zoom to the full extent of the map. Open the Layer Properties
dialog box for the Elevation layer and click the Display tab. In the
Transparency text box, highlight the default value of 0 and type 50.
The Elevation layer will be 50 percent transparent. Make sure your
dialog box matches the graphic, and then click OK.

Transparency can also be used effectively on vector layers. Transparency for all
features in vector layers is set on the Display tab. Alternatively, click the Advanced
button on the Symbology tab to set varying transparency according to field values.

The Hillshade layer is now visible below the Elevation layer, providing a realistic impression of the landscape.

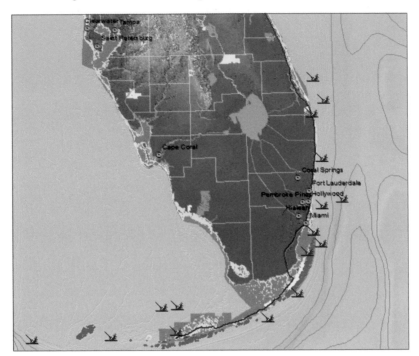

16 If you want to save your work, rename your map document and save to your MyData folder for chapter 7.

17 If you are continuing to the next chapter, leave ArcMap open. Otherwise, exit the application and click No if prompted to save your changes.

In chapter 8, you'll symbolize layers by quantitative attributes rather than categorical ones. You'll also learn how to classify data—that is, adjust the value ranges for applying symbols. And you'll work with symbology methods specifically for numeric data.

In chapter 9, you'll revisit the Florida data and add labels and annotation to complete your map.

Summary

In this chapter, you gained experience with the first step in mapmaking: symbolizing features. In chapter 9, you will learn how to create map labels, and in chapter 10, you will learn how to create a nice-looking map layout so you can print your map and present it to others. This chapter taught you the basics of symbolizing both vector and raster data. Realize that map symbology can get very sophisticated—consider the work of professional cartographers. Once you have mastered the basics, you can explore the Symbol Selector editor and advanced cartography tools to create more detailed symbology. Remember, to reuse or share symbology, create a layer file. The layer file stores the symbology and also points to the source data file, so when sharing, be sure to provide the source data as well as the layer file.

Terms

cartography

definition query

style

layer file

hillshade

topography

pyramid

Chapter 8
Classifying features

Classifying features by standard methods

Mapping density

Using graduated and chart symbols

Learning objectives

- *Explore what to consider when classifying quantitative attribute data*
- *Become familiar with classification methods*
- *Choose symbology based on classification type*
- *Create dot density maps*
- *Create graduated-color maps*

AN IMPORTANT ASPECT OF USING ARCMAP IS TO DECIDE WHAT DATA TO map and how to map it. You worked with qualitative data in chapter 7. Although a lot of data can be categorical in nature, data can be qualified quantitatively as well. Common examples of quantitative attributes are the population of a country, the area of a city block, and the length of hydroelectric transmission lines.

Quantitative maps show numeric values in relation to one another on a continuous scale. Although you can map individual values, it is often more useful to group or classify these values. This will help your readers to separate the data visually, so it can be understood more easily. Deciding how many classes are appropriate for the data and the method of classification used is called classifying data. Depending on the message you are trying to convey, ArcMap can create several types of quantitative maps: graduated color, graduated symbol, proportional symbol, dot density, and maps using charts.

A *graduated-color map* uses a change in color to represent data classifications. A ramp with a single color (called monochromatic) is suited to data that moves from small to large values. Sequential schemes such as population increase can be shown with light colors for smaller amounts of change and dark colors for larger amounts of change. A ramp with two colors can be used to represent a divergent scheme, such as population increase *and* decrease. The extreme values are shown in dark colors, and the middle values are shown in light colors.

A *graduated-symbol map* uses a symbol that varies in size to represent numeric attribute values. Points that increase in size can represent a city's growing population. Different line widths can be used to represent the increasing flow of a pipe or a river. Generally, three to seven graduated colors are used. Use more than that, and the difference between colors may not be distinguishable by the human eye.

A *proportional-symbol map* is similar to a graduated-symbol map, except that the symbols are drawn proportionally in size. A country with a population of 50 million people would have a symbol five times larger than a country with a population of 10 million people. These symbols are most appropriate for data that has a relatively narrow range of values.

A *dot density map* represents quantities using randomly placed dots to form a density pattern inside a polygon. Each dot is equal to a predetermined attribute value—the more dots, the greater the value of the polygon. However, the density pattern may be different from the actual distribution of values, and a common technique is to use a masking layer to exclude certain areas, such as a neighborhood or a ZIP Code boundary, to create a more realistic depiction of where dots should be placed.

A map uses charts to represent multiple attribute values for a single feature. ArcMap can create pie, bar, and stacked charts that can be similar in size or vary based on proportionality.

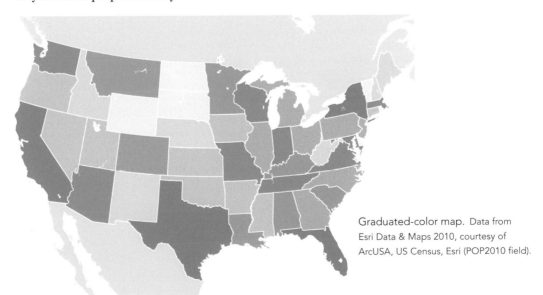

Graduated-color map. Data from Esri Data & Maps 2010, courtesy of ArcUSA, US Census, Esri (POP2010 field).

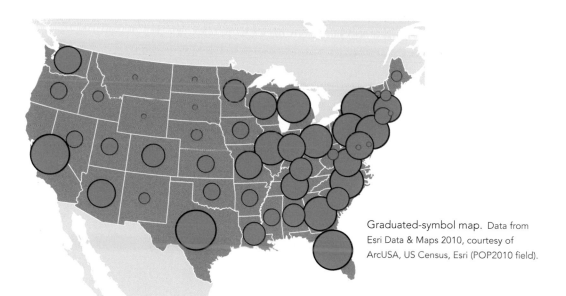

Graduated-symbol map. Data from Esri Data & Maps 2010, courtesy of ArcUSA, US Census, Esri (POP2010 field).

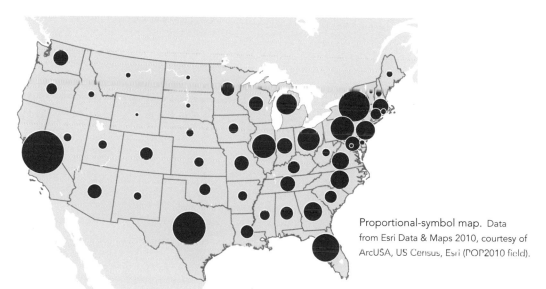

Proportional-symbol map. Data from Esri Data & Maps 2010, courtesy of ArcUSA, US Census, Esri (POP2010 field).

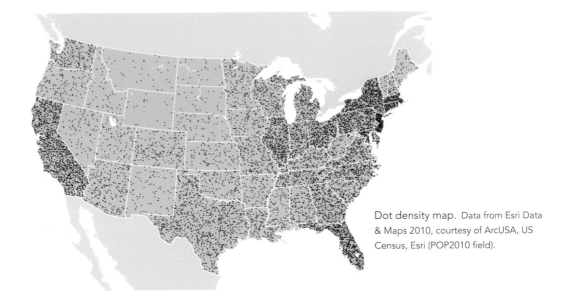

Dot density map. Data from Esri Data & Maps 2010, courtesy of ArcUSA, US Census, Esri (POP2010 field).

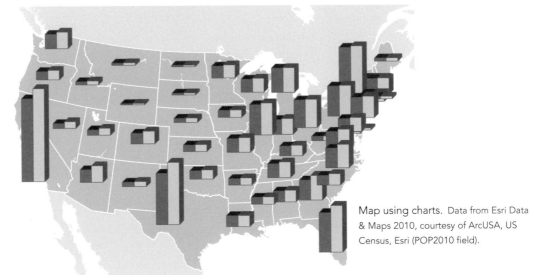

Map using charts. Data from Esri Data & Maps 2010, courtesy of ArcUSA, US Census, Esri (POP2010 field).

Data

This chapter uses the following dataset:

- Counties.shp—polygons of the counties within the state of Texas. It also contains crime data for the years 1994 and 2002. *(Source: ArcUSA, US Census, Esri, and National Archive of Criminal Justice Data)*

Exercise 8a
Classifying features by standard methods

When mapping quantitative data, you want to classify the data appropriately so the information is not skewed or misleading. ArcGIS has seven classification methods: *manual classification*, *equal interval*, *defined interval*, *quantile*, *natural breaks*, *geometrical interval*, and *standard deviation*.

Manual

With manual classification, you can set class breaks manually by typing them or by adjusting the histogram break bars in ArcMap.

Equal interval

The entire range is equally divided by the number of classes you choose. If the range of values is 1 to 1,000 and there are five classes, this method will create classes of 1–200, 201–400, 401–600, 601–800, and 801–1,000. This method is useful for highlighting changes in the extremes and is most appropriate for familiar data ranges, such as percentages or temperatures. With this method, it is also possible to have classes that contain no values.

Defined interval

A defined interval is similar to an equal interval. However, you get to define the interval size to determine how many classes there are. If the students in your class have a range of test scores from 50 to 90 percent, for example, and the interval size is 10, this method will create four classes, 50–60, 61–70, 71–80, and 81–90.

Quantile

With quantiles, all classes have the same number of features. This method is appropriate when data is linearly distributed. It can create a balanced-looking map because no classes have too many, too few, or no values, but the resultant map can be misleading. The key is to increase the number of classes to prevent data from being placed in a different class. When an appropriate number of classes is used, the data can be visually represented more accurately.

Natural breaks (Jenks)

Natural-breaks classes are based on natural groupings inherent in the data, and boundaries are set where there are relatively large gaps between values. Developed by Professor George Jenks of the University of Kansas, this method is useful for classifying unevenly distributed data such as population.

Geometrical interval

The geometrical-interval classification scheme creates class breaks based on class intervals that have a geometric series, such as a logarithmic distribution. This method was specifically designed to accommodate continuous data and minimize variance within classes to ensure that each class range has approximately the same number of values and that the change between intervals is fairly consistent.

Standard deviation

Data that you know is normally distributed (in a bell-shaped curve) can benefit from standard-deviation classification. This method creates classes according to a specified number of standard deviations from the mean value. A divergent color scheme is recommended for this type of classification.

In this exercise, you'll create thematic maps for a crime-mapping project by applying different classification methods and different numbers of classes.

1 **Start ArcMap. Open the ex08a map document from your C:\EsriPress\GTKArcGIS\Chapter08 folder.** The map is open in data view (see facing page). There is a single data frame called Texas that contains the Counties layer. You'll explore its attributes to gain an understanding of what you will be mapping.

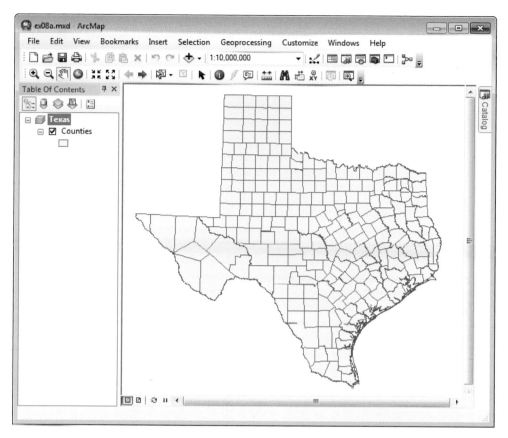

2 In the table of contents, right-click the Counties layer and click Open Attribute Table. Scroll to the right to see more of the table column headers. The table contains many attributes such as county name, population, and crime data from 1994 and 2002. You'll be using a few of these attributes for the mapping tasks in this chapter.

COUNTY	AGRASLT94	AGRASLT02	AYRPTARR94	AYRPTARR02	AYRPTCRM94	AYRPTCRM02	ARSONS94
Armstrong	5	3	1	1	1	1	1
Bandera	9	23	1	1	1	1	3
Borden	0	1	1	1	1	1	0
Briscoe	1	2	1	1	1	1	1
Clay	4	10	1	1	1	1	9
Cochran	12	6	1	1	1	1	0
Coke	3	5	1	1	1	1	1
Collingsworth	11	7	1	1	1	1	4
Cottle	3	2	1	1	1	1	1

(0 out of 254 Selected)

3 Close the attribute table. In the table of contents, double-click the Counties layer name to open the Layer Properties dialog box. Click the Symbology tab.

4 Under Show, click Quantities. The Graduated colors option of the Quantities method is highlighted, and the dialog box is changed. The Value list shows the available attributes—that is, numeric fields only.

5 In the Fields frame, click the Value arrow and click BURG02. The BURG02 attribute represents burglaries for each county in 2002. By default, the values in the BURG02 field are grouped into five classes using the Natural Breaks (Jenks) classification method.

6 If your color ramp is different from the one in the graphic above, click the Color Ramp arrow and click the Blue Light to Dark ramp. (To find it by name, right-click the color ramp color bar and click Graphic View to turn off that view.) Click OK to close the Layer Properties dialog box. The graduated-color symbology is applied to the map, and the table of contents is updated to show the values represented by each color. The light colors represent fewer burglaries, and the dark colors represent more burglaries. This may be useful to crime analysts who want to quickly see where the most burglaries are occurring. It doesn't account, however, for how many people live in a county or how many square miles are in a county. You'll explore these situations in exercise 8b. For now, you'll see how other classification methods affect the map.

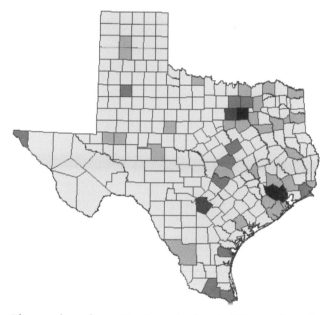

The number of counties in each class is different. The class with the highest values has just two members: Dallas and Harris Counties. The next highest class also has two members: Tarrant and Bexar Counties. The remaining counties belong to the three lowest classes.

On your own

Use the Identify tool to look at the attribute information of various counties.

7 In the table of contents, double-click the Counties layer to open the Layer Properties dialog box. Click the Symbology tab if it's not already active.

8 In the Classification frame, click the Classes arrow and click 3. The number of classes and their value ranges are adjusted. Click Apply and move the Layer Properties dialog box away from the map display. The map reflects the new classification. The highest burglaries class now contains four counties. Many counties that belonged to separate classes previously are now grouped together.

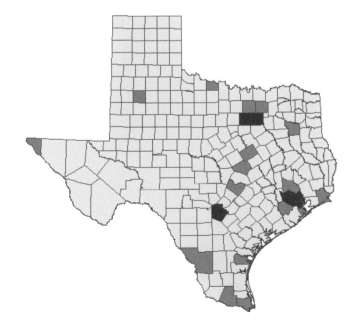

Why might these counties have a higher incidence of burglary than surrounding counties?

9 In the Classification frame, on the Symbology tab, click Classify. The Classification dialog box displays the current classification method and the number of classes. Under Classification Statistics, you can see the count, minimum, maximum, sum, mean, median, and standard deviation of the BURG02 field (you may need to scroll to the right to see the values). In the Break Values box, you can see the upper-threshold value that defines each class. The percentage (%) button displays decimals as percentages.

The large window is a *histogram*, or frequency distribution chart. The x-axis (horizontal) shows the range of values in the BURG02 field. The y-axis (vertical) is a count of features. Vertical blue lines are class breaks (also shown in the Break Values box). The gray columns represent the number of features that fall within a specific span of the total range of values. By default, 100 columns are drawn in the histogram, so each column represents 1 percent of the total range. In this case, values range from 0 to 39,272, so the total range of the histogram is 39,275. (This wouldn't be the case if the number of burglaries began with a number other than zero. Here, 39,275 divided by 100 is rounded to 393, so each column shows the number of features that fall within each span of 393 values.)

10 In the Break Values box, click the first break, at 1564.000000.

Many of the gray columns are grouped within the first class. This means there is a high count of counties with fewer than 1,564 burglaries—exactly 228 counties (the count is shown below the histogram window). Next, you'll take a closer look at the columns in the histogram.

How many elements (or features) are in the other classes?

11 Right-click the histogram near the origin of 0,0 (where the x-axis and y-axis intersect) and click Zoom In. Repeat the zoom in. If you zoomed in on another spot in the histogram by mistake, right-click and click Reset Zoom, and then try again. (Your zoom results may be a little different from what you see in the graphic.) By zooming in on the histogram, you can get a better sense of how the values are distributed. Many of the county burglaries actually fall within the bottom .5 percent of the value range (below an x-axis value of 196.5). The data is skewed toward the low end of the histogram, meaning that there are relatively few burglaries in these counties (the mean number of burglaries is 138). A gray column will not be drawn if no features fall within the value range it would represent. The number of columns can be set to anything from 10 to 100. With 10 columns, for instance, each one would represent 10 percent of the value range.

12 In the Columns box, change the default value of 100 to 25. Now each gray column represents one twenty-fifth (or 4 percent) of the value range.

13 **In the Classification frame, click the Method arrow and click Equal Interval.** The class breaks are now equally divided (still in three classes). This method places breaks so that each class has the same range of values. Almost all the values fall into the first class. Unlike the natural-breaks method, with equal-interval classification it is possible to have classes that have no values.

On your own

Try increasing the number of classes to see how the breaks are placed. When you are done, return the number of classes to 3.

14 **In the Classification frame, click the Method arrow and click Quantile.**
The quantile method creates classes that have an equal number of features.

On your own

Zoom in on the histogram twice to get a better picture of three classes with the same number of features. How many counties fall within each class?

15 Click OK to close the Classification dialog box. The current classification settings are updated to reflect the quantile method. Click Apply and move the Layer Properties dialog box away from the map display.

The map looks much different from the previous classification methods. Many more counties now fall within the highest burglaries class. Depending on how you want to depict the data, this can be misleading and may not accurately reflect the data. It also gives a false impression that a lot of burglaries occur in these counties, which may not be true. You'll have to adjust the classification method again to use natural breaks.

16 On the Symbology tab, click Classify. On the Classification dialog box, for Method, click Natural Breaks (Jenks) and change the number of classes to 7. Click OK.

Classification	

Classification
Method: Natural Breaks (Jenks)
Classes: 7

Data Exclusion
Exclusion ...   Sampling ...

Classification Statistics
Count: 2
Minimum: 0.0000(
Maximum: 39275.0000(
Sum: 211931.0000(
Mean: 834.3740
Median: 109.0000(

Columns: 100   ☐ Show Std. Dev.   ☐ Show Mean

Break Values   %
380.000000
1060.000000
2307.000000
3979.000000
8186.000000
17787.000000
39275.000000

☐ Snap breaks to data values

OK
Cancel

Before finalizing the map, you'll next make some small changes to the class labels to make them look better.

17 **In the lower section on the Symbology tab, click the Label column header and click Format Labels.** The Number Format dialog box has options to edit the look of the numeric labels. This does not affect the classification values, but only the way they appear. You can control the number of decimal places and significant digits, alignment, and whether to include comma separators, and you can even choose special formats listed in the Category box.

18 **On the Number Format dialog box, under Rounding, change the number of decimal places to 0. Also, select the Show thousands separators check box.**

19 Click OK to close the dialog box. See how your label changes are reflected on the Symbology tab, and then click OK.

20 On the menu bar, click View > Layout View.

21 On the Tools toolbar, click the Full Extent button. The layout includes a legend that shows the seven classes created using natural breaks.

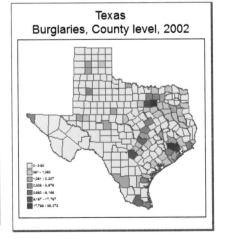

Texas
Burglaries, County level, 2002

On your own

Try classifying some of the other attributes using different methods to see how the data is distributed. Use the Identify tool to look at the COUNTY field in relation to the BURG02 field.

22 If you want to save your work, rename your map document and save to your MyData folder for chapter 8.

23 If you are continuing to the next exercise, leave ArcMap open. Otherwise, exit the application and click No if prompted to save your changes.

Exercise 8b
Mapping density

In this exercise, you'll continue to create thematic maps for a crime-mapping
project. In exercise 8a, you saw how crime can be mapped to show areas of
high and low burglary incidence by applying different classification methods.
However, you know that not all counties have the same population—for
example, Dallas County has over 2.2 million people, and El Paso County has
just over 650,000. But which of the two is more densely populated? To answer
that question, you would need to know the number of people per square unit
of area. As long as you have both a population attribute and an area attribute,
you simply divide the population by area. How about calculating crime density
per capita? Similarly, you divide a crime attribute by population. Dividing one
attribute by another to find the ratio is called *normalization*.

1 **Start ArcMap. Open the ex08b map document from your Chapter08 folder.** This opens the map in layout view. In the table of contents, there is a single active data frame called Motor Theft Per Capita. The Motor Theft Per Capita data frame has two identical Counties layers (coming from the same data source). You'll make a population density map that is overlaid by a dot density map of motor vehicle thefts.

2 On the menu bar, click View > Data View.

3 In the table of contents, right-click the Counties 2 layer and click Open Attribute Table.

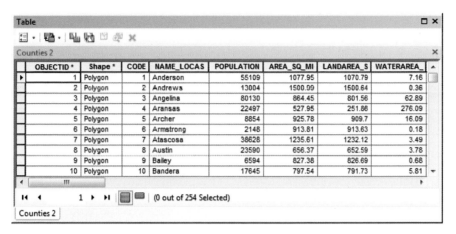

The POPULATION attribute contains the population count of each county. The AREA_SQ_MI attribute contains the size of each county in square miles. Next, you'll divide the POPULATION attribute by the AREA_SQ_MI attribute to create a ratio of population density for each county.

4 **Close the attribute table. Double-click the Counties 2 layer and click the Symbology tab. Under Show, click Quantities (Graduated colors is automatically selected). In the Fields frame, click the Value arrow and click POPULATION. Click the Normalization arrow and click AREA_SQ_MI. Set the color ramp to Red Light to Dark.** The Natural Breaks (Jenks) classification method is chosen, and five (5) classes are defined. In the Label column, the values are rounded, but the decimal places are not consistent.

5 Click the Label column header and click Format Labels. On the Number Format dialog box, select the Number of decimal places option and change the number to 2. Click OK to close the dialog box.

6 Click OK to close the Layer Properties dialog box. Select the check box next to Counties 1 to hide the layer.

The class with the greatest population density has three counties: Tarrant, Harris, and Dallas. (You can verify this using the Identify tool.) The next highest class contains four other counties. The remaining counties of Texas have low population density.

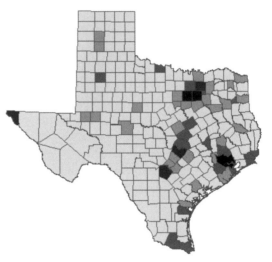

Graduated colors are one way to represent density. Another way is to use a dot density map, which you'll do next.

7 Select the check box next to Counties 1 to make the layer visible again.

8 Open the Counties 1 properties. On the Symbology tab, click Quantities > Dot density. In the Field Selection list, scroll down and click MVTHEFT02. Click the right-facing arrow (>) button to add it to the list on the right.

If you added the wrong attribute or added too many attributes, simply click the left-facing arrow (<) button to move an attribute back to the list on the left.

The box on the right in the Field Selection frame indicates that the incidence of motor vehicle theft in 2002 will be represented by the dot symbol. Next, you'll change the color of the dot so that it has higher contrast and can stand out from the population density graduated-color map.

9 Double-click the symbol next to MVTHEFT02 to open the Symbol Selector. Click the Color chip and click Black, and then click OK.

10 In the Background frame, click the Line button. In the Symbol Selector, click the Color chip and click No Color, and then click OK. This will prevent the county lines from appearing.

11 In the Densities Calculated frame (see next page), click the horizontal slider and move it left or right. The preview gives you an idea of what the density will look like. In the Dot Value box, change the value to 100, and then click OK.

A dot value of 100 means that each dot represents 100 motor vehicle thefts.

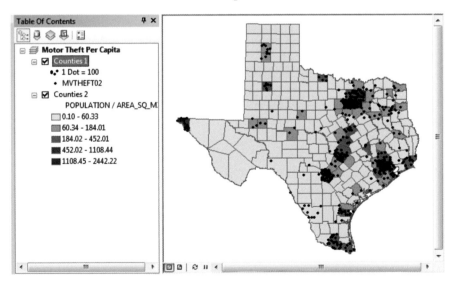

In ArcMap, you can also build query expressions to exclude dot density points or use other layers to act as a mask to exclude dots or place them in the specified area only. In the next few steps, you'll use a masking layer to hide dot density points.

12 Add the HarrisAdjacentCounties feature class from the TexasCounty geodatabase in your Data folder for chapter 8.

13 Zoom in on the newly added layer. It is located in the southeastern area of the state. (Your feature class may be colored differently from the one in the graphic below.)

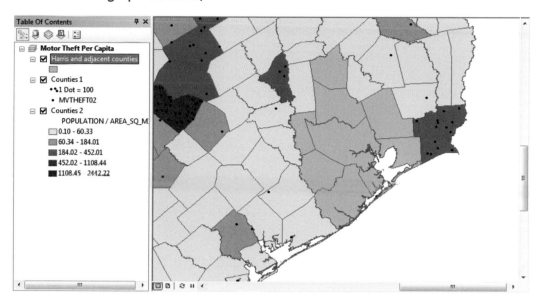

14 Open the Counties 1 properties. On the Symbology tab, click the Properties button. On the Dot Density Symbol Properties dialog box, select the Use Masking check box, click HarrisAdjacentCounties in the Control Layer list, and click the "Place dots only in these areas" option. Then click OK. The options on this dialog box allow you to use a mask to either hide or show dot density points. When using a mask to exclude, no points will be visible in the defined area. When using a mask to include, points in the defined area only will be visible.

15 Click OK again to close the Layer Properties dialog box. In the table of contents, hide the HarrisAdjacentCounties layer. When creating dot density maps, consider the area units (such as census tracts or block groups), dot size, dot value, and dot placement. The distribution of data can create variations in the pattern of a dot density map. The random distribution of dots can often be misleading. For example, the dots in a county are randomly distributed. However, in real life, the majority of the population is centered around urban areas. There is likely to be a trial-and-error period when choosing the most appropriate dot value and size—there is no exact method. Generally, a medium-size dot should be chosen. Small dots produce a widely dispersed dot pattern and may convey disproportionate map accuracy. Large dots produce overly dense dot patterns and hide any noticeable map patterns.

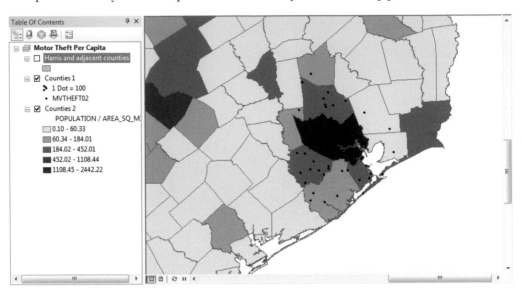

On your own

Try changing the dot value and the dot size to see how the density pattern changes. To turn off the masking layer, return to the Dot Density Symbol Properties dialog box and clear the Use Masking setting.

16 If you want to save your work, rename your map document and save to your MyData folder for chapter 8.

17 If you are continuing to the next exercise, leave ArcMap open. Otherwise, exit the application and click No if prompted to save your changes.

Exercise 8c
Using graduated and chart symbols

Graduated-symbol maps use proportionately sized symbols such as circles or squares to represent different ranges of values. This type of map allows you to visualize classifications by means of differently sized symbols, making an effective visual tool.

Charts are also an effective visual communication tool. A pie chart can show multiple sets of quantitative data at the same time: its overall size can be proportionate to an attribute value, and its divisions (or slices) can show the composition of several attribute values. It might be used to show how much of a country's population is divided into different age groups, for example.

Bar charts are useful in side-by-side comparisons, especially in cases where attribute values do not make up a whole. For example, bar charts might be used to compare census tract household income in 2000 versus 2010. Another variation of the bar chart is a stacked bar chart representing the sum of attributes. For example, if you wanted to display the marital status in a county, you could symbolize the values for three different attributes—single, married, and divorced—using a stacked bar chart.

In this exercise, you'll create three more maps for this crime-mapping project. The first will use graduated symbols to map aggravated assault. The second will use bar charts to compare two different years of larceny (theft). And the third will use pie charts to show the makeup of all the crime attributes included in the dataset.

6
7
8c
9
10

1 In ArcMap, open the ex08c map document from your Chapter08 folder. The map is open in data view. In the table of contents, there are three data frames: Assault, Larceny, and All Crime. Each data frame has a Counties layer (all from the same data source). The Assault data frame is active.

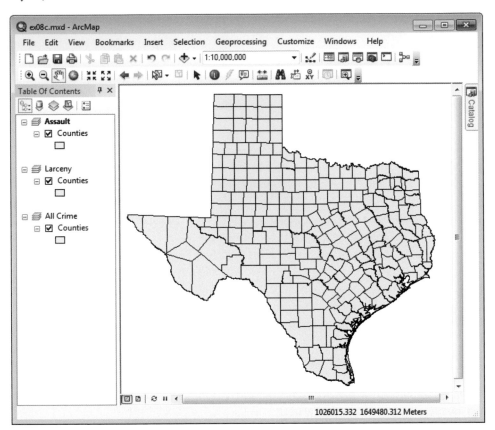

2 In the table of contents, right-click the Counties layer (of the Assault data frame) and open the attribute table (see facing page). Scroll to the right to the AGRASLT02 column header. Right-click the AGRASLT02 column header and click Sort Descending.

The AGRASLT02 attribute represents the number of aggravated assaults for each of the 254 counties in 2002. You'll use a natural-breaks classification to create five (5) classes.

3 Close the table. In the table of contents, double-click the same Counties layer to open its properties. On the Symbology tab, click Quantities > Graduated symbols. In the Fields frame, click the Value arrow and click AGRASLT02.

In the Symbol column, five differently sized symbols are shown representing five ranges. Next, you'll change the symbol template to something more appropriate.

4 Click the Template button to open the Symbol Selector.

5 In the Symbol Selector, click the Circle 1 symbol. Click the Color chip and click Poinsettia Red. Click OK to close the dialog box.

Next, you'll change the background color.

6 Click the Background color chip to open the Symbol Selector.

7 In the Symbol Selector (see facing page), scroll down the style library and click the Med Yellow symbol, and then click OK.

Next, you'll adjust the symbol size to make the difference between the smallest and largest symbols more apparent.

8 Change the Symbol Size from 4 and 18 to values of 3 and 20, respectively.

9 In the Label column, click 0 - 198 and type Less than 199. **For the remaining labels, change the hyphen to the word** to. **For example, change 199 - 604 to** 199 to 604, **and then click OK.**

The map now shows graduated symbols for each county. The table of contents shows the graduated-symbol settings you made. Similar to the density maps you created in exercise 8b, a greater number of incidents are located in counties with a higher population count.

You'll continue by creating bar charts to compare larceny between 1994 and 2002.

⓵⓪ In the table of contents, right-click the Larceny data frame and click Activate. Open the Counties attribute table. Scroll to the right to the LARCY94 and LARCY02 column headers.

MXCRM94	MXCRM02	INDXCRM02	LARCY94	LARCY02	Shape_Length	Shape_Area
1978	1721	1709	1149	986	362810.708823	2782536174.16016
510	359	357	357	226	256958.530282	3872520143.14844
2960	3253	3244	1850	1723	324423.028329	2230976753.73047
1166	1571	1564	608	1016	637430.8112	729743480.927581
116	134	132	43	56	195565.228866	2394913970.875
46	34	34	14	22	194434.295322	2362038343.85938
748	848	846	363	543	235091.824446	3146874670.30078
439	479	478	254	274	249878.302106	1694080722.15625
220	150	150	158	87	187438.105011	2137950567.92188
301	547	542	166	343	212441.762169	2058185277.12695

|◀ ◀   1 ▶ ▶| ▦ ▤ (0 out of 254 Selected)

Counties

These two layer attributes represent the number of larcenies for each of the 254 counties in 1994 and 2002. Next, you'll look at the attribute statistics to get a better sense of how the data is distributed.

11 Right-click the LARCY02 column header and click Statistics.

Many of the counties have low counts of larceny (the tallest blue bar in the Frequency Distribution chart). As you saw earlier, a majority of the incidents occur in highly populated counties. Instead of creating bar charts for each county, you'll exclude counties that have a low population.

12 Close both the Statistics of counties window and the attribute table.

13 In the table of contents, double-click the Counties layer (for the Larceny data frame) to open the Layer Properties dialog box. On the Symbology tab, under Show, click Charts > Bar/Column. Scroll down the Field Selection list, click LARCY94, and click the right-facing arrow (>) button to add it to the list on the right. Add LARCY02 as well. Your bar chart colors are randomly selected and may be different from the ones in the graphic below.

14 Click the Exclusion button to open the Data Exclusion Properties dialog box. Click the Query tab and in the Exclude clause list, double-click "POPULATION". Click the less than (<) button, and then press the SPACEBAR and type 500000 in the expression box below.

Do not use comma (,) separators—for instance, to separate digits—in query expressions. Commas will invalidate the query. The Verify button can check whether a query is valid or not.

You just constructed a query that *excludes* counties with a population less than 500,000.

Querying data is a fundamental technique when working with GIS. You'll learn more on how to query data in part 5, "Getting information about features."

15 On the Data Exclusion Properties dialog box, click the Legend tab. Select the Show symbol for excluded data check box. Click the Symbol color chip to open the Symbol Selector.

16 In the Symbol Selector, click the Fill Color chip, click Gray 20%, and click OK.

17 In the Label box, type Population less than 500,000 and click OK.
The "excluded" symbol and label will be added to the table of contents after you confirm your changes on the Layer Properties dialog box in step 19.

But first, you'll make a slight adjustment to the size of the bars in the chart.

18 Click the Properties button to open the Chart Symbol Editor dialog box. In the Bars frame, adjust the width to a value of 10, and then click OK.

19 **Click OK to close the Layer Properties dialog box.** The bar charts are drawn on the map for the counties that were not excluded (those with a population greater than 500,000.) These counties include Bexar, Travis, El Paso, Tarrant, Dallas, Harris, and Hidalgo.

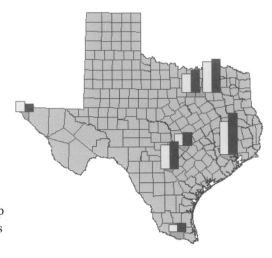

Next, you'll create the final map to show all the crime attributes included in the 2002 dataset.

20 **Activate the All Crime data frame. Open the properties of the data frame's Counties layer. On the Symbology tab, under Show, click Charts. The Pie option is highlighted. In the Field Selection list, click AGRASLT02 and click the right-facing arrow (>) button to add it to the list on the right. Continue to add the following attributes: ARSONS02, RAPE02, MVTHEFT02, MURD02, ROBB02, BURG02, and LARCY02.**

21 **Right-click the Color Scheme ramp and click Graphic View to turn off that view. Click the Color Scheme arrow and click Enamel.**

A symbol color is randomly assigned to each attribute using the selected color scheme. Similar to the bar charts, you'll exclude some values and change the size of the pie charts.

22 Click the Exclusion button to open the Data Exclusion Properties dialog box. Click the Query tab and in the Exclude clause list, double-click "POPULATION". Click the less than (<) button, and then press the SPACEBAR and type 500000 in the expression box below.

On your own

Change the color scheme to your liking.

23 Click the Legend tab. Select the Show symbol for excluded data check box. Click the Symbol color chip to open the Symbol Selector. Click the Fill Color chip, click Gray 20%, and click OK.

24 In the Label box, type Population less than 500,000 and click OK.

25 Next, click the Size button. On the Pie Chart Size dialog box, click the "Vary size using the sum of the field values" option. Adjust the symbol size to 20, select the Appearance Compensation (Flannery) option, and then click OK. The size of each pie chart will be based on several factors: the numeric size (in points), the variation due to the sum of the attribute values, and the appearance compensation option. When compensation is activated, ArcMap scales

up proportional circles (in this case, pie charts) so that the larger the circle, the greater the amount of scaling. The compensation of proportional circles was the major research focus of James Flannery, a professor of geography specializing in cartography, in the 1950s.

26 Click OK to accept your settings. The color scheme may be applied differently to your map than it is in the graphic below.

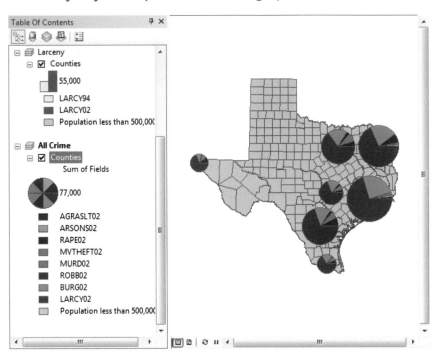

27 On the menu bar, click View > Layout View.

28 Your map scales may not be exact. On the page layout, click a data frame to activate it, click the Bookmarks menu, and then use the bookmark provided for each data frame.

On your own

Create layer files for several of your data layers.

29 If you like, rename your map document and save to your MyData folder for chapter 8.

30 If you are continuing to the next chapter, leave ArcMap open. Otherwise, exit the application and click No if prompted to save your changes.

Summary

In this chapter, you learned what to consider when classifying quantitative attributes. This was useful in determining which ArcMap classification method to use and when viewing a data histogram to see how data is distributed. You also learned how to create a visual summary with bar and pie charts. These are some of the most useful tools for displaying information. By now, you should be comfortable manipulating data to achieve different views using graduated colors, graduated symbols, proportional symbols, and dot density effects.

Terms

graduated-color map

graduated-symbol map

proportional-symbol map

dot density map

manual classification

equal interval

defined interval

quantile

natural breaks

geometrical interval

standard deviation

histogram

normalization

Chapter 9
Labeling features

Using dynamic labels
Setting rules for label placement
Creating graphic labels
Converting dynamic labels to annotation

Learning objectives

- *Create dynamic labels*
- *Modify the default appearance of dynamic labels*
- *Use predefined styles for labels*
- *Fine-tune the placement properties of dynamic labels*
- *Prioritize placement of dynamic labels*
- *Set a scale dependency for label display*
- *Create graphic labels using the text tools on the Draw toolbar*
- *Convert labels to annotation and modify placement*

BROADLY SPEAKING, A *label* IS ANY TEXT THAT NAMES OR DESCRIBES A FEATURE on a map, including proper names (Haiti, Buckingham Palace, Interstate 10), generic names (Lake, Tundra, Hospital), descriptions (Residential, Commercial, Hazardous), or numbers (7,000 placed on a mountain to indicate its elevation). In ArcMap, labels specifically represent values in a layer attribute table. You can add other bits of text to a map using the Draw toolbar, but, strictly speaking, these are not labels and do not behave quite the same way. Labels, for example, stay close to the features they are associated with as you zoom in or out. Text is not similarly sensitive. Its position may change markedly in relation to the features if you zoom or pan the map.

Default labels, *top*, are often sufficient, but you can modify the default labels, *bottom*, to improve the map's appearance by changing the font, color, size, character width and spacing; adding special effects such as halos; or specifying placement settings, such as curving river labels along the lines of features. You can also choose from many predefined label styles offered in ArcMap. Data from Esri Data & Maps 2010, courtesy of ArcWorld and ArcWorld Supplement.

As you have already seen in chapters 3 and 7, you can label all the features in a layer with the use of a single click. This process, called dynamic labeling, is fast, but it has its limitations. For one thing, ArcMap chooses the label positions. (You can set guidelines, but the actual placement is out of your control.) If the scale of your map changes, or if different layers are added or removed, ArcMap may change the position of labels. Still, dynamic labels may be useful for interactive Internet map publishing.

An alternative to dynamic labels is **annotation**. Annotation is static text or graphics that you can individually select, position, and modify. You can enter annotation manually or convert existing labels to annotation. Annotation features are stored either as part of a map; in an annotation group the same way as dynamic labels; or in a geodatabase, as an annotation feature class. If you are creating a printed map and want significant control over the appearance and placement of labels, annotation is your likely choice.

In the following exercises, your map display will be redrawn many times as ArcMap adds and moves labels. If your map fails to redraw as it should, click the Refresh button at the bottom left corner of the map display (see facing page).

Refresh button

Data

In this chapter, you will use most of the same Florida data you used in chapter 7. The feature classes, stored in the Florida geodatabase, are as follows:

- Bathymetry—line features that represent changes in ocean depth *(Source: Florida Fish and Wildlife Conservation Commission's Fish and Wildlife Research Institute and National Oceanic Atmospheric Administration)*
- Cities—major cities of Florida *(Source: National Atlas of the United States)*
- Coral—polygons representing coral populations off the coast of Florida *(Source: Florida Fish and Wildlife Conservation Commission's Fish and Wildlife Research Institute)*
- Counties—polygons representing Florida county boundaries *(Source: US Census Bureau)*
- Hwy1—a single line feature depicting a major state highway that traverses the Atlantic coastline and Florida Keys *(Source: Florida Department of Transportation)*
- Lakes—polygons representing Florida's major lakes *(Source: Florida Department of Environmental Protection)*
- Rivers—line features representing Florida's major rivers *(Source: Florida Department of Environmental Protection)*
- Shipwrecks—a selection of historical shipwrecks along the southern Atlantic coast of Florida and the Florida Keys *(Source: National Oceanic and Atmospheric Administration Office of Coast Survey's Automated Wreck and Construction Information System)*
- Southeast—a polygon layer that includes demographics *(Source: ArcUSA, US Census, Esri)*
- StParks—polygons representing Florida state parks *(Source: Florida Department of Environmental Protection)*
- florida_coast—an image of seafloor topography *(Source: Natural Earth)*

Exercise 9a
Using dynamic labels

Dynamic labels are easy to work with because they behave as a group. When you turn them on or off, change their symbol, or change the attribute value they express, it affects all the labels in the layer. The drawback to dynamic labeling is that ArcMap doesn't always put the labels just where you'd like them. For this reason, labels are often converted to annotation in the late stages of map production.

The position ArcMap chooses for dynamic labels depends on many factors, including the size of your application window. Your results may look slightly different from the graphics in the exercises for this chapter. For labeling, it's best to maximize your ArcMap window if possible.

1 **In ArcMap, open the ex09a map document from your C:\EsriPress\ GTKArcGIS\Chapter09 folder.** The cities already have default labels (see facing page), but you need additional labels to make the map more helpful.

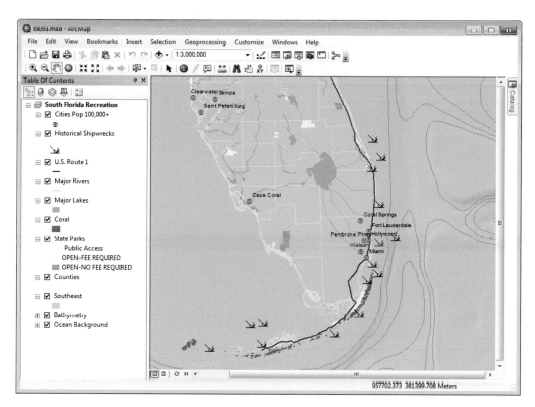

Recall that the map will be printed in tourism brochures. First, you will label the state of Florida and its water features.

2 In the table of contents, right-click the Southeast layer and click Label Features.

Each feature in the layer is labeled, but since this map shows only one state, you see only one label, for Florida, in the center of the map display. ArcMap tries to put the labels in the center of each polygon. Next, you'll change the font and color of the label.

3 In the table of contents, double-click the Southeast layer to open the Layer Properties dialog box. Click the Labels tab. In the Text String frame, the Label Field list shows which field in the attribute table is being used for labels (STATE_NAME). Below this, you see the current label symbology. The labels for this layer are drawn in 8 pt black Arial.

4 Click the color chip. On the color palette, click Gray 20%.

5 In the Text Symbol frame, in the size list, replace 8 with 14.

6 Next, click the Symbol button. In the Symbol Selector, click Edit Symbol.

7 On the Editor dialog box, click the Formatted Text tab. Next to Character Width, highlight 100.00 and enter 130. Under Text Case, click All Caps.

8 Next, click the Mask tab. Under Style, click Halo. Change the size to 1.

6

7

8

9a

10

9 **Click OK to close the Editor dialog box, and then OK again to close the Symbol Selector. Make sure your Layer Properties dialog box matches the graphic below, and then click OK.**

You have successfully created a label for the state of Florida that is well-balanced for its purpose: the halo ensures the text isn't lost on the map, but the light-gray color allows it to assume a lesser importance so it doesn't compete with other features or labels.

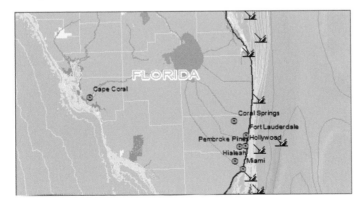

Next, you will use a predefined style to label lakes.

10 Open the properties of the Major Lakes layer and click the Labels tab. Select the Label features in this layer check box. Click the Symbol button to open the Symbol Selector.

11 In the Symbol Selector, scroll through the Esri style library and click the label style River, and then click OK. (There is not a unique style for lake labels, so the River style will work fine.)

AaBbYyZz
Sea

AaBbYyZz
River

AaBbYyZz
Stream

AaBbYyZz
Map Title E-Size

AaBbYyZz
Map Title D-Size

12 Make sure your Layer Properties dialog box matches the graphic below, and then click OK.

The lakes on the map are now labeled.

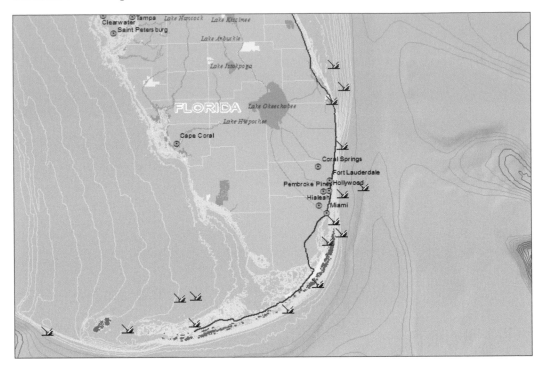

The Florida label moved position once the lakes were labeled. Dynamic labeling means labels will change position depending on the placement of other features and other labels and the size of the map scale.

13 If you want to save your work, rename your map document and save to your MyData folder for chapter 9.

14 If you are continuing to the next exercise, leave ArcMap open. Otherwise, exit the application and click No if prompted to save your changes.

ArcMap chooses label positions to minimize conflicts, but you can control the positioning somewhat, as you'll see in exercise 9b. If you want absolute control of the label positions, you must either create graphic labels, which you'll do in exercise 9c, or convert dynamic labels to annotation, which you'll do in exercise 9d.

Exercise 9b
Setting rules for label placement

When you place labels dynamically, ArcMap chooses label positions according to rules you can modify. Labels for points can occupy any of eight positions around the feature (above, below, above right, below right, and so on). You can rank these positions and prohibit certain ones as desired. Labels for line features can be placed above, below, or on the features and can be made to follow features' curves. Labels for polygons are always placed as close to the center of the polygon as possible, but you can choose to rotate the label vertically, or remove the label if it doesn't fit inside the polygon.

In addition, weights can be set for each layer on a map to prioritize and handle labeling conflicts. If there isn't room to label two nearby features in different layers, only the feature in the layer with the higher label weight will be labeled. Additionally, if you don't want certain features obscured by labels, you can set a higher feature weight to minimize the possibility of labels overlapping features.

In this exercise, you will modify the properties of the Cities labels, and then create labels for the Major Rivers layer with distinct rules for label placement. You will also change the layer priorities that determine which feature is labeled when there is a conflict.

1 In ArcMap, open the ex09b map document from your Chapter09 folder. The map looks as it did at the end of exercise 9a. If the cities labels were restricted to the left of the points, there would be less clutter in the southern Atlantic coast area.

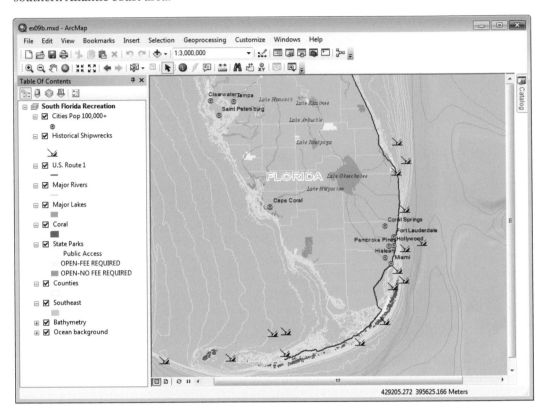

2 Open the properties of Cities Pop 100,000+, and on the Layer Properties dialog box, click the Labels tab.

3 Under Other Options, click Placement Properties. On the Placement Properties dialog box, make sure the Placement tab is active.

When labels are offset horizontally around points (the default), each possible position can be ranked. The graphic on the dialog box, in the Point Settings frame, shows the current setting. The feature is represented by the white square in the middle, and the numbers in the surrounding squares rank each position—1 is the most desirable and 3 the least. The number 0 means a label cannot occupy that position. In the current setting, ArcMap places labels above and to the right of features as long as they do not overlap other labels in any layer. All positions are allowed, although positions below and to the left of the feature are discouraged with this setting.

In this exercise, you want to center the labels to the left of the features.

4 On the Placement Properties dialog box, click Change Location to open the Initial point placement dialog box.

5 On the Initial point placement dialog box, scroll down and click the setting for Prefer Center Left, all allowed. Click OK.

6 Click OK to close the Placement Properties dialog box, and then click OK again to close the Layer Properties dialog box. This is better. Most of the city names should be located at the left side of the corresponding point. However, at this scale, many of the labels overlap the point symbols, and it's not clear which point is Hollywood and which is Pembroke Pines. And depending on the size of the ArcMap window, some city labels may still be in conflict with shipwreck features. You'll make these final adjustments later, when you convert the labels to annotation.

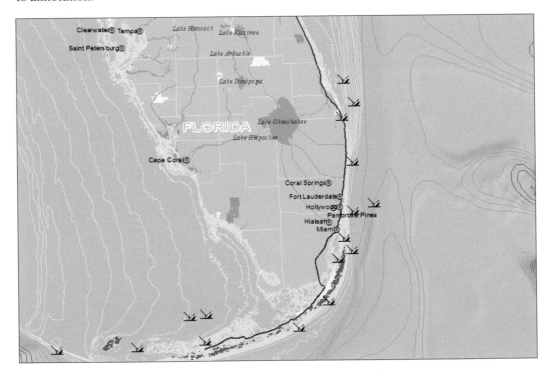

7 Open the properties of Major Rivers and click the Labels tab. Turn on the labels for the layer.

8 Accept the default style and click Apply.

9 On the Layer Properties dialog box, click Placement Properties. The Placement Properties dialog box is different for points, lines, and polygons. On the Placement tab, under Line Settings, click the Curved option. Make sure the Remove duplicate labels option is selected, and then click OK.

10 Click OK to close the Layer Properties dialog box and see how your changes look.

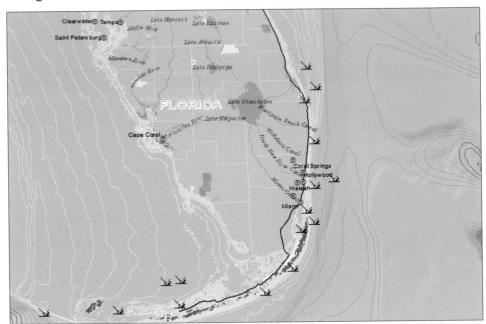

ArcMap tries to label as many features as possible without overlapping labels. You want to label most of the rivers and canals, but the map looks a little cluttered, and your city labels might look wrong again. So next, you will downgrade the priority level of rivers to help clean up the map.

Allowing labels to overlap

By default, ArcMap does not put labels on top of each other. This means that, depending on the scale and the part of the map you are looking at, some features in a layer may not be labeled. If you want to be sure every feature in a layer is labeled, you can tell ArcMap to place overlapping labels. You can also control label-on-label and label-on-feature overlaps for individual layers. For more information, in ArcGIS for Desktop Help, on the Contents tab, navigate to Mapping > Adding text to a map > Displaying labels > Placing labels with the Standard Label Engine > Placing overlapping labels.

---

11 Turn on the Labeling toolbar—on the Customize menu, click Toolbars > Labeling. The Labeling toolbar, with several buttons for managing labels, appears. The toolbar may be docked or floating.

12 On the Labeling toolbar, click the Label Priority Ranking button [⊿].

On the Label Priority Ranking dialog box, the layers of the map are listed in order of labeling priority. Next, you'll move the Major Rivers layer to the bottom of the list.

13 Click Major Rivers to highlight it, and then click the Bottom (not the Down) arrow.

14 Move the Label Priority Ranking dialog box out of the way and click Apply. Notice how the map changes—there are fewer labels for rivers and canals, and the map is easier to read.

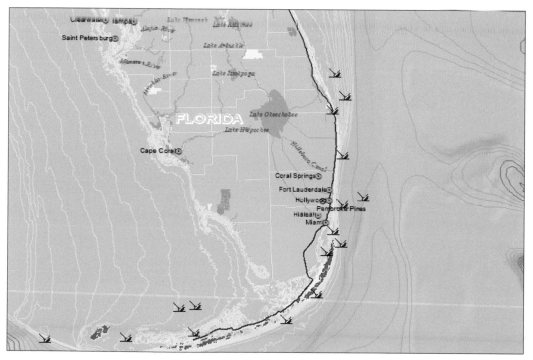

15 **Click OK to close the Label Priority Ranking dialog box.** Another way to reduce label clutter is to implement scale-dependent labels. You did this in chapter 3 and will do it here next.

On your own

Read the ToolTips for the other buttons on the Labeling toolbar.

Label Manager and setting label classes

You can use the Label Manager dialog box [icon] to set label properties and scale dependencies for all layers using one window rather than opening each layer's properties separately. You can also use the Label Manager dialog box to create different classes of labels based on symbology categories or other differentiating attributes. For instance, in the Florida map, you could choose one type of label for the "No fee" state parks and a different type of label (or no label) for the "Fee required" parks. Or if you have a roads feature class whose attribute table contains a field for road type, you could use a different label for each class—for example, symbol markers for highways, black text for main roads, and gray text for local roads. Creating label classes also allows you to have more than one label per feature—for example, if you wanted counties to be labeled with both their name and population, you could use two different labels. Once you create label classes, you can set the label properties for each class separately.

Maplex

For advanced label placement options, you can turn on the Maplex Label Engine, now accessible from the Labeling toolbar with any ArcGIS license. Maplex allows you to modify the density of labels and allows more control over label positioning for different data types. For more information on Maplex, in ArcGIS for Desktop Help, on the Contents tab, navigate to Mapping > Adding text to a map > Displaying labels > Working with the Maplex Label Engine.

16 Open the properties of Major Rivers. On the Labels tab, under Other Options, click Scale Range.

17 On the Scale Range dialog box, click the Don't show labels when zoomed option, and in the Out beyond text box, type 1:1,000,000. Click OK to close the Scale Range dialog box, and then OK again to close the Layer Properties dialog box. The Major Rivers labels will appear only at scales larger than 1:1,000,000. If the tourism brochure uses a large-scale local map in addition to the main map, rivers and canals will be labeled.

18 On the menu bar, click Bookmarks > Canals to see the scale-dependent labels.

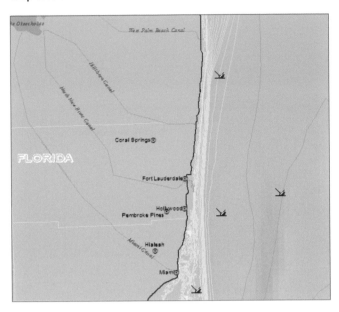

19 Zoom to the full extent of the map and close the Labeling toolbar.

20 If you want to save your work, rename your map document and save to your MyData folder for chapter 9.

21 If you are continuing to the next exercise, leave ArcMap open. Otherwise, exit the application and click No if prompted to save your changes.

Exercise 9c
Creating graphic labels

If you want to add stand-alone text to your map or if you just want to label a few features interactively, you can use the tools on the Draw toolbar. Most of the tools on the Draw toolbar are for creating stand-alone text, but there is also a Label tool that allows you to choose a label that is connected to a feature's attributes. The labels or text boxes created from the Draw toolbar act like graphics—they can be selected and moved wherever you like. In this exercise, you'll add text to mark the Gulf of Mexico and the Atlantic Ocean, and then create a graphic label for the US Route 1 feature.

This technique is meant for cases in which you have only a few features to label or if you want to label base layers that don't have attributes.

1
2
3
4
5
6
7
8
9c
10
11
12
13
14
15
16
17
18
19
20

1 **In ArcMap, open the ex09c map document from your Chapter09 folder.** As before, some of your dynamic labels may be in slightly different locations from what you see in the graphic.

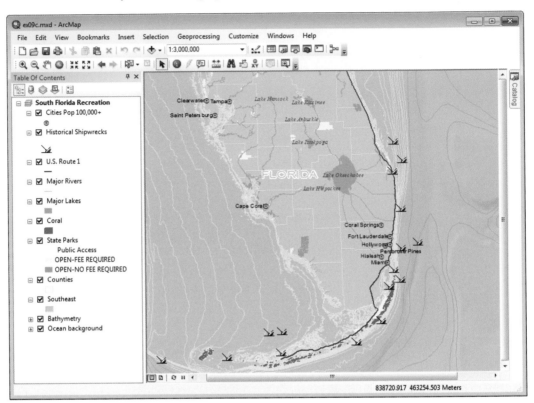

Map readers would be better oriented if you labeled the Gulf of Mexico and the Atlantic Ocean. However, you don't have an ocean layer with attributes to which you can attach labels. (The ocean background you see is a raster layer, which doesn't have text attributes.) Since there are no distinct Gulf of Mexico or Atlantic Ocean features, there is nothing to label. Instead, you'll next add a graphic label using the Text tool. Graphic labels are selectable and stand alone on the map—they are not connected to any feature.

2 **If it's not already available, display the Draw toolbar—on the Customize menu, click Toolbars > Draw.**

3 **On the Draw toolbar, click the Text tool** ☑️ .

4 Move the mouse pointer over the map. The pointer changes to a crosshair with an A (for text). Click in the Gulf of Mexico to add a text box.

5 In the text box, type Gulf of Mexico and press ENTER

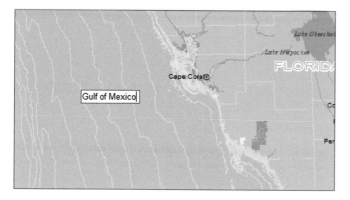

The text is drawn in 10 pt black Arial, as you can see on the Draw toolbar. The cyan dashed line indicates the text is selected. Next, you'll increase the font size.

6 Right-click the Gulf of Mexico label. On the shortcut menu, click Properties. On the Properties dialog box, click the Text tab, if not already active, and click Change Symbol.

This opens the Symbol Selector. Next, you will select a style, and then modify it according to your preferences.

7 In the Symbol Selector, scroll down until you see the predefined label style for Ocean and click it to select the color and spacing determined by that style. Next, click the font arrow and click Book Antigua (if it's installed on your computer) or another font of your preference.

8 Click OK to close the Symbol Selector, and then OK again to apply the new properties to the text. Now that it's larger, the text may have to be moved.

9 Move the mouse pointer over the text. The pointer becomes a four-headed arrow. Use the pointer to drag the text so that it approximately matches the graphic below, and then release the mouse button. Click in an empty blue area to clear the text selection.

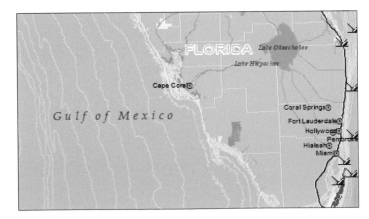

On your own

Repeat steps 3–9, but this time label the Atlantic Ocean.

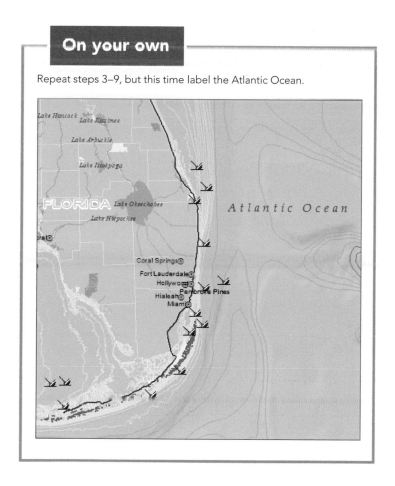

Finally, you will label US Route 1 using the Label tool.

10 On the Draw toolbar, click the Text tool arrow to see the other text tools available. Click the Label tool.

When you use any text tool, it will become the default tool shown on the Draw toolbar until you choose a different tool.

11 On the Label Tool Options dialog box, click the Place label at position clicked and Choose a style options. In the style library, click the first symbol, for U.S. Route.

12 Move your mouse pointer over the U.S. Route 1 line feature (the red line running down the Atlantic coast), just east of Lake Okeechobee. When you see the MapTip US1, click to place the label.

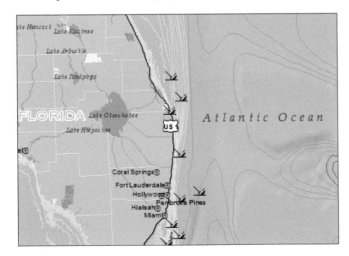

13 With the label selected—it should be outlined with a cyan dashed line—right-click it and click Properties. Change the symbol size to 6 pt black Arial.

14 Click OK on any open dialog boxes, click the Select Elements tool (on either the Tools toolbar or the Draw toolbar), and click anywhere on the map to clear the highway symbol selection.

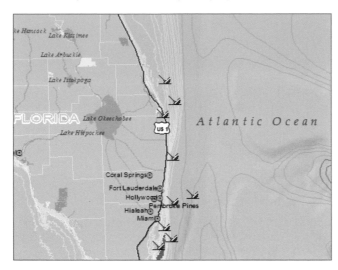

15 If you want to save your work, rename your map document and save to your MyData folder for chapter 9.

16 If you are continuing to the next exercise, leave ArcMap open. Otherwise, exit the application and click No if prompted to save your changes.

Exercise 9d
Converting dynamic labels to annotation

Although you have considerable control over dynamically placed labels, you can't specify exact positions or symbolize them individually. To do this, you must convert the labels to annotation, which means, essentially, that you tell ArcMap to turn over full responsibility for label placement and appearance to you. Annotation can be selected, moved, and modified piece by piece, like the text boxes you created in exercise 9c. The difference is, you can quickly create annotation for all features in a layer rather than one at a time. Typically, you convert labels to annotation when your map is nearly finished and you want to make final adjustments to the labels.

Annotation can be saved either within your map document in an annotation group, the same way as dynamic labels, or saved to an annotation feature class within a geodatabase. It then becomes much like any other spatial dataset—it can be copied, edited, and added to any map document.

For more information, in ArcGIS for Desktop Help, on the Contents tab, navigate to Mapping > Adding text to a map > Converting labels to annotation.

In this exercise, you will convert the cities labels to annotation so you can make adjustments to the label positions. Once converted to annotation, the labels will keep the same size and location relative to surrounding features.

1
2
3
4
5
6
7
8
9d
10

1 In ArcMap, open the ex09d map document from your Chapter09
folder.

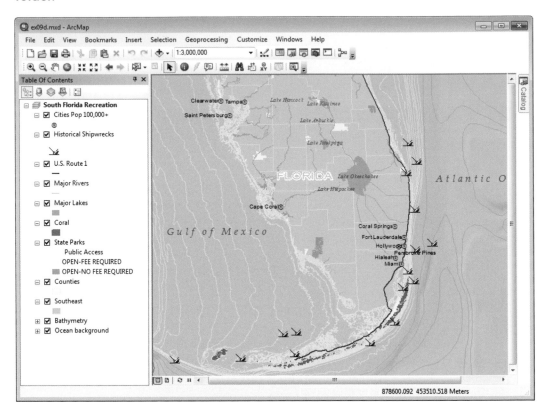

2 In the table of contents, right-click the Cities Pop 100,000+ layer
and click Convert Labels to Annotation. This opens the Convert
Labels to Annotation dialog box.

The default option for storing annotation is to store it in a geodatabase.
However, since you want to modify only a handful of labels, you will store the
annotation as an annotation group in the map document.

> **Map annotation or geodatabase annotation?**
>
> If you have a large number of labels, it is best to use geodatabase annotation, because it is faster and can handle a greater number of annotation features more efficiently. Creating annotation often takes a lot of time, so saving to a geodatabase can be desirable, allowing the work to be reused and shared. Geodatabase annotation can be either stand-alone or feature-linked. Stand-alone annotation is used for text that is not associated with a map feature. In feature-linked annotation, the labels are associated with another feature class, so if a feature is moved or deleted, the annotation is moved or deleted as well. You will learn more about geodatabases in chapter 11.

3 Click the option to store annotation in the map. Make sure your dialog box matches the graphic below, and then click Convert. The labels are converted to annotation. In the process, their positions may have changed slightly.

Next, you'll move the labels for Hollywood and Pembroke Pines so it's easy to see which feature goes with which label. You will also move the labels for the other cities just a bit, so the text doesn't touch the corresponding point symbols.

4 **Zoom to the Canals bookmark.** Notice how large the annotation labels are in the map display, compared with the dynamic labels for lakes and rivers. Unlike dynamic labels, annotation labels are static, meaning they act like a feature—just as a lake grows in appearance as you zoom in on a map, so does the annotation.

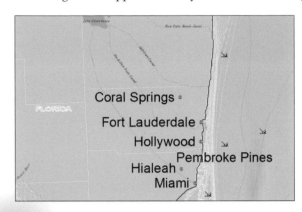

If you want your dynamic labels to act like static annotation when zooming in, go to the full extent of the map and set a reference scale for the data frame. (Right-click the data frame in the table of contents and click Reference Scale > Set Reference Scale.)

On your own

Use the Identify tool to determine which point represents Hollywood.

5 **On the Tools toolbar, click the Select Elements tool** [▶]. **Click the Hollywood label to select it (it will be outlined by a cyan dashed line).**

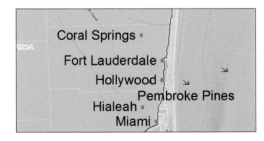

⑥ Drag the Hollywood label slightly north.

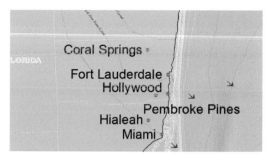

⑦ Click the Pembroke Pines label and drag it to the southwest of the point feature, as shown in the graphic. Click somewhere away from the label to clear the selection. Click Refresh if necessary.

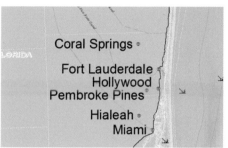

⑧ **Return to the full extent of the map to see if the readability of these labels has improved.** Much better   it's clear now which city is which.

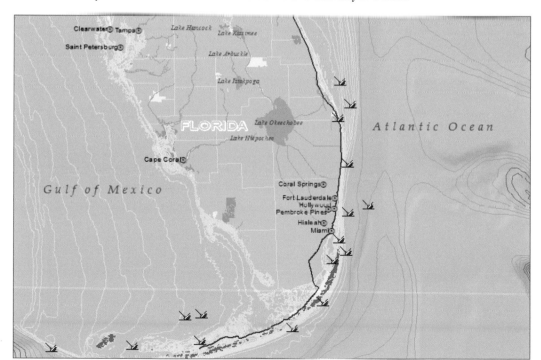

9 Zooming in as necessary, select each of the city labels and move them slightly farther away from their associated point features, to ensure that the text does not run into the point symbols.

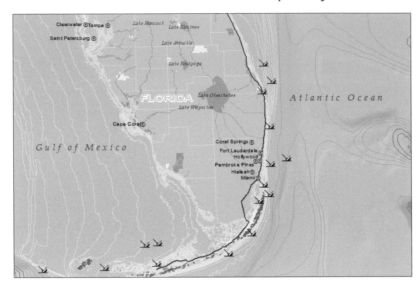

You have finished your Florida tourism map.

10 If you want to save your work, rename your map document and save to your MyData folder for chapter 9.

11 If you are continuing to the next chapter, leave ArcMap open. Otherwise, exit the application and click No if prompted to save your changes.

Summary

You now have a good foundation for creating nice-looking maps: in chapter 7, you learned how to symbolize basic GIS layers; in chapter 8, you learned how to create quantitative symbology; and in this chapter, you learned how to create labels using several methods. Remember, with dynamic labels, ArcMap places the labels (although you can influence that placement by setting priorities and weights), but with stand-alone text or annotation, you can select and move individual labels and give them a unique appearance. In chapter 10, you will continue building your mapmaking knowledge and learn how to create a layout.

Terms

label

annotation

Chapter 10
Making maps for presentation

Creating a layout
Adding titles and additional text
Adding standard map elements
Adding final touches and setting print options

Learning objectives

- *Become familiar with page layout tools*
- *Create effective map layouts*
- *Add text to a layout*
- *Create map elements such as a north arrow, scale bar, and legend*
- *Become familiar with print settings and options*

A MAP IS NOT COMPLETE WITHOUT SEVERAL CRITICAL FEATURES AND A well-composed page layout. A collection of map elements laid out and organized on a page is important to any map that is destined to be printed, or even for online (or onscreen) use. Common map elements that can be arranged in a layout include data frames (containing map layers), a legend, a scale bar, a north arrow, a map title, descriptive text, and additional graphics. Cartographic design should be informed by both aesthetics and the ability to communicate. Cartographic elements need to accurately convey information about the map without diverting attention from the map itself.

The measure of a good layout is how well it conveys information—how well it communicates to its readers—providing not only the facts, but also persuading opinion. A good map layout should contain all the surrounding elements needed to achieve this. When creating a layout, you will need to ask yourself some key questions:

- What is the purpose and intended use of the map?
- Who is the target audience?

- What physical (or digital) size and page orientation will the map have?
- Is the map stand-alone, or will it be part of a map series with maps that share a template?
- How will the map elements be organized on the page?
- Will the layout contain more than the common elements and include such things as graphs or reports?
- What is the intended map scale, and how large will the data frames be in the layout?

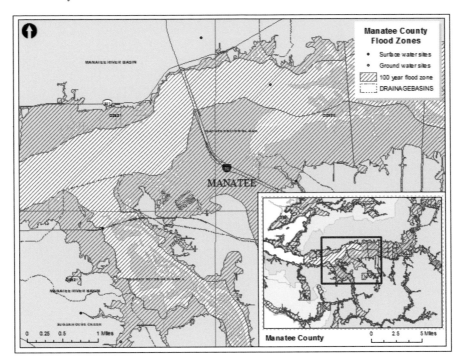

A layout created in ArcMap that shows the main map, an inset map, scale bars, a north arrow, and insert legend. Data from the Florida Geographic Data Library, courtesy of the University of Florida GeoPlan Center.

You may also want to think about how the layout of data frames and map elements affects the balance and composition of the layout. A well-planned page layout with properly sized data frames and a clear visual hierarchy of surrounding elements will allow readers to explore on a natural path without wandering to different sections of the layout. The focus will generally be on the map itself; however, with text, map elements, and other graphical objects surrounding it, you must be aware of how these elements may draw attention from the intended focal point. Finding the right balance between elements that should be in the foreground and those that are best suited to the background is one of the keys to good mapmaking.

In this chapter, you'll create the layout shown here of Queensland, Australia, bioregions and bio-subregions, which are researched areas of diverse biology and geographic landscape.

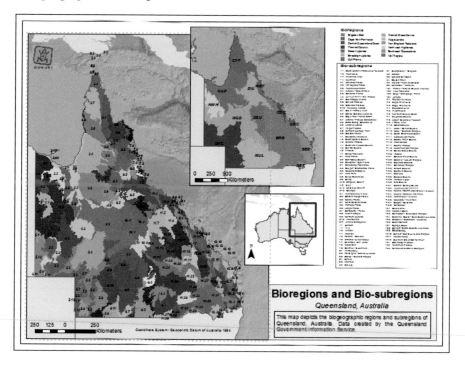

Data

The exercises in this chapter use the following datasets:

- australia.shp—polygons of the provinces of Australia *(Source: State of Queensland Department of Environment and Resource Management 2010)*
- bioregions.shp —polygons of bioregions in the province of Queensland, Australia *(Source: State of Queensland Department of Environment and Resource Management 2010)*
- biosubregions.shp—polygons of bio-subregions in the province of Queensland, Australia *(Source: State of Queensland Department of Environment and Resource Management 2010)*
- country_background.shp—polygons of surrounding countries, used as a background layer *(Source: ArcWorld Supplement)*
- ocean_background.tif—raster of the ocean, used as a background layer *(Source: Natural Earth)*

Exercise 10a
Creating a layout

The amount of space on which elements can be arranged may be anything from a letter-size sheet of paper to a wall poster. The orientation may be vertical (portrait) or horizontal (landscape). Choosing the dimensions and orientation of the map before you start makes the layout process easier. It will influence your decision on factors such as data frame size, map scale, text size, and other map elements.

ArcMap has rulers, guides, and a grid to help you arrange map elements on a page. You can also align, nudge (move incrementally), distribute (space evenly), order, rotate, and resize selected elements to place them where you want them.

For this exercise, you are a GIS student researching the biodiversity of Queensland, Australia, and have just finished mapping its 13 biogeographic regions and 133 biogeographic subregions (also known as bioregions and bio-subregions, respectively). Bioregions represent broad landscape patterns that are the result of the interplay among such factors as geology, climate, and biota. Bio-subregions are based on finer differences in geology, vegetation, and other biophysical attributes and are the basis for determining the major regional ecosystems. You will create a page layout as part of a presentation used to provide useful information to other researchers, students, and people interested in studying biodiversity.

1 **Start ArcMap and open the ex10a map document from your C:\EsriPress\GTKArcGIS\Chapter10 folder.** The map document contains three data frames: Bioregions, Bio-subregions, and Overview (see facing page). Data frames allow multiple maps to be shown on the page layout. The Bioregions data frame is currently active. Bioregions are classified and labeled based on their abbreviated names (for instance, Brigalow Belt is labeled BRB) and are symbolized using a color ramp with various shades of orange.

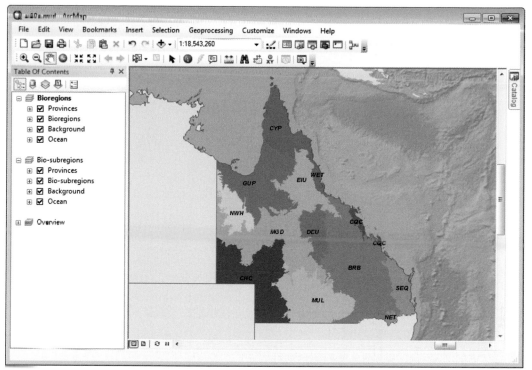

2 In the table of contents, expand the Bioregions layer to see the bioregion names and corresponding feature colors. When you are finished examining the feature symbols, collapse the layer.

What type of classification is used to classify these bioregions—categorical or quantitative?

All three data frames will be part of your page layout. You'll focus the attention on the bio-subregions map and position the bioregions map next to it. For reference, you'll include a map of Australia with an extent indicator showing where the state of Queensland is located. In ArcMap, an extent indicator is simply a rectangle that is placed in a data frame to reference the specific geographic location of another data frame.

3 **On the menu bar, click View > Layout View.**

In layout view, you see how the printed map will look. The data frames are stacked on the virtual page according to their order in the table of contents (although they can be reordered any way you want). Also included is an index of bio-subregions that consists of two text boxes and a rectangle. It will be easier to see what you're doing if you make the ArcMap application window as large as possible, so you'll do that next.

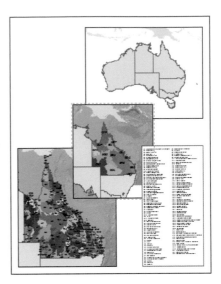

4 **On the ArcMap application title bar, click the middle button in the upper-right corner to maximize the application window. It is recommended that you maintain a large ArcMap window size for all the exercises in this chapter.** Notice that in layout view, the Layout toolbar automatically appears (if it wasn't already turned on) and is active. It may be floating or already docked to your ArcMap window. If it is docked, you will not see the title of the toolbar. To see it, you'll have to undock the toolbar.

5 **On the Layout toolbar, click the Zoom Whole Page button** 🔲 .

The virtual page is enlarged within the layout window. At the moment, the page orientation is portrait. Next, you'll adjust the page settings.

6 **On the menu bar, click File > Page and Print Setup.**

7 On the Page and Print Setup dialog box, in the Map Page Size frame (not to be confused with the Paper frame, in which the Landscape option will not change the orientation of the layout), set the orientation to Landscape, as shown in the graphic below. Click OK. (You may need to clear the Use Printer Paper Settings option.)

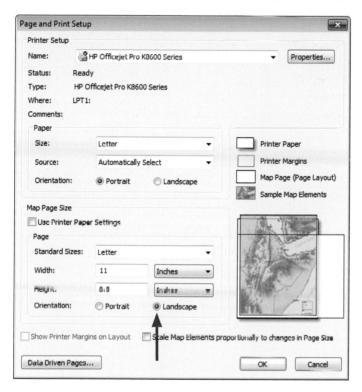

The orientation of the layout is changed. (You will also need to change the orientation of the printer paper, but you will do this later in the chapter.) Don't worry about the position of the overview map—you'll fix that next.

To help you place data frames exactly, you'll turn on page guides. Page guides are cyan-colored lines originating from arrows on the vertical and horizontal layout rulers. These guides help you align and position data frames and map elements but do not appear on the printed map. In this exercise, some guides have already been set for you.

8 **On the menu bar, click View > Guides.** The guides are displayed on the layout.

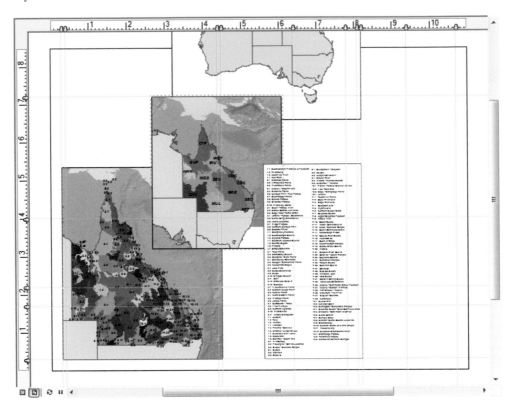

9 Place the mouse pointer over the vertical ruler on the left side of the display. Click to add a guide at 8 inches and another one at 8.2 inches, as shown in the graphic below.

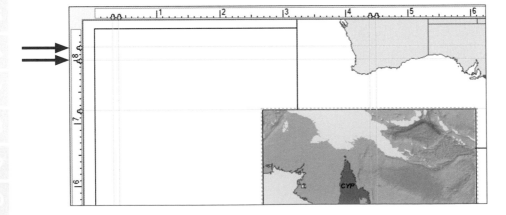

Don't worry if you didn't click exactly the right spot on the ruler. To move the guide, drag it to the correct position. The measurement is indicated in the upper-left corner of the display area.

The guides will be more useful if map elements snap (automatically align) to them, so you'll set the snapping next.

10 Right-click an empty part of the page. On the shortcut menu, click Guides > Snap to Guides.

Next, you'll resize and arrange the data frames on the page, beginning with Bio-subregions (the largest data frame).

11 In the display area, click the Bio-subregions data frame to select it. Once selected, it is outlined with a cyan dashed line and marked with cyan selection handles.

12 Move the mouse pointer over the data frame. When the pointer becomes a four-headed arrow, use the pointer to drag the frame so its upper-left corner snaps to the guides at 8.0 inches on the vertical ruler and 0.3 inch on the horizontal (top) ruler, as shown in the graphic below.

Next, you'll make the Bio-subregions data frame bigger by dragging a selection handle. When you resize a data frame, ArcMap automatically adjusts the *map extent* and the display scale.

If you want to make sure the map extent or the display scale stays the same, you can set a fixed extent or scale on the Data Frame tab on the Data Frame Properties dialog box.

13 Place the mouse pointer over the selection handle at the lower-right corner of the data frame. When the pointer becomes a double-headed arrow, use the pointer to drag the corner of the data frame until it snaps to the guides at 0.3 inch (vertical ruler) and 6.4 inches (horizontal ruler).

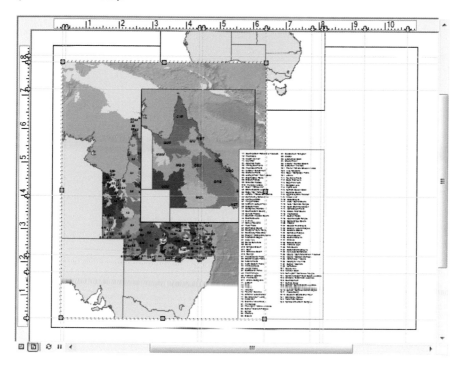

14 On the Standard toolbar, in the Map Scale box, replace the existing scale with 1:13,000,000.

1:13,000,000 ▾

The data frame is updated with an appropriate map scale to show the entire Bio-subregions layer.

Next, you'll move the Bioregions data frame.

15 On the virtual page, click the Bioregions data frame to select it. Drag the frame so its upper-left corner snaps to the guides at 8.2 inches (vertical ruler) and 4.4 inches (horizontal ruler). On the Standard toolbar in the Map Scale box, replace the existing scale with 1:25,000,000.

Next, you'll reposition the Bio-subregions index.

16 In the display area, click the index to select it. Drag the index so its lower-right corner snaps to the guides at 2.2 inches (vertical ruler) and 10.7 inches (horizontal ruler).

Next, you'll reposition, rescale, and change the layer order of the Overview data frame.

17 In the display area, click the Overview data frame to select it, and then right-click it. On the shortcut menu, click Order > Bring to Front.

18 Right-click the Overview data frame and click Properties. On the Data Frame Properties dialog box, on the Size and Position tab, in the Size frame, enter 2 in in the Width box and 1.8 in in the Height box. Then click OK.

Don't worry about the appearance or map scale of the overview map—you'll fix this in a moment.

19 Drag the Overview data frame so its lower-right corner snaps to the guides at 2.2 inches (vertical ruler) and 7.8 inches (horizontal ruler).

20 On the Standard toolbar, change the scale to 1:120,000,000. The data frames now have their final sizes and positions.

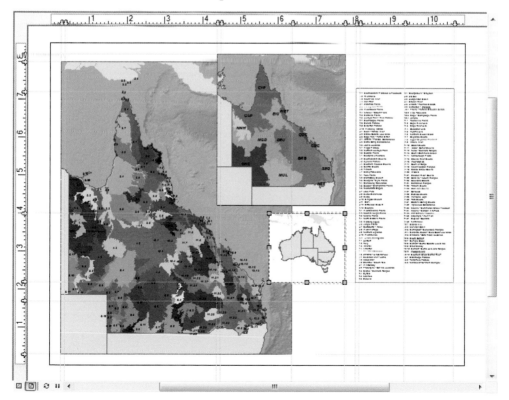

Next, you'll finish the Overview data frame by creating an extent indicator.

21 **Right-click the Overview data frame and click Properties. On the Data Frame Properties dialog box, on the Extent Indicators tab, click Bio-subregions in the Other data frames list and click the right-facing arrow (>) button to move the item to the "Show extent indicator for these data frames" list. Then click OK.** An extent indicator framed in red (see facing page) is drawn to show the spatial extent of the Bio-subregions data frame inside the Overview data frame.

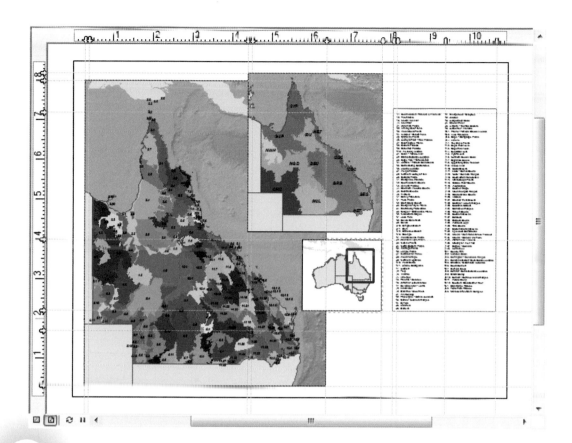

If your maps look slightly different, use the Pan tool on the Tools toolbar to position the maps better.

22 If you want to save your work, rename your map document and save to your MyData folder for chapter 10.

23 If you are continuing to the next exercise, leave ArcMap open. Otherwise, exit the application and click No if prompted to save your changes.

In exercise 10b, you'll add a title to the map.

Using map templates

One of the easiest ways to keep your maps consistent is to use a map template. Any map document (.mxd) can be made into a map template. Templates are layouts that contain layers, data frames, page layouts, and other common map elements. Using a template to create a map series can save time because you don't have to re-create each element of the map. Like other parts of ArcMap, templates can be shared to increase productivity within an organization.

Use a map template

When you open ArcMap or create a new document, the Getting Started dialog box appears and provides options to help you choose a recent map or a map template. When you create a new map document from a template, an untitled map appears based on the selected template. Provided templates are listed under the Templates node under New Maps. Types of templates range from blank, standard page-size templates to traditional layouts that include a basemap.

Continued on facing page

While working in a map document, you can change the current template to another one. On the Layout toolbar, click the Change Layout button, and then on the Select Template dialog box, choose a new template.

Create a map template

To create a new template that only you can see, save or copy a map document to a folder in your user profile (C:\Documents and Settings\<user profile>\ Application Data\ESRI\Desktop10.1\ArcMap\Templates). You can create folders to organize templates within this user profile folder. Templates stored in this location are listed under My Templates on the Getting Started dialog box.

Alternatively, to create a new template that all users of your system can use, you'll have to save or copy a map document to the default location in the ArcGIS installation directory (C:\Program Files\ArcGIS\Desktop10.1\MapTemplates). The default location can be changed using ArcMapAdvancedSettings.exe, located in the Utilities folder of the install directory. Similarly, new categories (folders) can be created to help you organize your templates.

Categories will appear on the Getting Started, New Map, and Change Layout dialog boxes only if a map document is present in the folder under the Templates or My Templates headings.

Exercise 10b
Adding titles and additional text

All maps have titles, and many have subtitles as well. A good title helps the reader understand what to look for on the map. Additional text might be a short description to further explain what the map is about.

Now that you have laid out the three data frames and index for the Queensland, Australia, bioregions and bio-subregions map in exercise 10a, you'll add a title, subtitle, and description.

1 In ArcMap, open the ex10b map document from your Chapter10 folder. If necessary, maximize the ArcMap window, and then click the Zoom Whole Page button on the Layout toolbar. The map is in layout view. It looks as it did at the end of exercise 10a.

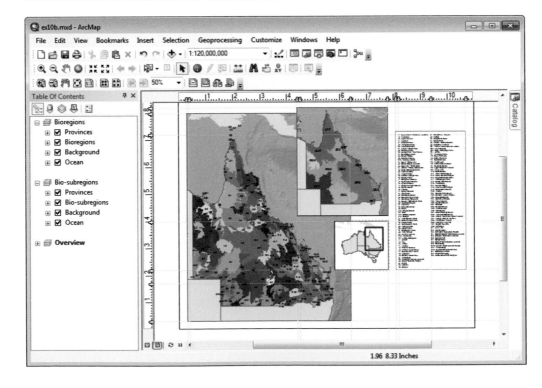

Next, you'll insert the title and position it in the space at the lower-right corner of the page.

2 **On the menu bar, click Insert > Title. On the Insert Title dialog box, in the text box, type Bioregions and Bio-subregions and click OK.**

If you make a mistake and need to change the text, double-click the title to open its text properties. You may notice that the Properties text box contains the following text: <dyn type="document" property="title"/>. This code dynamically calls the title entered, which is then populated on the Map Document Properties dialog box, in the Title text box (on the menu bar, go to File > Map Document Properties).

The title is centered near the top of the page. However, it is too small. You can change its font, size, style, or color on the Draw toolbar.

3 **Make sure the title is selected, indicated by a cyan dashed box surrounding it. Drag the title to the empty white area at the lower-right corner of the layout.**

4 **To turn on the Draw toolbar, on the Customize menu, click Toolbars > Draw. The toolbar may be floating or docked. On the Draw toolbar, for font size, highlight the current value and replace it with** 20, **and then press ENTER. Click the Bold button.** The title font is changed to 20 pt Bold Arial.

5 Snap the upper-right corner of the title to the guides at 1.7 inches (vertical ruler) and 10.7 inches (horizontal ruler).

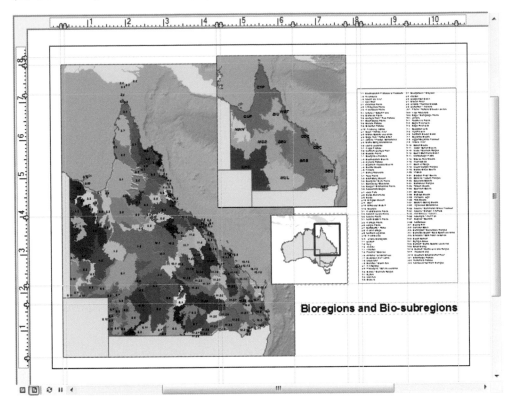

Next, you'll add a subtitle and position it below the title.

6 On the menu bar, click Insert > Title. A duplicate of the title is created at the top of the layout. Drag the new title just below the existing title in the lower-right corner of the map. Double-click the text box. On the Properties dialog box, replace the existing text with Queensland, Australia.

Next, you'll change the font and style of the subtitle.

7 On the Properties dialog box, click Change Symbol. In the Symbol Selector, change the font size to 14 and click the Italic button, and then click OK. Click OK again to close the Properties dialog box.

Next, you'll also add some descriptive text.

8 On the Draw toolbar, click the Text button arrow and click Rectangle Text.

9 **Press and drag the tool to draw a rectangle below the subtitle. Double-click the new text box. Delete the selected text and type** This map depicts the biogeographic regions and subregions of Queensland, Australia. Data created by the Queensland Government Information Service. **Next, click the Left Justify button.**

10 **On the Properties dialog box, click the Size and Position tab. In the Size frame, change the width to** 4 in **and the height to** 0.65 in, **and then click OK.**

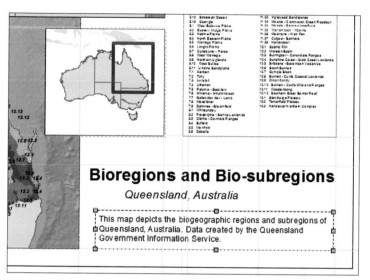

Next, you'll fine-tune the position of the three elements and group them together. The subtitle needs to be very close to the title to fit in the space prepared for it.

11 Right-click the subtitle. On the shortcut menu, click Nudge > Nudge Up. (You can also use the UP ARROW key to nudge it.) Keep nudging until the subtitle is as close to the title as possible without touching it. If you go too far, you can use Nudge Down (or press the DOWN ARROW key) to nudge the subtitle down. At any time, use the Layout Zoom In tool to zoom in if you have difficulty seeing these elements. You may also drag them into position using the Select tool.

12 Similarly, nudge the description rectangle up.

13 With the description rectangle selected, hold down CTRL and click the title and the subtitle. All three elements are selected. Right-click any element, and on the shortcut menu, click Align > Align Center.

14 **With the elements still selected, right-click any element, and on the shortcut menu, click Group.** Now anything you do to the grouped title, such as changing the font size or the color, will affect all the elements that are grouped together. If the right side of your grouped objects moved away from the vertical guide, simply nudge them or snap them back to the guide.

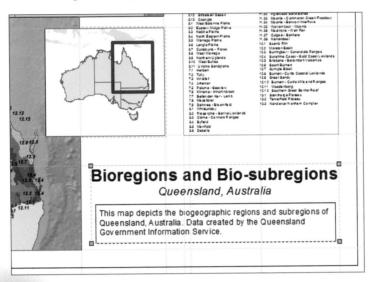

To ungroup an element, right-click the element, and on the shortcut menu, click Ungroup.

Next, you'll add a background rectangle to frame all the text elements.

15 **On the Draw toolbar, click the Rectangle tool. (If necessary, click the arrow and select the Rectangle tool.)**

16 Press and drag the tool to draw a rectangle that snaps to the guides around the title.

By default, the rectangle is pale yellow. Next, you'll change its color and move it behind the title.

17 Make sure the rectangle is selected. On the Draw toolbar, click the Fill Color arrow. On the color palette, click Sahara Sand. The rectangle color is changed to a light orange.

18 Right-click the selected rectangle and click Order > Send to Back.

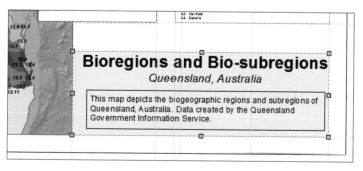

Finally, you'll center the title within the rectangle. Because ArcMap aligns elements in relation to the last element selected, you'll select the grouped text first.

19 Click the grouped element containing the title, subtitle, and description. Press and hold SHIFT and click the bottom of the colored rectangle, outside the grouped text elements selection box. Both the grouped title and the rectangle should be selected. The rectangle's selection color is cyan, indicating it is the last element selected. The grouped-text selection color is changed to green.

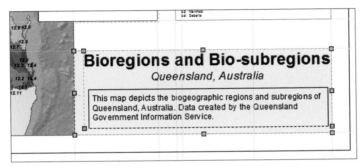

20 Right-click either selected element and click Align > Align Vertical Center.

21 Right-click either selected element and click Align > Align Center. This centers the title vertically and horizontally within the rectangle.

22 Right-click either of the selected elements once more and click Group. Then click an empty part of the virtual page to clear the grouped-element selection.

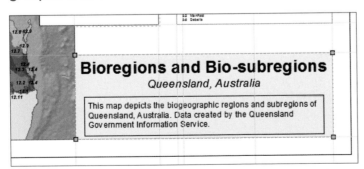

23 Zoom back to the extent of the full page by clicking the Zoom Whole Page button 🔳 .

24 If you want to save your work, rename your map document and save to your MyData folder for chapter 10.

25 If you are continuing to the next exercise, leave ArcMap open. Otherwise, exit the application and click No if prompted to save your changes.

In exercise 10c, you'll add a north arrow, scale bar, and legend to your page layout.

1
2
3
4
5
6
7
8
9
10b

Exercise 10c
Adding standard map elements

Map elements such as the north arrow, scale bar, and legend provide useful information to readers of your map. The **north arrow** provides an indication of true north (the direction toward the geographic North Pole) instead of magnetic north (the direction of a compass needle pointing toward the magnetic North Pole) for orientation. Displaying scale or distance with a **scale bar** is important and can be indicated using a graphical and numeric form. A **legend** lists the symbols used on a map and what they symbolize. ArcMap has many different styles for north arrows, scale bars, and legends and allows you to customize them.

These elements are dynamically associated with the data frame that is active when they are inserted. A scale bar changes when the display scale of the data frame changes. A north arrow rotates with the data frame. A legend is updated when layers in the data frame are deleted or resymbolized. Although these elements can be moved anywhere on a map layout, it is good practice to keep them near or within the data frame they belong to.

The people who view your presentation and your map may or may not be familiar with the bioregions and bio-subregions of Australia. Adding a legend is necessary to aid the interpretation of a map. In exercise 10a, you positioned a bio-subregions index in the layout. This can be considered a legend too because the values in it correspond to the values on the map. In this exercise, you'll begin by creating a north arrow and scale bar and finish by creating a legend for the bioregions layer.

1 In ArcMap, open the ex10c map document from your Chapter10 folder. Maximize the window as needed, and then on the Layout toolbar, click the Zoom Whole Page button. The map is in layout view (see facing page) and looks as it did at the end of exercise 10b. The Overview data frame is active.

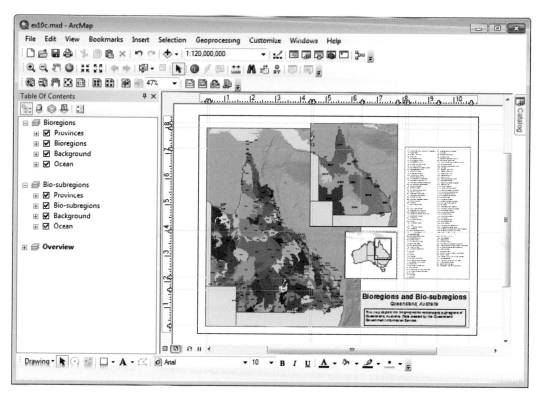

2 On the menu bar, click Insert > North Arrow. On the North Arrow Selector dialog box, click ESRI North 2 and click OK. The north arrow is added and placed near the center of the page.

Use the Zoom In tool on the Layout toolbar to zoom in on the page if you have difficulty seeing the north arrow. Using the Zoom In tool on the Standard toolbar will change the map scale of a data frame. Don't forget to click the Select Elements button on the Tools toolbar after zooming in so you can select features again.

3 Center the mouse pointer over the north arrow until the pointer becomes a four-headed arrow. Using the pointer, drag the north arrow to the lower-left corner of the Overview data frame. (Use the graphic as a guide. If necessary, turn off Snap to Guides for more control of placement.)

4 Double-click the north arrow. On the North Arrow Properties dialog box, click the color chip and click Gray 40%. In the Size box, type 40. Then click OK. When you're finished, click an empty part of the virtual page to clear the north arrow selection. (Click the Zoom Whole Page button if you zoomed in.)

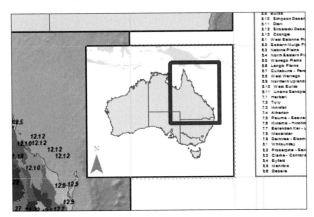

Next, you'll add a scale bar to the Bio-subregions data frame. Scale bars tell you how to convert measurements on the map into the real distances they represent on the ground.

5 In the layout, click the Bio-subregions data frame (the largest one) to activate it. Then, on the menu bar, click Insert > Scale Bar.

> You can also activate a data frame by right-clicking its name in the table of contents and clicking Activate. When a data frame is activated, it is shown in boldface in the table of contents.

6 On the Scale Bar Selector dialog box, scroll down the list and click Scale Line 1 Metric. Then click OK.

The scale bar is added and centered in the layout. Before moving it into position, you'll next change some of its properties

7 Right-click the scale bar and click Properties. (You can also double-click map elements to open their properties dialog box.) On the Scale Line Properties dialog box, click the Scale and Units tab. The scale bar currently has two divisions and four subdivisions (the scale line division has subdivisions for more precise measuring).

8 In the middle of the dialog box, click the When resizing arrow and click Adjust width. The resizing option will change the width of the scale bar if the display scale of the data frame is changed.

Adjusting scale bars

A scale bar changes as you zoom in or out of a data frame. If the scale bar is adjusted by division value, the distance represented by a scale bar division is variable. As you zoom in, it represents a shorter distance; as you zoom out, it represents a longer distance. If the scale bar is adjusted by the number of divisions, the distance represented by a division stays the same, but there are fewer divisions as you zoom in and more as you zoom out. If the scale bar is adjusted by width, the distance represented by a division and the number of divisions stays the same, but the entire scale bar grows wider as you zoom in and narrower as you zoom out. (If a scale bar is adjusted by width, you cannot widen it by dragging it; its width changes only with the display scale.)

---

9 In the Scale frame of the dialog box, in the Division value box, replace the current value with 250. Click the Number of divisions down arrow and change the value to 1. Leave the Number of subdivisions value as 4.

10 In the Units frame, in the Label box, change the label Kilometers to km.

Next, you'll make a slight change to the scale mark heights and text of the scale bar.

11 On the dialog box, click the Numbers and Marks tab. In the Marks frame, change Division Height to 5 pt and Subdivision Height to 4 pt.

12 **Click the Format tab. In the Text frame, change the size to 9, and then click OK to apply the format changes.** In the layout, the scale bar is updated to reflect the new properties you set.

13 **Press and drag the scale bar to the lower-left corner of the Bio-subregions data frame. (Zoom in on the layout to position the scale bar more accurately. Remember to click the Select Elements tool after zooming.)**

Next, you'll insert an additional scale bar for the Bioregions data frame. This time, to speed up the process, you'll copy the scale bar you just created and paste it when the Bioregions data frame is active. This is possible because the scale bar is dynamically linked to an active data frame. The values and width will adjust accordingly.

14 **With the scale bar selected, on the menu bar, click Edit > Copy. Activate the Bioregions data frame and click Edit > Paste.** A second scale bar is added to the layout and placed near the Bio-subregions scale bar. It doesn't look quite right because the Bioregions map scale (1:24,000,000) is different from the Bio-subregions map scale (1:13,000,000).

15 **Double-click the newly created scale bar designated for the Bioregions data frame.**

Notice that it has the same properties as the other scale bar. Next, you'll change this one slightly to a more appropriate width for its map scale.

16 **On the Scale Line Properties dialog box, click the Scale and Units tab. In the Scale frame, in the Division value box, replace the current value with 500. Change the Number of subdivisions value to 2, and then click OK.**

17 Zoom to the whole page if necessary, and drag the scale bar to the lower-left corner of the Bioregions data frame.

Next, you'll create a legend for the Bioregions data frame.

18 With the Bioregions data frame active, on the menu bar, click Insert > Legend to open the Legend Wizard.

By default, the legend includes all the layers on the map, and the number of legend columns is set to one (1). Since the only layer you want to show in the legend is Bioregions, you'll remove the other layers from the Legend Items list. You'll also increase the number of columns in the legend.

19 In the Legend Items list, click Provinces and click the left-facing arrow (<) button. Remove the Background and Ocean layers as well.

The layers are removed from the list, and only the Bioregions layer remains.

20 Change the "Set the number of columns in your legend" value to 2, and then click Next.

On the second panel of the wizard, you choose a title and font properties for the legend.

21 Replace the default legend title with Bioregions. **Click Next.**

On the third panel of the wizard, there are options to add a border. You'll skip this option.

22 Click Next.

Clicking Preview at any time displays the legend in its current form on the map and allows you to exit the wizard or return and make further changes.

On the fourth panel of the wizard, there are options to change the symbol patch, but you'll accept the default settings for now. You'll change the symbol patch size later.

23 Click Next.

6

7

8

9

10c

On the fifth panel of the wizard, you can change the spacing between elements, which you'll do next.

24 Change the spacing between Title and Legend Items to 1 and the Patches (vertically) value to 1, and then click Finish.

The legend is added to the page layout and centered.

Finally, you'll adjust a few more settings and move the legend.

25 Double-click the legend.

This opens the Legend Properties dialog box. On the dialog box, you can further
tweak and fine-tune how the legend looks.

26 On the General tab, in the Title frame, click the Symbol button. In
the Symbol Selector, change the font size to 8 pt, and then click OK.

27 On the Legend Properties dialog box, click the Items tab. In the
lower-left corner, click the Style button to open the Legend Item
Selector dialog box, and then click the Properties button.

28 On the General tab, in the Appearance frame, clear both the Show Layer Name and Show Heading check boxes. Select the Override default patch size check box and change Width to 8 pt and Height to 6 pt. Next, click Label Symbol. In the Symbol Selector, change the font size to 5 pt, and then click OK.

29 Click OK to close the Legend Item dialog box, and then OK again to close the Legend Item Selector. Finally, click OK to close the Legend Properties dialog box. The legend is updated and reflects all the changes you made.

30 Drag the legend above the Bio-subregions index. (Use the guides and snap to them if you wish.) When you're finished, click an empty part of the virtual page to clear the legend selection.

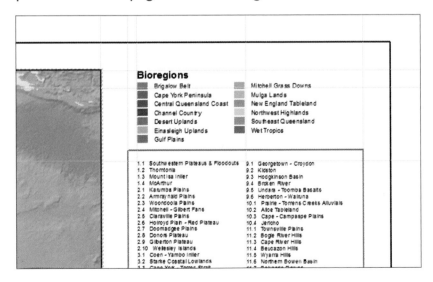

31 If you want to save your work, rename your map document and save to your MyData folder for chapter 10.

32 If you are continuing to the next exercise, leave ArcMap open. Otherwise, exit the application and click No if prompted to save your changes.

In exercise 10d, you'll add a few more map elements and prepare to print your map. Here's a look at the work you've done so far.

Exercise 10d
Adding final touches and setting print options

Before sending a map to print, you should check your page setup options and preview the map in ArcMap. It is very good practice to also check the overall composition of your layout, including the alignment of your data frames and map elements and the spelling of your titles and labels. Do an objective assessment: if your intended audience would benefit from more information (or in some cases, less), now is the time to make those final improvements.

The map is almost ready. In this exercise, you'll add a neatline, add more information about the map projection, add a title for the bio-subregions index, add a picture graphic, preview the map, and send it to a printer. Throughout the exercise, zoom in on the layout as much as you need to.

1 **In ArcMap, open the ex10d map document from your Chapter10 folder.** The map is in layout view (see facing page). It looks as it did at the end of exercise 10c. Maximize the ArcMap application window as needed, and then click the Zoom Whole Page button on the Layout toolbar.

2 Right-click anywhere outside the map display and click Guides > Snap to Guides to turn off snapping. (Your snapping may already be turned off, depending on how you worked with the layout in the previous exercise.)

3 On the menu bar, click Insert > Text. A text box appears in the middle of the layout with the word "Text" highlighted.

4 In the text box, type Geographic Coordinate System: Geocentric Datum of Australia 1994 and press ENTER. The text is selected. On the Draw toolbar, change the font size to 8.

5 Drag the text to the lower center of the Bio-subregions data frame and position it to the right of the scale bar. Clear the text selection.

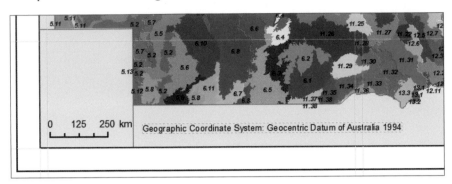

Next, you'll add a title for the Bio-subregions index and position it at the top of the index rectangle.

6 Insert a new text box. In the text box, type Bio-subregions and press ENTER. On the Draw toolbar, change the font size to 8, and then click the Bold button.

7 Drag the text to the right of the Bio-subregions index. (Don't worry if it is off the page—you'll reposition the index slightly to make room for the title.)

8. Use the zoom and pan tools on the Layout toolbar to zoom to the Bio-subregions index. Make sure you can see some white space below the index rectangle.

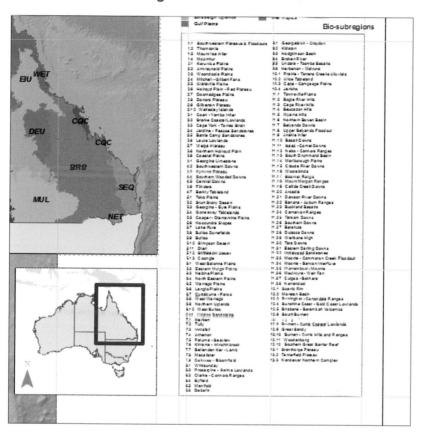

9. On the Tools toolbar, click the Select Elements tool. Click the Bio-subregions index to select it. Nudge the index rectangle down twice.

10. Right-click the selected index and click Ungroup (the rectangle and the text are grouped together). With the rectangle still selected, press and drag the upper-center selection handle so that it aligns with the horizontal guide just below the Bioregions legend.

11 Drag the Bio-subregions title to the space between the upper border of the rectangle and the top of the index text and align it with the left column. Change the drawing order so the text is visible.

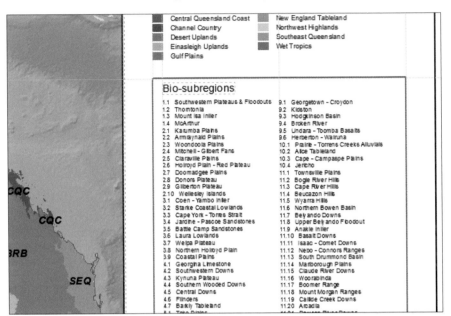

Next, you'll also add a small picture of your agency logo to the layout.

12 On the Layout toolbar, click the Zoom Whole Page button.

13 On the menu bar, click Insert > Picture. On the Open dialog box, navigate to \GTKArcGIS\Chapter10\Data. Click agency-logo.gif and click Open (see facing page).

14 Press and drag a corner selection handle to resize the graphic until it is approximately 0.5 inch wide (alternatively, right-click the image and change the width on the Size and Position tab). Drag the image to the upper-left corner of the Bio-subregions data frame. Other kinds of picture file formats can also be inserted, including JPEG, TIFF, EMF, and BMP.

The map is almost finished. The last element you will add is a *neatline*, a bounding line that frames the other elements.

15 Clear the logo selection.

16 On the menu bar, click Insert > Neatline.

17 On the Neatline dialog box, click the option to place the neatline around all elements. Click the Border arrow and click 1.5 Point. Click the Background arrow and click None. Then click OK.

18 Clear the neatline selection.

Before you print a map, you should make sure your map page size and printer setup specifications match. Otherwise, your map may be cut off or misaligned.

19 On the menu bar, click File > Page and Print Setup.

In exercise 10a, you changed the map page orientation from Portrait to Landscape. So next, you'll change your paper orientation, too.

20 On the Page and Print Setup dialog box, in the Paper frame, click the Landscape option. If necessary, click the Name arrow and click the printer you want to use. Accept the default paper size (Letter). Click OK.

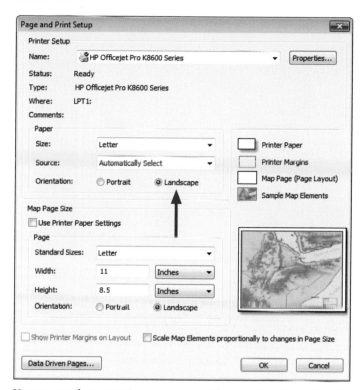

You are ready to preview your map. The preview shows how the map will look on the printed page, so you can correct any mistakes in advance.

21 On the menu bar, click File > Print Preview.

At this point, you should mainly be concerned with the alignment of the elements on the page. Your alignment will depend on the type of printer you are connected to and its configuration. If the map elements overlap the page edge, you'll correct this problem in a moment.

Don't worry that the quality of the logo graphic is poor in the preview—it will be fine when it is printed. (*Note:* Printing at this step is optional, in case you do not have access to a printer.)

22 On the top bar of the application, click Print to open the Print dialog box.

23 If the map looked good in the print preview, click OK to print it.

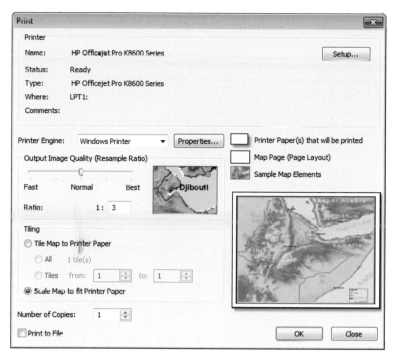

24 If the map went over the edge of the page, in the Tiling frame, click the Scale Map to fit Printer Paper option. Click OK if you want to print it. Click Close to return to the map.

The sample map in the lower-right corner of the Print dialog box shows the changes you make interactively. Printer settings vary depending on the default printer and its paper settings. There are many printer-specific options that may help you print your map.

Printing is not the only option for sharing your maps with others. Next, you'll explore other formats for sharing your map.

25 On the menu bar, click File > Export Map.

You can export your map to many types of file formats, each with its own unique options. A common format that can be easily opened and shared is the Adobe Acrobat PDF file type, which you'll use next.

26 Navigate to your MyData folder for chapter 10. Click PDF (*.pdf) in the Save as type list. The file name is automatically entered as ex10d.pdf. To see file format options, at the bottom of the dialog box, click Options to expand it. Change the resolution to 360 dpi. In the Options section, on the General tab, you can change both the resolution and the output image quality. Depending on the quality required, a resolution of 360 dpi (dots per inch) is sufficient for typical print jobs. Increasing the resolution will yield a slightly better image, but it will also increase the file size.

27 Under Options, click the Advanced tab. The Layers and Attributes option shows what will be exported to the PDF. Currently, it is set to export layers only, although ArcMap can also export all data attributes. Notice that the Export Map Georeference Information option is selected,

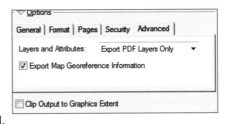

which allows the georeference and coordinate system information to be included with the exported PDF. Acrobat Reader contains Analysis tools that make the georeference information useful.

28 Click Save.

As the map is being exported, its progress is displayed at the bottom of the application window, on the ArcMap status bar.

29 When the export is complete, on the taskbar, click the Start button and click My Computer. Browse to your MyData folder for chapter 10 and open ex10d.pdf.

Adobe Acrobat contains tools to navigate your layout. Find and open the Layers navigation panel. It displays the layers of the maps similar to how they are in ArcMap. You can expand the layers and use the controls to hide and show layers. Depending on which Adobe Acrobat product you open the layout in, additional Analysis tools are available to find locations, measure distances, and add location markers.

You are now finished with the maps and page layout.

30 If you want to save your work, return to ArcMap, rename your map document, and save to your MyData folder for chapter 10.

31 Close the Draw toolbar. If you are continuing to the next chapter, leave ArcMap open. Otherwise, exit the application and click No if prompted to save your changes.

Summary

This chapter has given you a general overview of presenting data in ArcMap. You learned what to consider before placing and positioning elements on a page layout, including composition, data frame size, colors, titles, and text elements. With more practice, you should feel comfortable stylizing features, working with multiple data frames, and creating legends. You also learned other useful features, such as that guides can be created and snapped to so you can align data frames and map elements more easily; north arrows are placed in orientation to true north; and elements can be arranged in different stacking order, aligned, and nudged into position. You should now realize that ArcMap provides many tools for presenting GIS data effectively, and you should feel comfortable creating a map on your own.

Terms

map extent
north arrow
scale bar
legend
neatline

Creating and editing

data

Chapter 11
Building geodatabases

Creating a geodatabase
Creating feature classes
Adding fields and domains

Learning objectives

- *Create a new file geodatabase*
- *Map x,y data*
- *Convert coverages and shapefiles to geodatabase feature classes*
- *Create a new, empty geodatabase feature class and set its properties*
- *Add attribute fields to a new geodatabase feature class*
- *Establish a coded domain for attribute values and select a default value*

SPATIAL DATA COMES IN A VARIETY OF FORMATS THAT CAN BE MANAGED IN ArcCatalog and mapped in ArcMap. Many of these formats, including shapefiles, coverages, and CAD files, as well as the geodatabase, organize spatial data into feature classes. As you learned in chapter 3, a feature class is a group of points, lines, or polygons that represent geographic objects of the same kind, such as countries or rivers.

A shapefile is a single, stand-alone feature class. Cities.shp, for example, might be a point feature class representing cities. (Shapefiles are actually composed of several files with different extensions, although you will see only one item listed in ArcCatalog.) A geodatabase, by contrast, is a "container" in which sets of feature classes can be stored. For example, World.gdb might comprise a polygon feature class of world countries, a line feature class of world rivers, a point feature class of world cities, and more.

Most of the vector GIS datasets in use today are shapefiles or geodatabase feature classes. Although there are differences among spatial data formats, there is also a great deal of mutual compatibility, and you can usually convert one format to another. When you add spatial data to ArcMap, you see it and work with it as a layer, and usually aren't interested in the format of the source file.

You should know, however, that the geodatabase, as the latest and most sophisticated spatial data format, has certain advantages. One of them is the ability to store multiple datasets, including rasters, which makes the geodatabase a built-in tool for organizing data. Related feature classes can be grouped into a *feature dataset*—for example, sewer mains (lines) and manhole locations (points) could be grouped together in a Sewer feature dataset. When shapefiles are imported into a geodatabase, the size on disk is dramatically reduced, so the geodatabase is an ideal model for data sharing. Another advantage of using a geodatabase is the ability to create an *attribute domain*, which establishes and enforces valid values or ranges of values for an attribute field and minimizes data entry mistakes by prohibiting invalid values. For example, if "Open" and "Closed" were the domain values for a water valve status field, it would be impossible to enter any values in the attribute table other than Open or Closed.

Finally, the geodatabase has the ability to store behavior rules. Because of this, a geodatabase can support specialized types of data such as annotation feature classes, cartographic representations, topologies, and networks, which all have specific behaviors. These special data types are increasingly important to GIS, although they are beyond the scope of this book.

Once you've mastered the basics of ArcGIS, there are many resources available that can help you harness the extended capabilities of the geodatabase. For example, once you've completed this chapter and chapters 12 and 13, you might be interested in completing exercises 4, 5, 6, and 8 in ArcGIS for Desktop Help—on the Contents tab, navigate to Geodata > Geodatabases > An overview of the geodatabase >Building a geodatabase tutorial.

Geodatabases are sophisticated storage containers for feature classes, feature datasets, relationship classes, and rasters.

There are three types of geodatabases: file, personal, and the ArcSDE (or multiuser) geodatabase. In this chapter, you will create and work with a file geodatabase. A file geodatabase is designed to store datasets in a folder of files on your computer. It can store large datasets (up to one terabyte) and can be accessed by several users at the same time, although a feature class in a geodatabase can be edited by only one person at a time.

The original geodatabase introduced by Esri is the personal geodatabase (.mdb file extension). It uses the Microsoft Access data file structure and supports databases that are 2 GB or less. However, the effective database size is smaller, somewhere between 250 and 500 MB, after which database performance starts to slow down. For this reason, file geodatabases (.gdb) are recommended over personal geodatabases, particularly because datasets tend to grow larger over time.

ArcSDE geodatabases are scalable multiuser geodatabases that can accommodate large workgroups or enterprises. Using technology called versioning, ArcSDE geodatabases permit simultaneous editing of feature classes by different users. These geodatabases have no size limit in the top-tier ArcSDE product. Data is stored in a relational database management system (RDBMS), such as Microsoft SQL Server Express, SQL Server, Oracle, Informix, IBM DB2, or PostgreSQL.

For more information about geodatabase types, in ArcGIS for Desktop Help, on the Contents tab, navigate to Geodata > Geodatabases > An overview of the geodatabase > Types of geodatabases.

Data

In this chapter, you will use the following datasets:

- parcels—a coverage representing a selection of parcels for the city of Manhattan, Kansas *(Source: Riley County, Kansas)*
- FireHydrants.shp—a shapefile containing point features that represent a selection of city fire hydrants *(Source: City of Manhattan, Kansas)*
- WaterValves.csv—a spreadsheet containing x,y coordinate points that correspond to a selection of city water valves *(Source: City of Manhattan, Kansas)*

11a

Exercise 11a
Creating a geodatabase

You create new geodatabases using ArcCatalog—either the stand-alone application or the Catalog window in ArcMap. You can also create geodatabase feature classes, either from scratch or by importing a different data type into the geodatabase. In this exercise, you will create a file geodatabase and import coverage and shapefile data into it.

For this exercise, you work in the GIS department of a medium-size city in Kansas. You are part of a team that is deciding whether the city should convert its shapefile and coverage data to the geodatabase format. One question that needs to be answered is how difficult the process would be. To find out, you will create a file geodatabase and import a land parcels coverage, a fire hydrants shapefile, and a text file that contains coordinate points indicating the positions of water valves.

1 Start ArcMap and open a blank map session. Open the Catalog window. In the Catalog window, navigate to Folder Connections, and double-click C:\EsriPress\GTKArcGIS. (You should already have a folder connection established. If not, create one now.)

> You will be using the Catalog window exclusively in the exercises in this chapter. You may find it easiest to undock the window, expand it, and reduce the size of the ArcMap window. You could also perform the following steps using the stand-alone ArcCatalog application, although some instructions would need to be modified slightly.

2 **Double-click Chapter11. Right-click MyData and click New > File Geodatabase.**

A new file geodatabase is created in your MyData folder.

Next, you will give the geodatabase a more descriptive name.

3 In the Catalog window, right-click New File
Geodatabase.gdb and click Rename.
Type CityData.gdb and press ENTER.

Next, you'll import coverage polygons and shapefile
points into the new CityData geodatabase. You will run
both imports simultaneously from the same tool.

⊞ 🗀 Chapter06
⊞ 🗀 Chapter07
⊞ 🗀 Chapter08
⊞ 🗀 Chapter09
⊞ 🗀 Chapter10
⊟ 🗀 Chapter11
   ⊞ 🗀 Ex11aData
   ⊞ 🗀 Ex11bData
   ⊞ 🗀 Ex11cData
   ⊟ 🗀 MyData
      🗋 `CityData.gdb`
⊞ 🗀 Chapter12
⊞ 🗀 Chapter13
⊞ 🗀 Chapter14
⊞ 🗀 Chapter15
⊞ 🗀 Chapter16
⊞ 🗀 Chapter17

4 In the Catalog window, right-click CityData.gdb and click Import >
Feature Class (multiple). This opens the Feature Class to Geodatabase
(multiple) dialog box.

5 Move the dialog box away from the Catalog window (see facing
page). In the Catalog window, expand Ex11aData, and then click the
plus sign (+) symbol next to the parcels coverage. Drag the polygon
feature class to the Input Features box of the tool. The path of the
feature class is displayed in the Input Features box. In the Output Geodatabase
box, Catalog correctly assumes you want to import the data into the CityData
geodatabase. (You could change the output location if desired.)

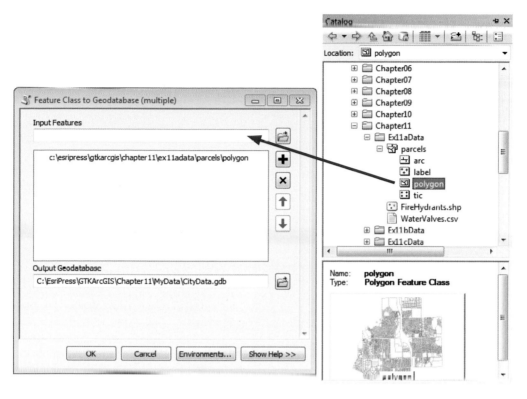

6 Repeat step 5, except drag the FireHydrants shapefile to the Input Features box. Make sure your dialog box matches the graphic below, and then click OK. When the operation is completed, a pop-up notification appears in the notification area. The coverage polygons and shapefile points are converted to feature classes in the CityData geodatabase.

11a

12

13

14

7 In the Catalog window, expand your MyData folder and double-click CityData.gdb to see the new feature classes.

⊞ 📁 Ex11bData
⊞ 📁 Ex11cData
⊟ 📁 MyData
  ⊟ 🗊 CityData.gdb
    📍 FireHydrants
    ▦ parcels_polygon
⊞ 📁 Chapter12

8 Rename parcels_polygon Parcels.

⊞ 📁 Ex11bData
⊞ 📁 Ex11cData
⊟ 📁 MyData
  ⊟ 🗊 CityData.gdb
    📍 FireHydrants
    ▦ Parcels
⊞ 📁 Chapter12

Next, you will preview the data to make sure it looks OK.

9 In the Catalog tree, right-click Parcels and click Item Description. In the Item Description window, click the Preview tab.

10 In the Catalog window, click FireHydrants. The Item Description window now previews fire hydrant points.

You still have a .csv (comma-separated variables) file to import into your geodatabase. It's common to find data in a nonspatial format such as this. In the case of Manhattan, Kansas, water valves, the x,y coordinates are stored in a spreadsheet. You will map the coordinates to create points that can be seen on the map.

11 Close the Item Description window. In the Catalog window, navigate to the Ex11aData folder, right-click WaterValves.csv, and click Create Feature Class > From XY Table. This opens the Create Feature Class From XY Table dialog box. The table already has an X field and a Y field designated.

12 Maintain the default input fields.

Next, you will specify the spatial reference, using the same coordinate system as the Parcels feature class (which is also the same as FireHydrants).

13 On the dialog box, click Coordinate System of Input Coordinates. On the Spatial Reference Properties dialog box, click the Add Coordinate System button arrow and click Import.

14 On the Browse for Datasets or Coordinate Systems dialog box, browse to \Chapter11\MyData\ CityData.gdb. Click Parcels and click Add. The spatial reference information is updated.

15 Click OK. Back on the Create Feature Class From XY Table dialog box, under Output, click the Browse button. On the Saving Data dialog box, for Save as type, click File and Personal Geodatabase feature classes. In the Look in box, navigate to CityData.gdb in your MyData folder for chapter 11. Name the feature class WaterValves and click Save. Click OK to create the feature class.

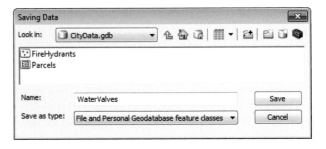

16 Preview the water valves.

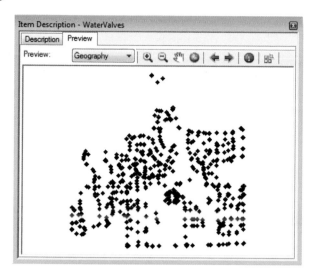

On your own

Create a description thumbnail for WaterValves.

When you create spatial data from a nonspatial source, such as tabular XY data or GPS points, it is good practice to create metadata, which will appear on the Description tab on the Item Description window. To create metadata, click the Edit button on the Description tab.

17 If you are continuing to the next exercise, leave ArcMap open. Otherwise, exit the application and click No if prompted to save your changes.

Exercise 11b
Creating feature classes

When you imported data in exercise 11a, you accepted the spatial properties of the existing data. When you converted nonspatial coordinates to a point feature class, you imported the spatial reference of another feature class. However, when you create a new feature class, you have to define these properties yourself. This means specifying the feature class geometry—point, line, or polygon—and the spatial reference.

You now know that you can convert the city's existing spatial data to a geodatabase format. You also need to know how to create new data in the geodatabase. In this exercise, you will make a feature class for water lines, although you won't put any features in it until chapter 12.

1 If you have completed exercise 11a, skip to step 2. Otherwise, start ArcMap and open the Catalog window. Expand your GTKArcGIS folder. Double-click Chapter11, and then double-click Ex11bData. Right-click CityData.gdb and click Copy. Right-click MyData and click Paste.

2 Open a blank map in ArcMap and open the Catalog window. In the Catalog tree (see facing page), expand your \GTKArcGIS\Chapter11\MyData folder. Right-click the CityData file geodatabase and click New > Feature Class. This opens the first panel of the New Feature Class wizard.

3 In the Name box of the wizard, type WaterLines.

By default, the Type property is set to Polygon Features. Since you are creating a feature class of water lines, you'll change it to Line Features.

11b

12

13

14

4 In the Type frame, click the "Type of features stored in this feature class" arrow and click Line Features.

5 Click Next.

Since no coordinate system has been selected, the "Current coordinate system" is unknown. However, if you completed exercise 11a, you will see the correct coordinate system for this dataset under Favorites. (If you do not see a state plane coordinate system under your favorites, skip step 6.)

6 Under Favorites, click NAD_1983_StatePlane_Kansas_North_ FIPS_1501_Feet. Then skip ahead to step 9.

7 Alternatively, click the Add Coordinate System button arrow and click Import.

8 On the Browse for Datasets or Coordinate Systems dialog box, browse to your MyData folder for chapter 11. Double-click CityData.gdb and click Parcels, and then click Add.

The Current coordinate system information is updated.

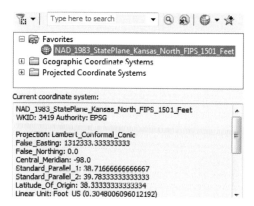

9 Click Next. The XY tolerance defines the minimum distance between two coordinates before they are considered equal. You will accept the default value.

10 **Click Next.** A database storage configuration allows you to fine-tune how data is stored in a file geodatabase. You will accept the default option.

11 **Click Next.** The fifth panel of the wizard displays the field names, data types, and field properties that belong to the new feature class. OBJECTID and SHAPE are required fields for geodatabase feature classes and are automatically added by the software. The OBJECTID field stores a unique ID number for each feature in the class. The SHAPE field stores each feature's shape and location.

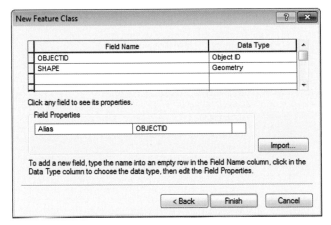

Depending on the type of feature class created, ArcGIS also creates and maintains measurement fields. A SHAPE_Length field is created for line feature classes. A SHAPE_Length and a SHAPE_Area field are created for polygon feature classes. (The SHAPE_Length attribute for polygons stores perimeter lengths.) These measurement fields appear only after the feature class has been created.

12 In the Field Name column, click SHAPE. The field properties for the SHAPE field are displayed. Notice that the geometry type is set to Line. You will add more attribute fields in exercise 11c.

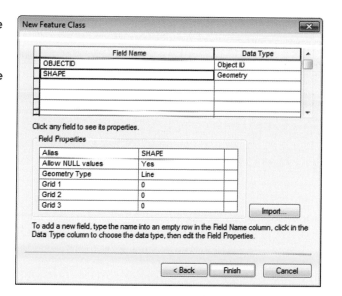

13 Click Finish.

14 In the Catalog tree, notice that a new feature class, WaterLines, has been added to CityData.gdb.

15 If you are continuing to the next exercise, leave ArcMap open. Otherwise, exit the application and click No if prompted to save your changes.

In exercise 11c, you will add attribute fields to the WaterLines feature class, and you will provide one of them with an attribute domain.

Exercise 11c
Adding fields and domains

When you create a feature class, you add fields to store attribute information. A field is defined by a name, a data type (for example, text or integer), and properties that vary according to both the data type and the spatial data format. One such property that can be defined for fields in geodatabases is an attribute domain.

An attribute domain is either a list or a range of valid values for an attribute. A city zoning attribute, for example, might have a dozen or so appropriate values, such as "residential," "industrial," and "commercial." By defining these values as a domain, you prevent any other values from being mistakenly added to a table during data entry or editing. Similarly, you might want to restrict the values for a numeric attribute to a particular range, such as 0 to 14 for a soil pH attribute.

Unlike other field properties, which apply exclusively to the field they define, a domain can be applied to more than one field in a feature class and to more than one feature class within a geodatabase.

In this exercise, you will add two fields to the WaterLines feature class you created in exercise 11b. One will store installation dates and the other the type of water line. There are three types of water lines: (1) mains, which run under streets; (2) domestic laterals, which run from mains to houses; and (3) hydrant laterals, which run from mains to fire hydrants.

After you add the fields, you will create a domain and apply it to the field for water line type. The domain will ensure that no values other than the three valid ones are entered in the attribute table. You will then specify one of these three as the default value.

1 If you have completed exercise 11b, skip to step 2. Otherwise, open a blank map in ArcMap, and then open the Catalog window and expand your \GTKArcGIS\Chapter11\Ex11cData folder. Right-click CityData.gdb and click Copy. Right-click MyData and click Paste.

2 Open a blank map in ArcMap. Open the Catalog window and expand your \GTKArcGIS\Chapter11\ MyData folder, and then expand CityData.gdb.

3 In the Catalog window, right-click WaterLines and click Properties. On the Feature Class Properties dialog box, click the Fields tab.

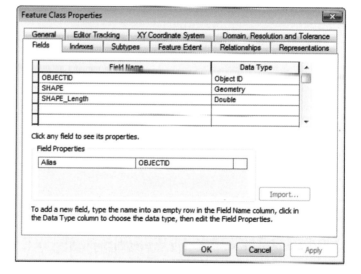

11c

12

13

14

4 **In the Field Name column, type** Install_Date **in the first empty row.**

Field names cannot contain spaces or special characters, such as hyphens, brackets, parentheses, or symbols such as $, #, %, or &.

Field Name	Data Type
OBJECTID	Object ID
SHAPE	Geometry
SHAPE_Length	Double
Install_Date	

5 **In the Data Type column, click the empty cell next to Install_Date and click Date.**

For more information about field data types, in ArcGIS for Desktop Help, on the Contents tab, navigate to Geodata > Geodatabases >Defining the properties of data in a geodatabase > Geodatabase table properties > ArcGIS field data types.

Field Name	Data Type
OBJECTID	Object ID
SHAPE	Geometry
SHAPE_Length	Double
Install_Date	Text ▼
	Short Integer
	Long Integer
	Float
	Double
	Text
	Date
	Blob
	Guid
	Raster

Subtypes

Another way to categorize geodatabase features is to use subtypes. With subtypes, you can group features into separate classes without having to create distinct feature classes. For example, you may want to classify a Roads feature class with the subtypes Main, Highway, and Utility. To create subtypes, you must have a short integer or long integer field in the feature class. The integer is a code that is associated with a description on the Subtypes tab on the Feature Class Properties dialog box. Attribute domains may be established for each subtype—for example, you may want to establish a domain range for the speed limit attribute, which will vary by road type.

6 In the Field Name column, in the next empty row, type Line_Type. **For Data Type, click Text. Then click OK.**

Now that you have added the two attribute fields, you will next create a domain and apply it to the Line_Type field.

7 In the Catalog window, right-click CityData.gdb (in your MyData folder) and click Properties to display the Database Properties dialog box. Click the Domains tab.

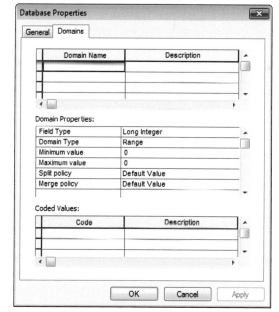

8 In the Domain Name column, in the first cell, type WaterLineType. **Click the first cell in the Description column and type** Type of water line installed.

11c

12

13

14

In the Domain Properties list, the default Field Type value is Long Integer and the Domain Type value is Range. The field type must match the data type of the feature class it is being applied to. Since you made Line_Type a text field in step 6, you must do the same here and change the field type to text.

9 **Under Domain Properties, click Long Integer and click Text.**

Domain Properties:

Field Type	Text ▼	▲
Domain Type	Short Integer	
Split policy	Long Integer	
Merge policy	Float	
	Double	
	Text	
	Date	▼

The domain type is changed automatically from Range to Coded Values.

Domain Properties:

Field Type	Text	▲
Domain Type	Coded Values	
Split policy	Default Value	
Merge policy	Default Value	
		▼

A Coded Values domain is used when the attribute values you want to enforce are best defined in a list. (The values may still be numeric because numbers are commonly used as codes for descriptions.) A Range domain is used when the values you want to enforce are quantities with upper and lower limits.

10 **In the Coded Values list in the lower portion of the dialog box, click the first empty cell in the Code column. Type M. Press the TAB key or click the first cell in the Description column and type Main.**

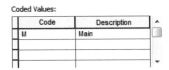

Coded Values:

Code	Description	▲
M	Main	
		▼

11 **In the next empty cell in the Code column, type DL (see facing page). For Description, type Domestic Lateral.**

12 In the next empty Code column cell, type HL. **For Description, type** Hydrant Lateral. **Then click OK.** After a coded value domain has been applied to a feature class, you will be picking from a list of these descriptions when you enter attribute values. The coded values are stored in the geodatabase, but you will see the descriptions in the attribute table only.

Now that you have created the attribute domain, you will next apply it to the Line_Type field of the WaterLines feature class.

13 In the Catalog window, right-click the WaterLines feature class (in your MyData folder) and click Properties. On the Feature Class Properties dialog box, click the Fields tab.

14 In the Field Name column, click Line_Type. Its properties are displayed under Field Properties.

11c

12

13

14

15 In the Field Properties frame, click the empty cell next to Domain and click **WaterLineType.** You have now created an attribute domain and applied it to the Line_Type field.

Field Properties		
Alias	Line_Type	
Allow NULL values	Yes	
Default Value		
Domain	WaterLineType ▼	
Length	50	

Finally, you will assign the field a default value. Mains are the most common type of water lines, so you'll set "Main" as the default. Whenever a water line is added to the feature class, it will automatically get this value in the Line_Type field. As you'll see in chapter 12, you can easily change the value if the line is a lateral instead.

16 In the Field Properties frame, click the empty cell next to Default Value and type **M,** the code for mains. Then click **OK.** You can also set a default value for a field without creating a domain. If a domain is applied, however, the default value must be a valid domain value.

17 Restore the Catalog window to the position you prefer and exit the application. Click **No** if prompted to save your changes.

Summary

In this chapter, you learned how simple it is to create a geodatabase container, as well as to convert other data types to geodatabase feature classes. Both tasks can be completed with the click of a button. Creating a feature class from x,y tabular data takes only one additional step—selecting a coordinate system. Setting up a new feature class from scratch takes a few more steps, since you have to set properties, determine the spatial reference information, create attribute fields, and establish attribute domains, if desired.

Terms

feature dataset

attribute domain

Chapter 12
Creating features

Drawing features
Using feature construction tools

Learning objectives

- *Open an edit session in ArcMap*
- *Set the snapping environment in an edit session*
- *Create geodatabase features using various construction tools*
- *Use commands that facilitate the feature creation process*
- *Change a default attribute value of a new feature*

MUCH OF THE VECTOR SPATIAL DATA USED IN A GIS COMES DIGITIZED FROM paper maps and aerial or satellite photographs. *Digitizing* data involves placing a map or a photograph on a digitizing tablet (a drawing table connected to a computer) and tracing features with a puck, which is a device similar to a mouse. In a variation called "heads-up digitizing," features are drawn directly on the computer screen with a mouse by tracing an aerial photograph, a scanned map, or other spatial data (see illustration on the next page).

Italy's Lago di Bolsena digitized "heads up" from a satellite image. Map and data from i-cubed, information integration & imaging LLC, distributed through i-cubed's DataDoors Archive Management at www.datadoors.net.

Features can also be digitized without tracing. ArcGIS has several tools for creating circles, rectangles, curves, and other shapes of exact dimensions.

Points are the simplest features to digitize—all it takes is a single click. Lines are more complicated, because they start, end, and may change direction. Polygons, from the point of view of digitizing, are just special cases of lines—they are lines that return to their origin.

Points that define the shape of a line or polygon are called **vertices** (plural of **vertex**). Points that mark the ends of a line segment are called **endpoints**. The segments between vertices are called **edges**.

Features that can be digitized in ArcGIS.

When you look at line or polygon features on a map, you don't see them as edges, vertices, and endpoints, although you do when you create or edit features. When you see a feature in this mode, you are looking at an **edit sketch**. A new feature exists only as an edit sketch until it is saved. After it is saved, it becomes a feature.

In ArcGIS, all digitizing—whether it is to create new features or modify existing ones—is done in the course of an **edit session**. An edit session begins when you click the Start Editing command on the Editor toolbar and ends when you click the Stop Editing command. After an edit session is started, features are created through the use of feature templates. Feature templates define all the information required to create a feature: the layer where the feature will be stored, the attributes the feature is created with, and the default tool used to create that feature.

Editor toolbar.

Editor menu

Create features

The Create Features window is the central place for creating and managing feature templates. The window has three main components: a toolbar to manage your templates and their properties, a list of templates used to create new features, and a set of construction tools used to define the shape of the features. You'll learn how to use this window and its options in this chapter.

Create Features window.

11
12
13
14

When you digitize features, you'll sometimes make mistakes. You can undo mistakes with the Undo button 🔄 on the Standard toolbar.

Data

In this chapter, you will use the same data as in chapter 11 (parcels, fire hydrants, and water valves for Manhattan, Kansas). The original datasets have now been converted to the geodatabase format. Additionally, the data for this chapter includes a raster layer, as follows:

- Subdivision21.TIF—an aerial image of the study area *(Source: Riley County, Kansas)*

Exercise 12a
Drawing features

In this exercise, you will create water lines for the feature class you created in the CityData geodatabase in exercise 11b.

New features are often connected to existing features. For example, the boundaries of new land parcels may be adjacent to boundaries that have already been digitized. New streets intersect or extend to existing ones. New power lines run to and from existing poles. Making sure that features are connected would be difficult if it depended entirely on eyesight.

ArcMap uses a technique called *snapping* to automatically connect features placed within a certain distance of each other. The rules that specify which features, and which feature parts, snap to others make up the snapping environment. The distance at which snapping occurs is called the snapping tolerance.

For this exercise, in a subdivision of your growing city, a contractor has laid new water lines, including an extension to a main and a line connecting a main to a fire hydrant. During construction, the city's project inspector recorded the locations of water valves and hydrants using a Global Positioning System (GPS) device. These locations have been imported to your geodatabase as point feature classes. As the city's GIS technician, you will use this point data and the ArcMap editing tools to digitize the new water lines.

11 12a 13 14 15 16 17 18 19 20

1 **Start ArcMap. Open the ex12a map document from your C:\EsriPress\GTKArcGIS\Chapter12 folder.**

The map shows a subdivision under construction. Layers for water valves, fire hydrants, water lines, and parcels are listed in the table of contents. The WaterLines feature class you created in exercise 11b has no features yet. Before starting an edit session, you will zoom to the area where you are going to digitize water lines.

2 On the menu bar, click Bookmarks > New Water Lines.

The display zooms to an area where there are several water valves and a fire hydrant. The field crew has laid three new water lines: from valve 763 to valve 813, from 813 to 831, and from 813 to the nearest fire hydrant. You will next digitize these three lines.

First, you'll add the Editor toolbar and start an edit session.

3 On the Standard toolbar, click the Editor Toolbar button 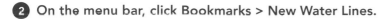 . This opens the Editor toolbar. The toolbar appears dimmed, meaning the tools are unavailable.

4 If necessary, move the Editor toolbar away from the map display or dock it. On the Editor toolbar, click Editor > Start Editing. The editing tools are now activated.

5 On the Editor toolbar, click the Create Features button 𝔼 to open the Create Features window. Move or dock the window as you prefer.

By default, no feature template is selected. You need to change the template to the water lines layer, which you'll do next.

11

12a

13

14

6 In the Create Features window, click the WaterLines template.

Because the water lines must connect (snap) to each other and to other features, such as valves and hydrants, you'll next set the snapping environment, before you start digitizing.

7 On the Editor toolbar, click the Editor menu and click Snapping > Snapping Toolbar. This opens the Snapping toolbar.

Snapping settings

The Snapping toolbar contains buttons for the main snapping settings, although additional ones are available on the Snapping menu. As you digitize a new feature, the cursor snaps to existing features according to the buttons that are clicked. (Clicking either the Vertex Snapping or the Edge Snapping button for a point layer, for example, makes the cursor snap to point features.) The cursor snaps when it comes within a specified distance of a feature. To set this snapping tolerance, on the Editor menu, click Options.

The Snap To Sketch command on the Snapping menu sets the snapping rules for an edit sketch. It is useful, for example, when you are creating a feature that intersects or connects to itself, such as a boundary line that forms a closed loop. For more information, in ArcGIS for Desktop Help, on the Contents tab, navigate to Editing > Fundamentals of editing > Using snapping > About snapping.

8 On the Snapping toolbar, click the End Snapping, Vertex Snapping, and Edge Snapping buttons to turn them off. Only the Point Snapping button is left active (the circle icon).

Remember, to be sure you are using the correct tool, move your mouse pointer over a tool button to see the ToolTip.

Next, you need to digitize the water lines.

9 In the Create Features window, in the Construction Tools panel, click Line, if not already selected.

10 Move the mouse pointer over the map display. The pointer changes to a crosshair. Move the pointer toward valve 763. When you come within the snapping tolerance of the valve, a circle appears and a WaterValve: Point label identifies it.

Away from feature.   Near and snapping to feature.

Next, you will digitize the first water line, from valve 763 to valve 813. To draw a line feature, click once to begin the line, click again for each vertex, and double-click to finish the line. Because the water lines don't change direction, you won't need any vertices.

11 With the cursor snapped to valve 763, click to start the first water line. Move the mouse pointer toward valve 813. The edit sketch of the line moves with the pointer (an effect known as "rubberbanding").

12 Move your pointer over valve 813. When the cursor snaps to it, double-click to end the line. (If you make a mistake, click the Undo button on the Standard toolbar and repeat steps 11 and 12.)

The line is created. Next, you will add a second line to connect valve 813 to valve 831.

13 Start with the cursor snapped to valve 813 and click to begin a new line. Move your pointer over valve 831. When the cursor snaps to it, double-click to end the line.

The first line changes from cyan (the default color of edit sketches) to Cretan Blue (the color symbolizing water lines on the map). Next, you will digitize the third and final water line, from valve 813 to the fire hydrant.

14 **Move your pointer over valve 813 so that the cursor snaps to it. Click to start the line. Move your pointer over the fire hydrant. When the cursor snaps to it, double-click to end the line.**

Next, you will view the attributes of the individual water lines you just created.

15 **With the water line to the fire hydrant still selected, on the Editor toolbar, click the Attributes button** ▣ **.** This opens the Attributes window. If the Create Features window is undocked, the Attributes window appears in the same space—tabs allow you to switch between windows.

The Attributes window can show one or more features at a time. Use the Edit tool on the Editor toolbar to select features to see their attributes. Alternatively, open the attribute table of the WaterLines layer in the table of contents to view all the records without selecting them.

The length of each line has been calculated in the SHAPE_Length field in feet. (Length and area attributes maintained by the software accurately measure features in the geodatabase, but they do not have legal authority. An official utility database would include an "as-built" length attribute containing the field results of actual surveys.)

The Install_Date field is null because no attribute values have been entered for it. The Line_Type attribute values have been set to Main for all three water lines, in accordance with the default value for water lines you specified in exercise 11c. The default value isn't right for the last line you digitized, which connects a valve to a hydrant, so you'll change that value next.

16 **In the Line_Type field, click Main. In the drop-down list that appears, click Hydrant Lateral.** You have finished digitizing for now and can save your work.

OBJECTID	3
SHAPE_Length	68.67425
Install_Date	<Null>
Line_Type	Hydrant Lateral ▾

17 **Close the attribute table. On the Editor toolbar, click Editor > Stop Editing. Click Yes to save your edits. Close the Snapping toolbar and close the Editor toolbar.**

If you docked a toolbar, right-click in an empty area near the toolbar and turn off the toolbar from the shortcut menu.

There is no need to save your map document. Your edits are saved to the feature class in the geodatabase. Any map document that contains the WaterLines layer will include the new features.

18 **If you are continuing to the next exercise, leave ArcMap open. Otherwise, exit the application and click No when prompted to save your changes.**

Exercise 12b
Using feature construction tools

In this exercise, you will digitize features representing land parcels. The basic procedure—starting an edit session and setting the snapping environment—is the same as in exercise 12a, but you will be using different tools. ArcMap can set the lengths and angles of edit sketch edges to precise specifications. It also has tools to automatically complete some operations, such as finishing a square, once you provide the necessary information.

For this exercise, you have left your job with the city and gone to work for a developer who is planning to subdivide a large tract of land. Having obtained parcel data from the city's GIS department, you will be using preliminary field measurements to digitize some new parcels in accordance with the developer's proposal. Eventually, your map will go to the city's planning department for review. If the design is approved, the developer will commission a licensed survey crew to conduct a legal survey of the parcel boundaries.

11

12b

13

14

1 **In ArcMap, open the ex12b map document from your Chapter12 folder.** The map shows the parcels displayed against a background photo that lets you see what has been built on the parcels. Some parcels have finished houses, some have houses under construction, and some are vacant.

You will zoom in and digitize a new parcel next to parcels 2731 and 2726.

2 On the menu bar, click Bookmarks > First Parcel Site.

3 On the Standard toolbar, click the Editor Toolbar button.

4 Start an edit session.

First, you will set the snapping environment so that boundary lines for the new parcels snap to the vertices of existing parcels.

5 Open the Snapping toolbar (on the Editor menu, click Editor > Snapping > Snapping Toolbar). On the Snapping toolbar, turn off the Point Snapping button and turn on the Vertex button, as shown in the graphic below. Close the Snapping toolbar.

The new parcel will share two sides with existing parcels, as shown in the graphic. A task called Auto Complete Polygon is used to digitize features that share boundaries with existing features. You will digitize the two new sides, and ArcMap will finish the polygon.

6 On the Editor toolbar, click the Create Features button.

7 In the Create Features window, the Parcels feature template is shown, because there is only one feature class in the map document. In the window, click Parcels, and then, in the Construction Tools panel, click Auto Complete Polygon. (You may need to scroll down if you've resized the window.)

8 In the table of contents, turn off the Subdivision Photo layer to see the parcels more clearly.

9 Move your pointer over the northernmost corner of parcel 2731 so that the cursor snaps to its vertex (the ToolTip for Parcels: Vertex will appear). Click to begin the polygon. The two boundaries you'll digitize must be aligned with the boundaries of the existing parcels. Your field notes say that the new parcel's northwest boundary should be parallel to the northwest boundary of parcel 2731 and that it should be 118 feet long.

10 Right-click anywhere on the northwest boundary of parcel 2731 to open the shortcut menu. On the shortcut menu, click Parallel. (If you left-click by mistake and add a vertex, click the Undo button on the Standard toolbar.)

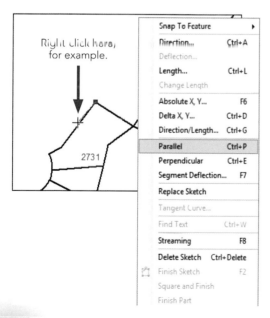

You can press ESC on the keyboard to close the shortcut menu without clicking a command. Pressing ESC will also cancel the effect of a menu command.

11 Move the pointer over the map display without clicking. The line is constrained parallel to the line segment you just clicked.

12 With the pointer located anywhere in the display, right-click. On the shortcut menu, click Length. In the Length pop-up box, delete the existing value and type 118, and then press ENTER. (The graphic shows both the shortcut menu and the pop-up box, although in reality, the menu disappears when the box is displayed.) The first side of the new parcel is finished. It is 118 feet long and runs parallel to the boundary of parcel 2731. The cursor moves to anticipate the next line segment.

13 Move your pointer over the northernmost corner of parcel 2726 so the cursor snaps to its vertex. Double-click to end the line segment.

If you single-click and move your pointer away from the vertex, right-click anywhere and click Finish Sketch on the shortcut menu.

The third and fourth sides of the polygon are completed automatically, using the existing parcel boundaries.

Next, you will add a second parcel to a different part of the subdivision.

14 In the table of contents, turn on the Subdivision Photo layer. Zoom to the Second Parcel Site bookmark. You will digitize the new parcel in the area shown by the yellow box in the graphic. (The box appears only in the book, not on your screen.)

This time you don't have existing boundaries to work from, so in the Construction Tools panel, you'll change the task to Polygon. Your notes tell you that the new parcel should have a corner at a point exactly 61.1 feet from a corner of parcel 2719 and 119.4 feet from a corner of parcel 2707. You will use the Distance-Distance tool, on the Trace tool palette, to choose the start point for the new parcel.

15 In the Create Features window, click Parcels. Polygon is automatically selected in the Construction Tools panel.

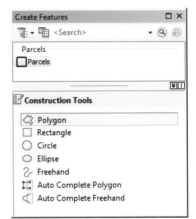

16 On the Editor toolbar, click the Trace tool arrow and click the Distance-Distance tool.

17 In the table of contents, turn off the Subdivision Photo layer.

18 Move your pointer over the northwest corner of parcel 2719 and click when the cursor snaps to its vertex. Press D on the keyboard once. In the Distance pop-up box that appears, type 61.1, and then press ENTER. A circle appears on the map. Its center is the vertex you clicked, and its radius is 61.1 feet.

19 Move your pointer over the northeast corner of parcel 2707 and click when the cursor snaps to its vertex. Press D once. In the Distance pop-up box, type 119.4, as shown in the graphic, and then press ENTER. A second circle is drawn, with a radius of 119.4 feet. The circles intersect at two points.

20 In the area of intersection, move your pointer left and right without clicking. The highlighted dot is constrained to move from one point of intersection to the other. Theoretically, either point could be the correct start point. It so happens, however, that the point on the left would start the parcel in the middle of an unbuilt road. The point on the right is the one you want.

21 Move your pointer to the point on the right and click. The circles disappear. Move your pointer away slightly. A red square marks the spot.

You are now ready to digitize the parcel, using field notes that detail the length and angle of the parcel sides.

22 On the Editor toolbar, click the Straight Segment tool ✎ .

23 Right-click anywhere on the map display. On the shortcut menu, click Direction. In the Direction pop-up box, replace the current value with 73.4, and then press ENTER. The line is constrained to the angular direction you set. The Direction command uses 0 degrees as east and measures positive angles counterclockwise.

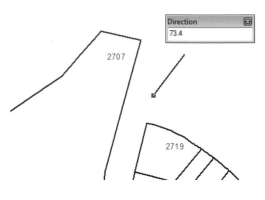

If your Direction pop-up box looks different from the one shown in step 23, your setting for Direction Type may be incorrect. To change the setting, on the Editor toolbar, click Editor > Options. On the Editing Options dialog box, click the Units tab. Change Direction Type to Polar. Make sure your settings match those shown in the graphic, and then click OK.

24 Right-click anywhere above the start point. On the shortcut menu, click Length. In the Length pop-up box, replace the current value with 54.87, and then press ENTER. The first side of the parcel is complete. The start point changes to a green square, and the current vertex becomes a red square.

Next, you will set a direction and a length—this time using a single command—to create the second line segment.

25 Right-click anywhere to the right of the red vertex. On the shortcut menu, click Direction/Length. In the Direction/Length pop-up box, replace the first value with −15.72 (note the minus sign). Replace the second value, the length, with 117, and then press ENTER.

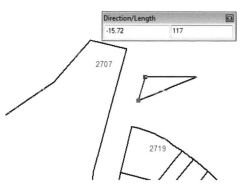

The second side of the parcel is done. To complete the last two sides, you'll next use the Square and Finish command, which completes a polygon by drawing a right angle.

26 Right-click anywhere below the red vertex. On the shortcut menu, click Square and Finish. The parcel is drawn to your specifications.

27 On the Editor toolbar, click Editor > Stop Editing. Click Yes to save your edits. In the table of contents, turn on the Subdivision Photo layer.

On your own

Start an edit session and experiment with the editing tools you didn't use in this chapter. Don't forget about the Undo button, and you probably don't want to save your edits—just practice editing to figure out how some of the tools work.

You have added two proposed parcels to the subdivision. The developer has more in mind, but you won't have to digitize them in this book. You don't need to save your map document. Your edits were made directly to the feature class in the geodatabase. When you saved them, the geodatabase was updated.

28 **Close the Editor toolbar and the Create Features window. If you are continuing to the next chapter, leave ArcMap open. Otherwise, exit the application and click No when prompted to save your changes.**

Summary

Although you didn't get to use all the construction tools ArcMap has to offer, this chapter taught you the basics of feature creation: you learned how to start an edit session, choose options in the Create Features window, use various edit commands to facilitate feature creation, and use various sketch tools that were logical choices for the features you were creating. Thankfully, it's not necessary to create all new features freehand, although that is an option, too. With these basic skills, you should feel more comfortable trying out new tools the next time you have to create geographic features.

Terms

digitizing
vertex (vertices)
endpoints
edge
edit sketch
edit session
snapping

Chapter 13
Editing features and attributes

Deleting and modifying features
Splitting and merging features
Editing feature attribute values

Learning objectives

- *Delete features in an edit session*
- *Modify a feature's geometry*
- *Create a topology to enforce geometry rules*
- *Split a single feature into two features*
- *Merge multiple features into one*
- *Edit attribute values*
- *Add a new attribute field and calculate its values*

CREATING FEATURES IS ONE PART OF MAINTAINING A GEODATABASE. ANOTHER part, equally important, is making changes to existing features.

Features can be edited in several ways. They can be deleted, moved, split, merged, resized, reshaped, and buffered. In addition to the tools on the Editor toolbar, there are also commands on the Editor menu for changing features.

When features are edited, their attributes often need to be updated as well. In a geodatabase, the pre-existing length and area attributes are updated automatically by ArcMap, but other attributes must be updated manually. Sometimes attribute values need to be edited, even when features don't change—for example, when a land parcel is sold, the owner name needs to be changed. And sometimes new attributes are added to a table.

With an ArcGIS for Desktop Basic software license, you can edit shapefile features and those found in file and personal geodatabases. With a Standard or Advanced license, you can also edit multiuser geodatabase features, as well as relationship classes and other behaviors. To edit raster layers, you need to use the ArcGIS Spatial Analyst for Desktop extension. You will be introduced to the ArcGIS Spatial Analyst extension in chapter 20, although editing rasters is beyond the scope of this book.

Data

In this chapter, you will work with the same parcel features from chapters 11 and 12, overlaid on the image raster you saw in chapter 12.

Exercise 13a
Deleting and modifying features

To modify a feature's geometry, you add, delete, or move vertices on the feature's edit sketch. If a feature shares an edge or vertex with another feature, you might want to ensure that no gap or overlap results. You can make features with coincident geometry behave as if they are connected by creating a *map topology* within your edit session.

For this exercise, you are the GIS specialist for the city's planning department. To prepare for a meeting of the City Commission, you will make some changes proposed for parcels in a new subdivision. Because the changes are recommendations only, you will not work with the city's live geodatabase, but with a copy designated for planning use.

Your first task is to delete two parcels on land the city is negotiating to buy back from the subdivision developer. Your second task is to enlarge the boundaries of two parcels to make room for a pond to catch water runoff from a nearby road and sidewalk. The map is simply for discussion, so your edits don't have to be exact. If the proposals are implemented, a licensed survey crew will resurvey the parcels.

11

12

13a

14

1 Start ArcMap. Open the ex13a map document from your C:\EsriPress\GTKArcGIS\Chapter13 folder.

The parcels are outlined in black against an aerial photograph of the subdivision. You will first zoom in on the parcels to be deleted.

2 On the menu bar, click Bookmarks > First Parcel Site.

The parcels are labeled with their ID numbers. (The labels are set to display at scales of 1:4,000 and larger.) You will delete parcels 2712 and 2718 in the middle of the display.

③ **Display the Editor toolbar, if not already available, and start an edit session (on the Editor toolbar, click Editor > Start Editing). Make sure the Edit tool** ▶ **is active.**

④ **On the Editor toolbar, click the Create Features button.** This opens the Create Features window. The template layer is Parcels, the only editable layer on the map.

⑤ **Close the Create Features window if you wish. On the map, click anywhere on parcel 2712. It is outlined in cyan to show that it is selected. Press and hold SHIFT and click parcel 2718 to add it to the selection.** The small blue *x* in the center of the selected parcels is called a selection anchor. Every selection in an edit session, whether of one feature or many, has one. (Selection anchors are used to rotate and scale features, which you won't do in this exercise.)

6 **With parcels 2712 and 2718 selected, click the Delete button ⊠ on the Standard toolbar.** The selected features are deleted.

7 **Save your edits (on the Editor toolbar, click Editor > Save Edits).**

Next, you will modify the shapes of two parcels to accommodate the proposed detention pond. You need to move the parcel edges to connect to a nearby sidewalk. You will add a vertex to one parcel's edit sketch and move another vertex that is shared by two parcels.

8 **Zoom to the Second Parcel Site bookmark.** The proposed pond is outlined in yellow in the graphic (but not on your screen).

The parcels to be edited are 2768 and 2845. The parcels will be extended to the sidewalk in the lower portion of the display. The sidewalk appears on the raster layer as a thin white line that runs east–west and lies just north of the road. To make the edits, you will zoom in closer.

9 **Zoom to the Proposed Drainage Change bookmark.**

You will make four edits: (1) you'll move an existing vertex in parcel 2768 to the sidewalk; (2) you'll add a new vertex to parcel 2768; (3) you'll move the new vertex to the point where the sidewalk jogs; and (4) you'll move a vertex shared by parcels 2768 and 2845.

10 **Double-click anywhere on parcel 2768 to select it and display its edit sketch.** On the edit sketch, the vertices are green squares. (Only the last digitized vertex— which you don't see because you are zoomed in on a corner of the parcel—is a red square.) ArcMap has automatically activated the Edit Vertices toolbar, which can be started from the Editor toolbar by the Edit Vertices button, which is highlighted.

11 Place your pointer over the first vertex to be moved. The pointer becomes a four-headed arrow.

12 Use the pointer to drag the vertex down to the sidewalk and release the mouse button. The black line shows the new shape of the feature.

The highlighted cyan line remains in place as a reference. It will disappear when you clear the selection or select another feature.

Next, you'll add a new vertex and move it to accommodate the jog in the sidewalk.

13 On the Edit Vertices toolbar, click the Add Vertex tool ⊠ .

11 Click the black line about where the arrow is in the graphic.

A new vertex is added where you clicked.

15 Place your pointer over the new vertex and drag the vertex to where the sidewalk changes direction. Release the mouse button.

11

12

13a

14

16 Double-click somewhere outside the polygon to complete the edit and clear the selection.

17 Save your edits.

Next, you'll move a vertex shared by parcels 2768 and 2845 to the sidewalk. This "shared" vertex (circled in yellow in the graphic) is actually two vertices (one belonging to each polygon) that lie at exactly the same spot.

You could use the Edit tool to move each vertex separately. Since you want the parcels to remain adjacent, this would mean setting the snapping environment, and then moving one parcel's vertex, moving the second parcel's vertex, and snapping that vertex to the first. This wouldn't be too hard this time, but if you had to do it several times, or if you had four or five polygons sharing a vertex, it might start to get tedious.

A better approach is to create a map topology. Map topology makes features aware of their coincidence to other features so that common vertices and edges can be moved together.

In what situations can you imagine map topology would be useful?

18 On the Customize menu, click Toolbars > Topology. (You may need to scroll down in the list of toolbars.) The Topology toolbar appears.

19 Click the Select Topology button ⊟. This opens the Select Topology dialog box.

On this dialog box, you can choose the layers for the map topology (in this case, Parcels is the only choice). You can also set a cluster tolerance, which is the distance at which ArcMap considers nearby edges or vertices to be coincident, or shared.

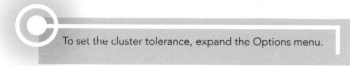

To set the cluster tolerance, expand the Options menu.

20 On the Select Topology dialog box, in the Select the layer box, select the Parcels check box and leave the cluster tolerance at its default value. Then click OK.

21 On the Topology toolbar, click the Topology Edit Tool 🖾 .

22 Place your pointer over the map display and click the location of the shared vertex. The vertex is highlighted in purple. (If you highlight a polygon edge by mistake, click outside the parcel and try again.)

11

12

13a

14

23 Place your pointer over the purple vertex. The pointer changes to a four-headed arrow.

24 Right-click the shared vertex and select Stretch Topology Proportionately. This command allows the shared vertex to move while its adjoining edges remain where they are.

	Build Topology Cache
⊞	Shared Features
	Select Shared Features
⊞	Clear Selected Topology Elements
🖺	Stretch Topology Proportionately
⊠	Modify Edge

25 Use the pointer to drag the purple vertex to the point where the sidewalk changes direction again. As you drag, both polygons are reshaped simultaneously. Release the mouse button.

While dragging the vertex.

After releasing the mouse button.

26 Close the Topology toolbar.

On your own

Examine the sketch tools available on the Editor
toolbar and the Advanced Editing toolbar.

27 On the Editor toolbar, click Editor > Stop Editing. Click Yes when
prompted to save your edits.

The map is ready for the City Commission meeting. You don't need to save your
map document because your edits were made directly to the feature class in
the geodatabase.

Map topology and geodatabase topology

Map topology can be created for shapefiles and geodatabase feature classes. It lasts
only for the duration of an edit session and allows you to edit features (within the
same layer) as if they were connected to each other.

With an ArcGIS for Desktop Standard or Advanced license, you can create a more
sophisticated topology for geodatabase feature classes, called
geodatabase topology. Geodatabase topology lets you enforce more than 30
spatial rules within and between feature classes, helping to guarantee the integrity of
your data. For example, you can specify that polygons may not overlap or have gaps,
that lines must be connected to other lines, or that points in one feature class must
lie within polygons of another feature class. Geodatabase topology is created within
a geodatabase feature dataset; it is independent of a particular map document or
edit session.

For more information, in ArcGIS for Desktop Help, on the Contents tab, navigate to
the following topics:

- Editing > Editing topology > Editing shared geometry with topology >
  Creating a map topology

- Geodata > Data types > Topologies > Creating a topology

---

28 If you are continuing to the next exercise, leave ArcMap open.
Otherwise, exit the application and click No when prompted to save
your changes.

11
12
13a
14

Exercise 13b
Splitting and merging features

To split a polygon or a line (create two features from one), you digitize a line across the feature where you want to split it. You can split several features at once.

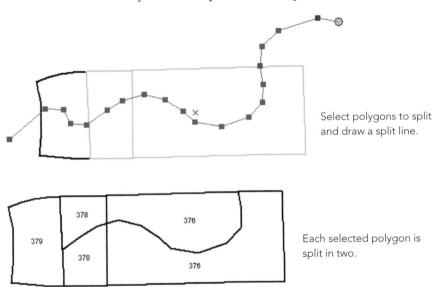

Select polygons to split and draw a split line.

Each selected polygon is split in two.

To merge polygons or lines (create one feature from two or more), you select them and click the Merge command on the Editor menu (see facing page). The merged features do not have to be contiguous. For example, you could merge a large polygon representing a country and smaller polygons representing its offshore islands. The country and islands would then exist as a multipart feature with a single record in the attribute table.

Select polygons to merge.

The three polygons are merged into one.

When features in geodatabase feature classes are split or merged, ArcMap updates their Shape_Length and Shape_Area attributes. This is not done for features in other data formats, such as shapefiles. For all feature classes, ArcMap updates the new feature's identifier attribute, which keeps track of the number of features. Changes to other attributes, including user-defined Ids, such as parcel numbers or legal measurements of length and area, must be made manually.

For this exercise, the planning department has been notified of two pending property changes. In one case, an owner wants to subdivide her land so that she can sell a piece of it. In the other, the owner of half a duplex has bought the other half. Because the duplex had two owners, there was a parcel boundary dividing the house. Now the boundary needs to be removed.

11

12

13b

14

1 **In ArcMap, open the ex13b map document from your Chapter13 folder.**

The map displays the familiar subdivision. You'll zoom in on the parcel to be split.

2 **Zoom to the First Parcel Site bookmark.**

The display zooms in on parcel 2707. The yellow line (which appears in the graphic but not on your screen) shows where the parcel is to be split. When the parcel is split, ArcMap will update the attribute table with the correct area of each new parcel. To see this for yourself, you'll check the area of parcel 2707 next, before you split it.

3 On the Tools toolbar, click the Identify tool 🛈 and click parcel 2707. The Shape_Area field shows a value of 107,248 square feet, before the decimal place.

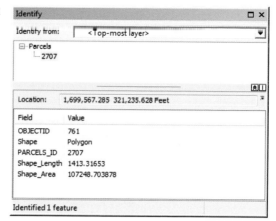

4 Close the Identify window. In the table of contents, turn off the Subdivision Photo layer.

5 Display the Editor toolbar, if not already available. Start an edit session and open the Create Features window, if not already displayed.

6 In the Create Features window, click Parcels, the only editable layer. Close the window if you wish.

7 On the Editor toolbar, click the Edit tool ▶ , if not already active.

8 Double-click parcel 2707 to select it and display its edit sketch. You will draw the split line between the vertices indicated by the two red arrows in the graphic. When you double-click to display a feature's edit sketch, the Edit Vertices toolbar appears.

Before digitizing the split line, you will next set the snapping settings so that the line snaps to vertices.

9 Display the Snapping toolbar (on the Editor toolbar, click Snapping > Snapping toolbar). On the Snapping toolbar, click Vertex Snapping, if not already turned on, and turn off any other snapping options that may be activated. When you are finished, close the toolbar.

10 On the Editor toolbar, click the Cut Polygons tool ⊞ . Although the parcel vertices are no longer displayed when the task is changed, the split line you digitize will still snap to them.

11 On the display, the pointer changes to a crosshair. Move the pointer to the position shown in the graphic. When the cursor snaps to the vertex, click it.

12 Move your pointer over the second vertex. When the cursor snaps to it, double-click to end the split line. If you moved your pointer away from the vertex, press F2 to finish the sketch. The parcel is split in two. Each parcel is labeled with the same PARCELS_ID of 2707. In exercise 13c, you'll update the PARCELS_ID attribute for the smaller parcel.

13 On the menu bar, click Selection > Clear Selected Features, and then save your edits.

⑭ On the Tools toolbar, click the Identify tool. Click each of the two new parcels and examine the Shape_Area values. The sum of the areas of the two new parcels equals the area of the original parcel. The PARCELS_ID number, a user-defined attribute, does not change. However, the new parcel is assigned a new OBJECTID number.

Your next task is to merge two parcels where the owner of half a duplex has bought the other half.

⑮ Close the Identify window. Zoom to the Second Parcel Site bookmark.

⑯ Turn on the Subdivision Photo layer.

In the center of the map display, the two parcels you need to work with—2855 and 2861—are located on a cul-de-sac. The parcel boundary crossing the duplex needs to be removed, so you'll do that next.

11

12

13b

14

17 Activate the Edit tool, and then click anywhere on parcel 2855 to select it. Press and hold SHIFT and click parcel 2861 to add it to the selection.

18 On the Editor toolbar, click Editor > Merge. On the Merge dialog box, click 2855 (Parcels), and then click OK.

The two parcels are now one. The merged parcel retains the attributes of parcel 2855.

19 In the table of contents, right-click the Parcels layer and click Open Attribute Table. Click the Show selected records button at the bottom of the table to display the selected records. The record that remains is for the feature you selected on the Merge dialog box.

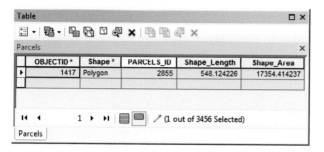

What happened to the Shape_Length and Shape_Area values for parcel 2855?

20 Close the table. On the Editor toolbar, click Editor > Stop Editing. Click Yes to save your edits. You don't need to save your map document, because your edits were made directly to the Parcels feature class in the geodatabase.

21 If you are continuing to the next exercise, leave ArcMap open. Otherwise, exit the application and click No when prompted to save your changes.

11

12

13b

14

1
2
3
4
5
6
7
8
9
10
11
12
13c
14
15
16
17
18
19
20

Exercise 13c
Editing feature attribute values

In ArcMap, you can change attribute values one at a time by typing new values in the table during an edit session. You can also change values for a selected set of records or for all records in a table using the Field Calculator. The Field Calculator assigns values to a field according to an expression, which may simply be a value, such as 101 or "Residential," or an expression that uses math operators. The Field Calculator can be used in or out of an edit session.

Besides editing attribute values, you can also add and delete records or add new fields to a table and define properties for those fields. Records are added and deleted during an edit session. Fields are added and deleted outside of an edit session.

Having split parcel 2707 in exercise 13b, you need to assign a new PARCELS_ID number to one of the new parcels. After that, you will add an attribute to the parcels feature class. In a previous meeting with the City Commission, you found that parcel sizes were often given in acres rather than in square feet. The Parcels feature class doesn't have an acreage attribute. You will add one, and then have ArcMap calculate values for it, converting square feet to acres.

1 In ArcMap, open the ex13c map document from your Chapter13 folder.

The map shows the parcels you split in exercise 13b. Both have an ID of 2707. You will next assign a new ID number to the smaller parcel.

2 Display the Editor toolbar, if it's not already turned on, and start an edit session.

3 Open the Create Features window. Again, click Parcels.

Even in a map with only one editable layer, it is necessary to choose the correct feature template in the Create Features window to activate specific editing tools.

4 On the Editor toolbar, click the Edit tool. Click the small parcel 2707 to select it.

5 Open the Parcels attribute table. Click the Show selected records button at the bottom of the table. Field names on gray backgrounds in the table indicate fields that are maintained by ArcMap. They cannot be edited. Names on white backgrounds indicate fields that you can edit. (Note that with some display schemes, the color difference may be hard to see.)

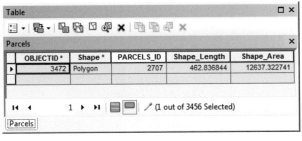

6 Click in the cell under the PARCELS_ID field. Highlight and delete the existing value of 2707. Type 4001 and press ENTER. The value 4001 does not duplicate an existing parcel ID. This is the only value you need to edit. If there were other attributes, such as an address or owner name, you would update those as well.

7 On the Editor toolbar, click **Editor > Stop Editing. Click Yes** to save your edits. Then move the table away from the map display. When the edit session is finished, there are no longer selected records in the table or selected features on the map. On the map, the smaller parcel is labeled with its new PARCEL_ID.

Next, you will add a field to the table to store the parcel size in acres.

8 At the bottom of the Parcels attribute table, click the Show all records button.

Currently, the Shape_Area field contains the square footage of each parcel. After adding the new field, you will write an expression in the Field Calculator to convert square feet to acres.

OBJECTID *	Shape *	PARCELS_ID	Shape_Length	Shape_Area
1	Polygon	219	771.272726	34756.355064
2	Polygon	220	951.705862	37783.615298
3	Polygon	222	1094.739692	74870.416578
4	Polygon	221	3841.477939	367909.42156
5	Polygon	223	953.156376	21397.130912
6	Polygon	221	575.000011	17000.00010
7	Polygon	225	714.166729	31394.712776
8	Polygon	241	1309.803923	96184.958257
9	Polygon	226	5870.145899	1830162.19609
10	Polygon	242	1212.419567	81143.275191
11	Polygon	227	1609.435725	161872.373062

(1 out of 3456 Selected)

Parcels

What expression do you anticipate writing to perform this conversion?

9 At the top of the table, click the Table Options arrow and click Add Field.

			Shape_Length
🔍	Find and Replace...		
	Select By Attributes...	9	771.272726
	Clear Selection	0	951.705862
	Switch Selection	2	1094.739692
	Select All	1	3841.477939
		3	953.156376
	Add Field...	4	575.903911

If the Add Field command is not available, you may not have concluded the edit session in step 7.

The Add Field dialog box is where you name the new field, specify its data type, and set its field properties. (This is equivalent to what you did in chapter 11 when you added fields to the WaterLines feature class in Catalog.)

Adding fields from ArcCatalog

New fields can be added from ArcCatalog or ArcMap. With an ArcGIS for Desktop Basic license, fields can be added to file and personal geodatabases or shapefile feature classes only. With a Standard or Advanced license, fields can also be added to feature classes in coverages and ArcSDE geodatabases. For more information, in ArcGIS for Desktop Help, on the Contents tab, navigate to Geodata > Data types > Tables > Creating and editing tables > Adding fields.

---

10 On the Add Field dialog box, in the Name box, type Acres. Click the Type arrow and click Float (Float is a numeric field that allows decimal points). Click OK.

The new field is added to the table. Its values are set to Null.

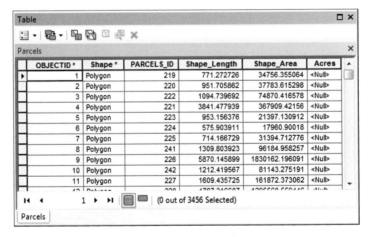

Next, you will calculate values for the Acres field, which you can do whether you are in an edit session or not. Field calculations made in an edit session take longer but have an Undo option. Calculations made outside an edit session are faster but cannot be undone. (To replace an incorrect calculation, you can make a new one or delete the field and start over.) For the sake of speed, you will make your calculation without starting an edit session.

11 In the attribute table, right-click the Acres field name and click Field Calculator.

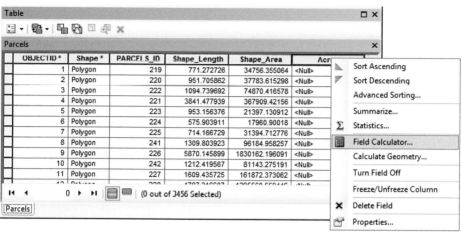

12 If you see a warning about calculating outside an edit session, click Yes to dismiss it. (You may also select the check box to no longer show this warning.) The Field Calculator appears.

Next, you'll create an expression to convert feet to acreage. Since one acre equals 43,560 square feet, dividing the values in the Shape_Area field by 43,560 will give each parcel's acreage.

13 In the Fields box of the Field Calculator, double-click Shape_Area to add it to the expression box below. Click the division (/) button. Press the SPACEBAR and type 43560. Make sure your dialog box matches the graphic, and then click OK.

11
12
13c
14

The values are calculated and written to the Acres field of the attribute table.

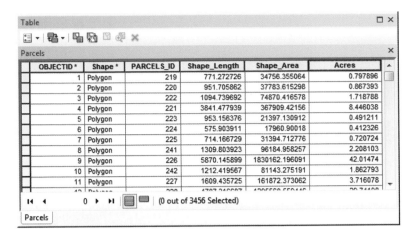

When you edit a feature in a geodatabase feature class, ArcMap automatically updates its Shape_Length and Shape_Area values. ArcMap does not, however, manage fields that you have added yourself. Therefore, when you make changes to parcels that affect their size, you'll have to recalculate the Acres field.

You don't need to save your map document. Your edits were made directly to the feature class attributes in the geodatabase.

14 Close the attribute table. Close the Editor toolbar. If you are continuing to the next chapter, leave ArcMap open. Otherwise, exit the application and click No when prompted to save your changes.

Summary

Editing GIS data is an important skill. Features change, new information is received, and in some cases, data needs to be simplified. This chapter taught basic editing skills that should provide a basis for learning more advanced editing techniques, which are available in ArcMap.

Terms

map topology

geodatabase topology

Chapter 14
Geocoding addresses

Creating an address locator
Matching addresses
Rematching addresses

Learning objectives

- *Create and manage address locators*
- *Learn how addresses are eligible for geocoding*
- *Manually match unmatched addresses*

IN CHAPTER 12, YOU LEARNED THAT A LOT OF SPATIAL DATA WAS ORIGINALLY digitized from paper maps. It's also possible to create spatial data—specifically, point features—from information that describes or names a location. The most common kind of information that is used for this is an address. The process of creating map features from addresses, place-names, and similar information is called *geocoding*.

The value of geocoding is that it lets you map locations from data that is readily available. If you own a business, you can't buy an atlas with a ready-made map of your customers. With a list of your customers' addresses, however, you can create this map yourself.

Geocoding addresses requires that you have an address table—a list of addresses stored as a database table or a text file. You also need a set of reference data, such as streets, on which the addresses can be located.

	FID	Shape	Address	City	State	ZIP
▶	0	Point	737 Redman Ave	Haddonfield	NJ	0803
	1	Point	45 Treaty Elm Lane	Haddonfield	NJ	0803
	2	Point	130 Chestnut St	Haddonfield	NJ	0803
	3	Point	412 Kings Hwy E	Haddonfield	NJ	0803
	4	Point	601 Grove St	Haddonfield	NJ	0803

Address table.

Reference data.

ArcGIS uses address information in the attribute table of the reference data to figure out where to locate address points. The more detailed the reference data, the more accurately addresses can be located.

The same table of addresses is geocoded using two sets of reference data. *Left*, on the street map of Haddonfield, New Jersey, address points are placed in the correct location on the correct street. *Right*, on the ZIP Code map of the United States (zoomed to western New Jersey), the same address points are placed in the center of the correct ZIP Code. Because all five points have the same location, they look like one point.

Street-level reference data is commercially available. Its attributes include street name, street type (Avenue, Boulevard, and the like), and directional prefixes and suffixes needed to avoid ambiguity in address locations. Each street feature is divided into segments that have starting and ending addresses, such as seen on neighborhood street signs. This makes it possible to estimate the position of an address along the length of a street segment. There may be separate address ranges for each side of the street, allowing an address to be placed on the correct side of the street as well.

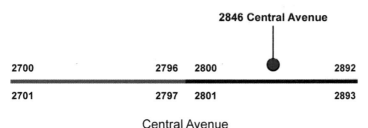

Central Avenue

To geocode the address 2846 Central Avenue, ArcGIS examines the reference data and finds the street segment of Central Avenue with the correct address range. Because 2846 is midway between the range of 2800 to 2892 for Central Avenue, the address point is placed in the middle of the street segment, on the even-numbered side of the street.

In addition to an address table and reference data, geocoding requires an *address locator*. An address locator is a file that contains the reference data (the relevant attributes of the address table, such as street name and ZIP Code) and various geocoding rules and tolerances.

The output of the geocoding process is either a shapefile or a geodatabase feature class of points. The geocoded data has all the attributes of the address table, some of the attributes of the reference data, and optionally, some new attributes, such as the x,y coordinates of each point. For more information on what fields are included in the output feature class for geocoding results, in ArcGIS for Desktop Help, on the Contents tab, navigate to Guide Books > Geocoding > Locating addresses > About geocoding a table of addresses.

Data

The exercises in this chapter use the following datasets:

- Interstates.shp—US interstate highways *(Source: Tele Atlas)*
- streets.shp—Atlanta street data *(Source: Tele Atlas)*
- zipcodes.shp—several ZIP Codes used as a background layer *(Source: Tele Atlas, US Census, Esri (Pop2010 field))*
- Customers.csv—a table of customer addresses *(Source: Esri)*

Exercise 14a
Creating an address locator

Geocoding starts with creating an address locator. Address locators come in different styles, depending on the attributes of the reference data. For example, the style called "Single Field" is used with reference data that contains only a single attribute of geographic information. In a reference map of the United States, for instance, this would probably be a state name field. Using this type of locator, the data to be geocoded could be located in the correct state, but no more precisely than that.

The address locator style ZIP 5 Digit is used with reference data that has a ZIP Code attribute. A table of US addresses could be located within the correct ZIP Codes using this type of locator.

The US Address – Dual Ranges style is used with street reference data that contains a street name attribute, and starting and ending address ranges for each side of the street. A table of US addresses could thus be geocoded to their approximate positions along the length of a street and placed on the correct side of the street as well.

Each address locator style requires that certain attributes be present in the address table. For more information about address locator styles, in ArcGIS for Desktop Help, on the Contents tab, navigate to Guide Books > Geocoding > Preparing for geocoding > Understanding address locator styles, and also Commonly used address locator styles.

For this exercise, you work for an online business directory that sells advertising—everything from listings and banners to interactive storefronts. A new service you plan to offer is a digital map that displays the location of clients' businesses. Your job is not to create the Internet map service itself but rather to test the technology to convert customer addresses to map points. Being inexperienced with geocoding, you'll work with a small sample of your potential customer base.

1 **Start ArcMap and open the ex14a map document from your C:\EsriPress\GTKArcGIS\Chapter14 folder.** The map on the facing page shows a portion of downtown Atlanta, Georgia, with layers of interstate highways, streets, and ZIP Codes.

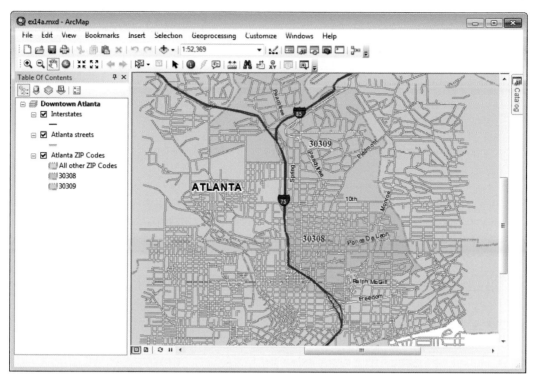

2 **At the top of the table of contents, click the List By Source button.**
In the List By Source view, the table of contents shows nonspatial tables, such as dBASE files, that belong to the map document. In this case, there is a table of customers, which contains a sample of your customer database. This is the address table you'll geocode in exercise 14b.

11

12

13

14a

It may be helpful to use spreadsheet software, such as Microsoft Excel, to prepare data for geocoding. This is useful in formatting your tables so that ArcMap will recognize the data.

The address locator style you choose depends on the attributes in your reference layer (the Atlanta streets layer) and in your address table. Next, you'll open the two tables to look at this information.

3 **In the table of contents, right-click the Customers table and click Open.** The attributes include customer ID, name, street address, five-digit ZIP Code, and business type. For geocoding, the relevant attributes are ADDRESS and ZIP.

CUST_ID	NAME	ADDRESS	ZIP	TYPE
840367	Accessories Inc	725 LAKEVIEW AVE NE	3030	Retail Clothing
291063	Antiques Inc	491 BISHOP ST NE	3030	Retail
520746	Damar Sales	388 7TH ST NE	3030	Car Dealer
197302	Eastern Graphics	150 6TH ST NE	3030	Printing
914582	Gelroy Associates	762 ARGONNE AVE NE	3030	Real Estate
107892	Homes For Sale	500 RANKIN ST NE	3030	Real Estate
308574	Industrial Supply Co	751 JUNIPER	3030	Wholesale
489041	Leonard Associates	555 10TH ST NE	3030	Stockbroker
734173	Lomer-Lockler Beauty College	220 6TH ST NE	3030	Trade School
227309	Oar Clothing Co	400 7TH ST NE	3030	Retail Clothing
197842	Ron's Food Mart	844 MYTLE	3030	Retail Food

(0 out of 16 Selected)

Customers

4 **Close the Customers table. In the table of contents, right-click the Atlanta streets layer and click Open Attribute Table (see facing page).** The four address range attributes are L_F_ADD (left from address), L_T_ADD (left to address), R_F_ADD (right from address), and R_T_ADD (right to address). These are the starting and ending street numbers (to and from) on each side (left or right) of a street segment.

There are several other address attributes, as follows:

- PREFIX is a direction that precedes a street name, such as the "N" in 139 N. Larchmont Blvd.
- PRE_TYPE is a street type that precedes a street name, such as "Avenue" in 225 Avenue D.
- ZIPL and ZIPR are the ZIP Codes for the left and right sides of a street segment, respectively.
- CITYL and CITYR are the city names for the left and right sides of a street segment, respectively.
- NAME, TYPE, and SUFFIX direction are the familiar attributes of a street address.

The attributes in the Atlanta streets table most closely match the US Address – Dual Ranges address locator style.

5 Close the table. On the Standard toolbar, click the Catalog button. (You may dock your Catalog window or leave it floating.) Or if the Catalog tab is already available, click it to show the Catalog window.

11

12

13

14a

6 **In the Catalog tree, navigate to \GTKArcGIS\Chapter14\MyData. Right-click My Data and click New > Address Locator.** The Create Address Locator dialog box appears.

Folder
File Geodatabase
Personal Geodatabase

Database Connection...
ArcGIS Server Connection...

Layer...
Group Layer

Python Toolbox
Shapefile...

Turn Feature Class...

Toolbox
dBASE Table
LAS Dataset
Address Locator...
Composite Address Locator...

XML Document

You may have noticed the Composite Address Locator command. A composite address locator consists of two or more individual address locators that allow addresses to be matched against either locator to find the best match. If an address does not match one of the locators, it has a chance of matching the other. For more information, in ArcGIS for Desktop Help, on the Contents tab, navigate to Guide Books > Geocoding > Building an address locator > Creating a composite address locator.

7 Click the Browse button next to the Address Locator Style box. On the Select Address Locator Style dialog box, click US Address – Dual Ranges and click OK. The Dual Ranges style is used for address ranges that have data for both sides of a street segment.

Notice that an error message appears in the upper portion of the Create Address Locator dialog box. Scroll down and click the warning icon next to the Field Map label. The error message, which reads "Field(s) required by this style are missing," simply means that field names need to be matched to alias names from the reference data.

8 Click the Browse button next to the Reference Data box. Browse to the \Chapter14\Ex14aData folder, click the streets shapefile, and click Add. The reference data is added to the list and has the role of Primary Table (you may have to expand the window slightly to see the full path). Below it, the Field Map section is populated with the fields required to use the Dual Ranges locator style.

9 **Scroll down to view the Field Map list.** The Field Map list shows the types of address information contained in the reference data that the address locator style can use. Required information for geocoding is signified by an asterisk (*) in front of the field name. The values in the corresponding lists under Alias Name show the reference data attributes that contain this information. In each case, ArcMap has already selected the correct field (alias) name. (If it hadn't, you could use the lists to choose the corresponding fields.)

10 **In the lower portion of the Create Address Locator dialog box, click the Browse button next to the Output Address Locator box. Browse to \GTKArcGIS\Chapter14\ MyData, and in the Name box, type** Atlanta customers. **Then click Save.**

11 **Click OK to create the address locator.** When the Create Address Locator process is completed, a pop-up notification appears in the notification area, at the far right of the taskbar. Click the notification link if you want to view the results.

ArcMap creates the new address locator in the output folder specified.

12 **If you're continuing to the next exercise, leave ArcMap open. Otherwise, exit the application and click No if prompted to save your changes.**

In exercise 14b, you'll add the Atlanta customers address locator to ArcMap and use it to match addresses.

Exercise 14b
Matching addresses

Now that you've created an address locator in exercise 14a, you can compare the table of addresses to the reference data attributes and look for matches.

The first step in address matching is to standardize the addresses in the address table—that is, divide them into their component parts. This is done by ArcGIS in the address locator. For example, the address "500 Rankin Street NE" has four parts: the street number (500), the street name (Rankin), the street type (Street), and a suffix (NE). The reference data is already standardized—each address part is listed as a separate attribute.

ArcGIS takes each standardized address from the address table and looks for features in the reference data that have matching parts. For each address, it generates a list of probable or possible matching locations, called candidates, in the reference data. Each candidate gets a score that rates its likelihood of being the correct location. For ArcGIS to make a match and create a point feature, a candidate's score must reach a certain level. You can choose both the score that defines a location as a candidate and the score that defines a candidate as a match. By default, the minimum candidate score is 10, on a scale of 1 to 100, and the minimum match score is 60.

When ArcGIS has made as many matches as it can, it creates a new dataset, which has a point feature for each address, whether it is matched or not. Features that have unmatched addresses are not assigned spatial coordinates, and are therefore not displayed. At that point, you can adjust the geocoding options to try to find matches for unmatched addresses.

Most of the time, geocoding is done from a table of addresses, but as long as you have created an address locator, you can also geocode by typing address information in the Find dialog box. Geocoding in this manner does not create new features, but it will flash a matching location on the map if there is one or place a graphic there. This is useful in situations where you need to locate a single address or find a few addresses quickly—for example, to make immediate deliveries.

For this exercise, before geocoding your customers table, you'll test the address locator by locating the address of the Ace Market, a downtown Atlanta business.

1 **In ArcMap, open the ex14b map document from your Chapter14 folder, and at the top of the table of contents, click the List By Source button.** The map shows the layers of interstate highways, Atlanta streets, and ZIP Codes. The Customers table is at the bottom of the table of contents.

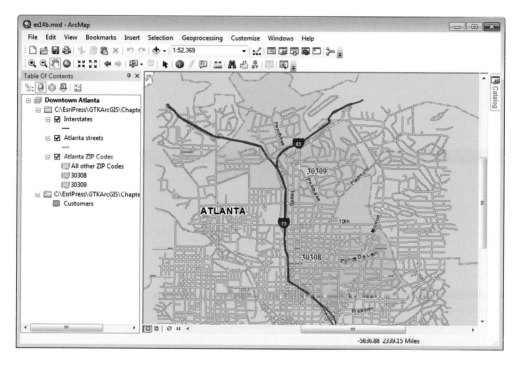

2 **On the Customize menu, click Toolbars > Geocoding.** The Geocoding toolbar appears.

Next, you'll add the address locator to the map document.

3 **On the Geocoding toolbar, click the Select Address Locators arrow and click <Manage Address Locators...>.**

The Address Locator Manager dialog box appears.

④ **On the Address Locator Manager dialog box, click Add. On the Address Manager Locator dialog box, navigate to your \GTKArcGIS\ Chapter14\MyData folder. Click the Atlanta customers address locator you created in exercise 14a and click Add.** The Atlanta customers address locator and a description of the style appear on the Address Locator Manager dialog box.

⑤ **Close the Address Locator Manager dialog box.**

Now that the Atlanta customers address locator has been added to the map document, you can find addresses on the map, which you'll do next.

⑥ **On the Tools toolbar, click the Find button. On the Find dialog box, click the Locations tab. The Atlanta customers address locator is displayed in the Choose a locator list. If it isn't, click the list arrow and click Atlanta customers.**

7 **In the Full Address box, type**
1171 Piedmont Ave, **the address of
the Ace Market, and then click Find.**
One candidate street segment appears in
the lower section of the dialog box. Its
score is 90.97. By default, only the
candidate with the highest score is
displayed.

On your own

Display additional candidates by
selecting the Show all candidates check
box, and then explore the address
information for each one.

8 **Scroll across the dialog box and look at the candidate's address
information.** The address range for odd street numbers, FromAddr and
ToAddr, is correct because it includes the street number 1171. The street name,
street type, and suffix direction are also correct.

A score of 90.97 is high, but to be sure that you have the right one, you will
include more information for the Ace Market address—specifically, the suffix
and ZIP Code.

9 **On the Locations tab, click the
Full Address box and type**
1171 Piedmont Ave NE 30309,
and then click Find. The candidate
now gets a perfect score of 100.

11

12

13

14b

10 If necessary, move the Find dialog box so you can see the main ArcMap window. Click the first candidate to highlight it. Notice that a green point flashes at the location of Ace Market. In the Candidates box, right-click anywhere on the highlighted candidate. On the shortcut menu, click Add Point. Close the Find dialog box. ArcMap places a black dot on the map at the location of the candidate.

You geocoded the Ace Market to verify that the address locator was working. This point is a graphic (much like a point that can be drawn using the Draw toolbar) and not a feature associated with a shapefile or a feature class. Next, you can delete the graphic.

11 On the Tools toolbar, click the Select Elements tool. Click the graphic to select it (cyan selection handles appear). Press DELETE.

Next, you'll geocode the customers table.

12 On the Geocoding toolbar, click the Geocode Addresses tool.

You can also right-click the Customers table in the table of contents and click Geocode Addresses. You may want to add the table to ArcMap first, however, to ensure the fields are recognized.

On the "Choose an Address Locator to use" dialog box, the Atlanta customers address locator is highlighted.

ArcGIS Online geocoding services are also available. These are geocoding services based on ArcGIS 10.1 geocoding technology. ArcGIS 10.1 geocoding services allow you to find addresses using a single-line input. You can also use it to find an address with a spatial offset or by its latitude and longitude coordinates.

11

12

13

14b

13 **Click OK to open the Geocode Addresses dialog box.** The City or Placename and the State lists are designated as None because the Customers table does not contain this information.

Geocoding creates a point feature for every address. The points can be saved as a shapefile or as a geodatabase feature class. Your Atlanta data is in a shapefile format, so you'll save the address points as a shapefile as well.

14 **Click the Browse button next to the Output shapefile or feature class box.**

15 **On the Saving Data dialog box, browse to \GTKArcGIS\Chapter14\ MyData and name the output** Geocode1.shp. **Make sure the Save as type list is set to Shapefile and click Save.** The Geocode Addresses dialog box is updated with the new path.

16 **Click OK to begin the geocoding process.** ArcMap geocodes the addresses and creates the output shapefile. When the process is completed, a Geocoding Addresses results window appears.

Of the 16 addresses in the Customers table, 13 are matched, and three addresses are unmatched. You'll match the unmatched addresses in exercise 14c. For now, you'll look at the attributes of the geocoded shapefile.

17 **Close the Geocoding Addresses results window.** The Geocoding Result: Geocode1 layer is listed at the top of the table of contents. Meanwhile, 13 point features are displayed on the map, representing the locations of the matched addresses.

18 In the table of contents, right-click the Geocoding Result: Geocode1 layer and click Open Attribute Table.

The table contains 16 records, one for each address in the Customers table. The Status attribute tells you whether each address was matched (M), unmatched (U), or tied (T) with another address for the highest score. (In this case, none of the addresses scored a tied match.) Point geometry is created for records with status M only. The Score attribute contains the match score, and the Side attribute tells you whether an address is on the right side or left side of the street.

The next two attributes that are used, ARC_Street and ARC_ZIP, contain the information from the ADDRESS and ZIP fields, respectively, in the Customers table. The remaining attributes are the same as those in the Customers table.

Although the ARC_Street and ARC_ZIP attributes contain the same values as ADDRESS and ZIP, they serve a different purpose. When you rematch unmatched addresses, you can edit the values in ARC_Street and ARC_ZIP to make matches. You would do this, for example, if you found mistakes in the Customers table. The ADDRESS and ZIP attributes, on the other hand, preserve the original information from the Customers table and should not be edited.

19 Close the table. You have already saved the geocoded shapefile. There is no need to save your map document.

20 If you are continuing to the next exercise, leave ArcMap open. Otherwise, exit the application and click No when prompted to save your changes.

Exercise 14c
Rematching addresses

When you geocode at the street address level, it's common to have addresses that aren't matched. Occasionally, this is the result of reference data that has errors or is incomplete. For instance, an address may belong to a subdivision that is newer than the reference data. More often, there is a mistake, such as a spelling error, in the address table. In some cases, the problem is with the way ArcGIS has standardized an ambiguous address.

Unmatched addresses can be rematched automatically (ArcGIS does all of them at once) or interactively (you do them one at a time). If you choose to rematch automatically, you need to adjust the ArcMap geocoding matching options, particularly the minimum match score and the spelling sensitivity, or your results won't change. If you rematch interactively, you have additional options. You can edit the way ArcGIS has standardized an address, and you can also match a candidate with a score below the minimum match score.

So far, the address locator has found matches for 13 addresses in the Customers table, leaving three addresses unmatched. In this exercise, you'll match the remaining three. First, you'll modify the address locator settings to generate more candidates. Then you'll interactively match each address.

1 In ArcMap, open the **ex14c map document from your Chapter14 folder.** The map layers include the familiar interstate highways, streets, and ZIP Codes of downtown Atlanta. If necessary, click the List By Source button at the top of the table of contents.

Next, you'll add the layer of geocoded addresses you created in exercise 14b.

2 On the Standard toolbar, click the Add Data button. On the Add Data dialog box, navigate to your \GTKArcGIS\Chapter14\MyData folder. Click Geocode1.shp, and then click Add. The Geocode1 layer is added to the table of contents. It is symbolized with the default ArcMap point symbol in a random color.

On your own

If you want, you can change the symbol to make it easier to see. In the following graphics, the point symbol has been set to Circle 2, Fire Red, and size 7 pt. In addition, you can change the symbology of the other layers however you wish.

Next, you'll rematch the unmatched addresses.

3 In the table of contents, make sure the Geocode1 layer is
highlighted. On the Geocoding toolbar, click the Review/Rematch
Addresses button 🔤 . (If the Geocoding toolbar is not activated, on
the Customize menu, click Toolbars > Geocoding.) The Interactive
Rematch dialog box appears. (If needed, resize the dialog box to make it larger.)

In the Statistics panel in the upper-right corner, you see the results from exercise 14b. The Show results option at the upper left allows you to specify which addresses you want to rematch, so you'll do that next.

④ Click the Show results arrow and click Unmatched Addresses.

Show results: Unmatched Addresses ▾ | Manage result sets... | Refresh | Rematch Automatically |

	FID	Shape	Status	Score	Match_type	
▶	5	Point	U	0	A	
	8	Point	U	0	A	
	14	Point	U	0	A	

◄   III

|◄ ◄   1 ► ►| (of 3)

The table displays the three unmatched records with scores of 0. (This is where you may need to enlarge the dialog box to see more records.) The first one is highlighted in cyan. Below the table and to the left, in the Address panel, the Street or Intersection box and the ZIP Code box contain the current address information for the highlighted record. Edits to this information will be written to the ARC_Street and ARC_ZIP fields of the Geocode1 table. You will not use the City or Placename or the State boxes.

The large box in the lower midsection displays 14 match candidates for the highlighted record. There are no match candidates for the second and third records.

⑤ In the address table (not the Candidates table), use the navigation arrows to scroll through the three unmatched addresses. Scroll to the right and notice that the ZIP Code for the third record, 30309, is different from the ZIP Codes for the other records.

It is possible that the wrong ZIP Code was entered in the Customers table. So next, you will edit the ZIP Code to see if this produces any candidates.

6 In the address table, select and highlight the third record. In the Address panel below it, highlight the ZIP Code value 30309 and type 30308. **Then press ENTER.** Six match candidates appear in the list. The top candidate has a perfect score of 100.

7 Make sure the first candidate is highlighted. Under the Candidates table, click Match. The candidate is matched. In the address table, the record's Status value is changed from U to M, the Score value is changed from 0 to 100, and its ARC_ZIP value is changed from 30309 to 30308 (you may need to scroll left or right to see these changes). In the Statistics panel, 14 addresses are now reported as matched.

Next, you'll match the other two addresses.

8 At the top of the address table, click the gray tab to the left of the first record to select it. There are 14 match candidates. Notice that the top match score is 78.32 for 751 Juniper St NE (see facing page). Not all candidates require a score of 100 to be matched. You can see that this record closely resembles the first candidate. You can also verify that the address falls within the proper range. This is the correct candidate for matching the address.

9 **Make sure the first candidate is highlighted and click Match.** The candidate is matched. Fifteen addresses are now reported as matched.

10 **In the address table, click the gray tab to the left of the second record to select it.**

	FID	Shape	Status	Score	Match_type	
	5	Point	M	78.32	M	751 Juniper St NE, Atlanta, GA, 30308
▶	8	Point	U	0	A	
	14	Point	M	100	M	725 Lakeview Ave NE, Atlanta, GA, 30308

Show results: Unmatched Addresses ▼ Manage result sets... Refresh Rematch Automatically

I◀ ◀ 2 ▶ ▶I (of 3)

There are 12 match candidates. Notice the spelling of the address is MYTLE, and the top candidate score is only 60.26. Next, you'll edit the spelling to see if any candidate achieves a better score.

11 **In the Address panel, edit the Street or Intersection box entry to** 844 MYRTLE **and press ENTER.**

The first of the 12 candidates for 844 MYRTLE has a score of 78.32. It has the correct address range, because 844 falls within the range of FromAddr 806 to

ToAddr 870. For the other candidates with lower scores, their address range doesn't match. It looks, however, as if the street type and direction suffix are missing. So next, you'll take a closer look at the addresses on the map.

12 In the lower-right corner of the Interactive Rematch dialog box, click Zoom To Candidates. If necessary, move the dialog box so you can see the map. The map is zoomed to the extent of the 12 candidates.

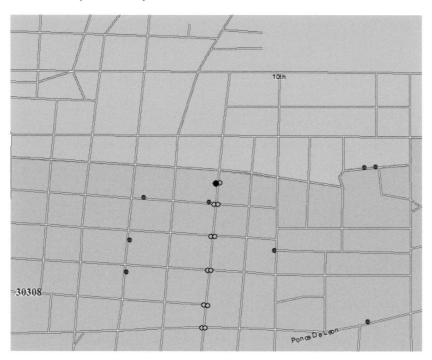

The candidate that is highlighted in the dialog box is in black, and the other candidates are in cyan. Common sense suggests that the first candidate is the correct address location and that the omitted street type and direction suffix was an error when this address was entered in the Customers table. Next, you'll add the missing information to see if it improves the candidate score.

13 In the Address panel, in the Street or Intersection box, type ST NE after 844 MYRTLE. The scores of the original candidates are adjusted. The first candidate now has a score of 100.

14 Make sure the top candidate is highlighted and click Match. The record's status is changed from U to M, and its name is changed in the ARC_Street field.

Geocoding Options

Geocoding Options can be accessed from the Address Locator Manager, Interactive Rematch, or Geocoding Addresses dialog boxes. These options allow you to fine-tune the interactive-matching experience. You can adjust spelling sensitivity, minimum candidate score, and minimum match score to control how closely an address must match a candidate to score a match. For more information, in ArcGIS for Desktop Help, on the Contents tab, navigate to Guide Books > Geocoding > Building an address locator > Modifying address locator settings > Settings in the Address Locator Properties dialog box > Geocoding options.

You've matched all the addresses in your customers table. The Statistics panel shows that 16 addresses were matched.

15 **Close the Interactive Rematch dialog box.** On the map, there are now 16 geocoded points.

16 In the table of contents, right-click the Geocode1 layer and click Zoom To Layer. Right-click the layer again and click Label Features.

The points are labeled with the names of your customers' businesses.

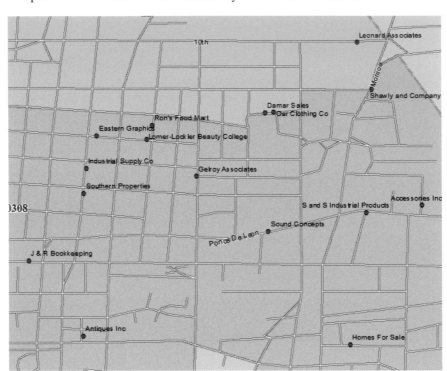

17 In the table of contents, right-click the Geocode1 layer and click Open Attribute Table. All the records have the value M in the Status field. Edits that you made during rematching ("MYTLE" to "MYRTLE ST NE," for example) are reflected in the ARC_Street and ARC_ZIP fields.

18 Close the table. Return the table of contents to the List By Drawing Order view and close the Geocoding toolbar.

19 If you want to save your work, rename your map document and save to your MyData folder for chapter 14.

20 If you are continuing to the next chapter, leave ArcMap open. Otherwise, exit the application and click No if prompted to save your changes.

Now that you know how to geocode addresses, the project to offer digital maps as one of your company's advertising services can move forward.

Summary

In this chapter, you learned how to create an address locator in ArcCatalog by choosing an address locator style and using street data as the primary reference. Then in ArcMap, you were able to use the locator to geocode your customers table and create a new point layer. You also used geocoding options to search for unmatched addresses and to geocode. Geocoding is a common GIS task that has many practical applications. As with many real-world applications, the data may not always be perfect, so you will have to make sure that the quality of the data coming in results in quality data going out.

Terms

geocoding

address locator

Getting information about

features

Chapter 15
Querying data

Getting information about features
Selecting features by attribute
Creating reports

Learning objectives

- *Identify and search for features*
- *Become familiar with various querying methods*
- *Create feature hyperlinks that point to a photo*
- *Select features by using SQL queries based on attribute values*
- *Generate useful reports*

A GREAT STRENGTH OF GIS IS THAT IT DOESN'T JUST SHOW YOU WHERE THINGS are, it also tells you a lot about them. If you're curious about a road leading to the next town—for example, US Interstate 70—you may be able to find out such things as its direction, length, speed limit, and the number of lanes it has. If that road leads to a place such as Topeka, Kansas, you may want to know the city's population (127,473, based on the census of 2010), the median household income ($35,928), the median home value ($95,400), and many other things, in addition to the city's geographic attributes.

A lot of information about features can be conveyed by the way they are symbolized and labeled. But for most features, it isn't possible to display everything that is known about them on a single map.

In ArcMap, there are several ways to retrieve unseen information about features. As you may have already done, you can click features with the Identify tool to display their attributes. This is called identifying features. You can click features on the map with the Select Elements tool to highlight them and look at their records in the layer attribute table. This is called selecting features interactively. You can write a query that automatically selects features that meet specific qualitative and quantitative criteria combined (for example,

15

16

three-bedroom houses with swimming pools). This is called selecting features by attribute. Or you can provide ArcMap with a piece of information, such as a name, and see which feature it belongs to. This is called finding features.

Attributes are not the only kind of information that can belong to a feature. Pictures, text documents, and web pages can also be associated with features through hyperlinks. Clicking a hyperlinked feature opens a file on disk, points your web browser to a URL, or runs a macro (a sequence of automated commands).

Data

This chapter uses the following datasets:

- cabbagetown.shp—polygons of the postal code areas that make up the Cabbagetown neighborhood in Toronto, Ontario, Canada *(Source: Generation 5 VA, Statistics Canada)*
- toronto_background.shp—polygons of the postal code areas that make up an area of Toronto, Ontario, Canada *(Source: Canada Post)*
- toronto_streets.shp—lines of select streets around the Cabbagetown neighborhood in Toronto, Ontario, Canada *(Source: City of Toronto)*

Exercise 15a
Getting information about features

The fastest way to get information about a single feature is to identify it. If you want to compare information about several features, the best way to do it is to select the features on the map and look at their records in the layer attribute table. When you have a piece of information—a place-name or address, for instance—but don't know which feature it belongs to, you can use that information to search the map for the feature.

Hyperlinks link features to things that can't be stored as attribute values. A point feature representing a government office, for example, could be linked to the office website.

For this exercise, you are a marketing analyst in Toronto, Ontario, Canada. Many of your clients are major retail chains located in the area that also operate online e-commerce websites. You have access to marketing data about neighborhoods throughout the city, and you have decided to use ArcGIS to help you identify current market profiles. A market profile attempts to characterize a neighborhood based on its demographics and socioeconomic composition.

Currently, you're working with a large retail chain that sells everything from household products to banking services. Your client has asked you to look at online sales for potential customers of goods because it wants to concentrate its marketing and advertising efforts on consumers who meet a certain criteria (which you will see in the exercise). The target area is in a neighborhood on the east side of downtown Toronto called Cabbagetown, which got its name from an influx of Irish immigrants who populated the area in the 1840s and were known to grow cabbage on their properties. Today, the neighborhood is known for its Victorian row homes, beautiful gardens, and mixture of residents of various income levels. You are going to explore the composition of the neighborhood first before you analyze its marketing statistics.

15a

16

1 **Start ArcMap. Open the ex15a map document from your C:\EsriPress\GTKArcGIS\Chapter15 folder.** The map shows target postal code areas (in light purple) in the Cabbagetown neighborhood, surrounding postal code areas (in light brown), and local streets. You may need to widen your table of contents to see layer names in their entirety (drag the table of contents divider).

2 **On the Tools toolbar, click the HTML Popup tool** 💬 **. The mouse pointer becomes a hand and pointer finger.**

15a

16

3 **On the map, click a postal code feature, such as the one in the graphic.** The HTML Popup window displays the feature's attribute values. The bold title at the top tells you which postal code feature is being identified (in this case, M4X1R1). Included is some demographic and socioeconomic data, such as age range (for example, 30_to_34) and total number of households (Total_Hous). The Canadian postal code format is similar to the United States ZIP+4 format in that it uses a base code, plus additional digits to identify a geographic segment such as a city block (M4X represents a larger area and 1R1 identifies a subunit of this area).

15a

16

The HTML Popup window can be customized as a table of the visible fields (as shown in the graphic), as a URL on the Internet, or as a formatted page based on a spreadsheet template. To learn more about HTML Popups, in ArcGIS for Desktop Help, on the Contents tab, navigate to Mapping > Working with layers > Interacting with layer contents > Setting HTML pop-up properties for feature layers.

4 Click a few more postal code areas. A new HTML Popup window appears each time you click a feature. Arrange the HTML Popup windows side by side so you can see each one.

M4X1P8	
M4X1P8	
FID	164
POSTALCODE	M4X1P8
Population	72
Total_Hous	20
15_to_19	2
20_to_24	2
25_to_29	5
30_to_34	5
35_to_39	5
40_to_44	6
AvgHHIncom	77660
GROCERY	0.03
GRCYL250	0.02
GRCYG250	0.01
HOMEGARDEN	0.08
HMGRL250	0.05
HMGRG250	0.03
VACATRAV	0.33
V100_249	0
V250_499	0.04
V500_999	0.1
V_G1000	0.16
IMAGE	

M4X1R1	
M4X1R1	
FID	208
POSTALCODE	M4X1R1
Population	12
Total_Hous	5
15_to_19	0
20_to_24	2
25_to_29	1
30_to_34	2
35_to_39	1
40_to_44	1
AvgHHIncom	34205
GROCERY	0.07
GRCYL250	0.06
GRCYG250	0.01
HOMEGARDEN	0.05
HMGRL250	0.03
HMGRG250	0.02
VACATRAV	0.35
V100_249	0.01
V250_499	0.06
V500_999	0.03
V_G1000	0.19
IMAGE	

Using the HTML Popup tool is a quick way to view feature attributes. As you've seen in previous chapters, using the Identify tool is also a fast way to get information about features. Neither method, however, is a convenient way to compare the attributes of several features. To do this, you'll next select features on the map, and then look at their records in the layer attribute table.

5 Close the HTML Popup windows. In the table of contents, click the List By Selection button.

By default, all layers are selectable. You want to select features from the Cabbagetown Postal Code Areas layer only, so you'll adjust for that next.

6 For both the Roads and Surrounding Postal Code Area layers, click the toggle key to make them not selectable.

Both the Roads and Surrounding Postal Code Area layers are now in the Not Selectable category.

7 On the menu bar, click Selection > Selection Options.

On the Selection Options dialog box, the selection tolerance is set to 3 pixels. Because you are working with small features that have many features adjacent to them, you'll next reduce the tolerance. This ensures that when you click a location, you won't be selecting more than one feature at a time.

8 In the Selection tolerance box, replace the value of 3 with 0, and then click OK.

You can also change selection color using the Selection Options dialog box.

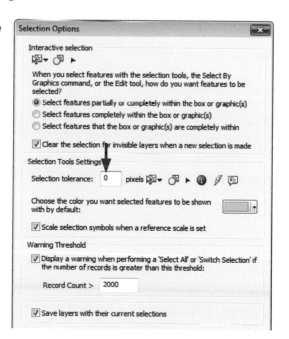

15a

16

9 On the Selection menu, click Selection Method > Add to Current Selection. The Add to Current Selection option allows you to click features without the need to press and hold SHIFT. When the Create New Selection option is chosen, you must press and hold SHIFT to select multiple features.

10 On the Tools toolbar, click the Select Features tool 🔲 .

11 On the map, click the postal code areas shown in the graphic, one at a time. Each area you click is added to the selection set. The table of contents shows how many postal codes are selected and the name of each postal code.

If you select a postal code area by mistake, on the Selection menu, click Interactive Selection Method > Remove From Current Selection, and then click the selected postal code.

The eight selected postal code areas are outlined in cyan (or a different color if you changed the selection color). In the attribute table of the layer, the records corresponding to the selected postal codes will also be highlighted.

12 **In the table of contents, right-click Cabbagetown Postal Code Areas and click Open Attribute Table.** The records for selected postal code areas are highlighted but not grouped together.

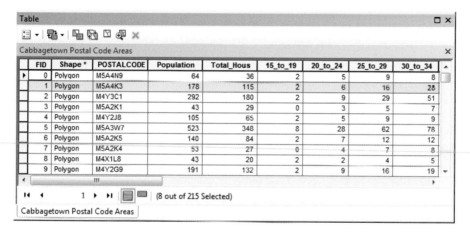

FID	Shape *	POSTALCODE	Population	Total_Hous	15_to_19	20_to_24	25_to_29	30_to_34
0	Polygon	M5A4N9	64	36	2	5	9	8
1	Polygon	M5A4K3	178	115	2	6	16	28
2	Polygon	M4Y3C1	292	180	2	9	29	51
3	Polygon	M5A2K1	43	29	0	3	5	7
4	Polygon	M4Y2J8	105	65	2	5	9	9
5	Polygon	M5A3W7	523	348	8	28	62	78
6	Polygon	M5A2K5	140	84	2	7	12	12
7	Polygon	M5A2K4	53	27	0	4	7	8
8	Polygon	M4X1L8	43	20	2	2	4	5
9	Polygon	M4Y2G9	191	132	2	9	16	19

(8 out of 215 Selected)

Cabbagetown Postal Code Areas

13 **At the bottom of the attribute table, click the Show selected records button to show the selected postal code records only.**

(8 out of 215 Selected)

15a

16

14 Scroll to the right in the table to see its attributes. You can now compare the attributes of the selected features for such values as average household income, total population, and age ranges. Right-click the Total_Hous field name. On the shortcut menu, click Sort Descending. Now the records are sorted in order, from the most households per postal code to the fewest. The first record shows a value of 115 (that is, 115 households for postal code M5A4K3).

FID	Shape *	POSTALCODE	Population	Total_Hous	15_to_19	20_to_24	25_to_29	30_to_34	3
1	Polygon	M5A4K3	178	115	2	6	16	28	
45	Polygon	M5A2L1	217	108	1	11	22	23	
86	Polygon	M4X1M7	85	40	4	4	6	7	
172	Polygon	M4X1E1	84	39	2	2	8	8	
20	Polygon	M5A3A1	40	21	0	2	4	5	
174	Polygon	M4X1V5	27	13	0	2	2	2	
106	Polygon	M5A3P9	10	4	0	2	0	1	
51	Polygon	M4Y2G8	5	3	0	0	2	1	

(8 out of 215 Selected)

Cabbagetown Postal Code Areas

15 Click the gray tab to the left of the first record (postal code M5A4K3). The record is highlighted in yellow (because it has already been selected in cyan), as is the corresponding map feature. You may need to move the attribute table to see the highlighted feature.

16 Click the gray tabs for the other records to familiarize yourself with the location and attributes of each postal code. When you're done, close the attribute table. The yellow highlighting on the map features disappears. On the menu bar, click Selection > Clear Selected Features.

> You can change the selection color for the features in each layer on the Selection tab of the layer properties. Or you can change the selection color for *all* attribute tables by modifying your ArcMap options (on the Customize menu, click ArcMap Options, and then the Tables tab).

Attribute information is useful, but a photograph can convey information that table attributes can't—such as the appearance of a neighborhood. Pictures, documents, web pages, and macros can be connected to features by a hyperlink.

The attribute table of the Cabbagetown Postal Code Areas layer has a field called IMAGE that contains paths of photographs that represent what neighborhoods look like. Next, you will hyperlink the features to the photos, and then view them.

17 In the table of contents, switch back to the List By Drawing Order view and double-click Cabbagetown Postal Code Areas. On the Layer Properties dialog box, click the Display tab. In the Hyperlinks frame, select the Support Hyperlinks using field check box. Click the Field list arrow and click IMAGE. Below the list, make sure the Document option is selected, and then click OK.

15a

16

Next, you're ready to see photos of the area to get a better sense of what kind of buildings and residential homes make up the Cabbagetown neighborhood.

18 **On the Tools toolbar, click the Hyperlink tool** ⚡ . On the map, hyperlinked features are outlined in blue. Pausing the mouse pointer over a feature that is hyperlinked displays the path of the document (in this case, a photograph).

19 **Click any hyperlinked feature to display a photograph of the property (see facing page). The photograph appears in your default image-browsing software. After you look at a few more photographs, close the application that opened it.**

15a

16

17

18

19

20

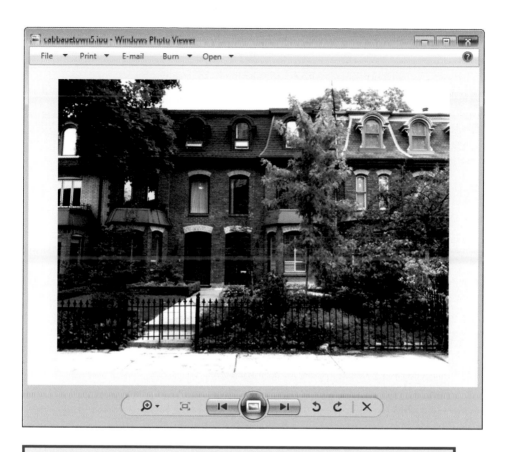

Creating hyperlinks

There are two ways to create hyperlinks. One is to add document paths or URL addresses to a field in the layer attribute table. This is efficient, if you are hyperlinking many features. The other way is to click a feature using the Identify tool, and then right-click the feature in the Identify dialog box and click Add Hyperlink. This is easier, if you are setting just a few hyperlinks. It is also the only way that lets you hyperlink a feature to more than one object. For more information, in ArcGIS for Desktop Help, on the Contents tab, navigate to Mapping > Working with layers > Interacting with layer contents > Using Hyperlinks.

15a

16

During a research meeting prior to your work in ArcGIS, your co-worker had suggested a postal code (M4X1C5) for further investigation of its average household income. So next, you'll use the Find tool to locate the area on the map and display its attributes.

20 On the Tools toolbar, click the Find tool 🔍 to open the Find dialog box. On the Features tab, in the Find box, type M4X1C5.

To speed things up, you can specify a layer, and even a field, to search in, which you'll do next.

21 Click the In arrow and click Cabbagetown Postal Code Areas. Under Search, click the In field option, and then click the arrow below and click POSTALCODE. Then click Find.

ArcMap searches the POSTALCODE field of the Cabbagetown Postal Code Areas attribute table for M4X1C5 and displays the matching record.

Now that you've found the feature, you can locate it on the map and get its attributes, so you'll do that next.

22 Move the Find dialog box away from the map but so you can still see it. In the lower portion of the dialog box, right-click M4X1C5. On the shortcut menu, click Identify. The postal code flashes briefly in the middle of the map, and the Identify window appears. (If the Identify window covers the postal code, move it and repeat step 22.)

23 In the Identify window, scroll down to view the average household income. This postal code area has a high household income, of $143,870. Similarly, if you identify the immediate surrounding areas, you'll see that they also have high household incomes.

24 Scan the other attribute values. When you're finished, close the Identify window and the Find window.

25 If you want to save your work, rename your map document and save to your MyData folder for chapter 15.

26 If you are continuing to the next exercise, leave ArcMap open. Otherwise, exit the application and click No if prompted to save your changes.

The area of Cabbagetown you just looked at may meet the target criteria for customers your client is hoping to advertise to. You'll look at your client's specific target criteria in exercise 15b.

15a

16

Exercise 15b
Selecting features by attribute

Your large retail chain client has asked you to target potential customers in the Cabbagetown neighborhood who meet specific criteria. The client will target households in these postal code areas for customized marketing campaigns focused on three different products sold online (groceries, home and garden products, and vacation packages). By targeting these postal codes, you will be able to provide your client with information about the neighborhood so that advertisement efforts can be focused on customer needs.

Interactive selection works when you can see what you're looking for on the map—the largest postal code, for instance. To select features according to criteria you can't see on the map (such as households that have an average income of at least $100,000), you will need to write a query.

A valid query results in a selected set of features that meet the specified conditions. The simplest *query expression* consists of an attribute (such as income), a value (such as $100,000), and a relationship between them (such as "greater than or equal to"). Complex queries can be created by connecting simple queries with terms such as "and" and "or." Relationship terms and connecting terms are called *operators*.

Queries are not written in ordinary English but in Structured Query Language (SQL). This is not as daunting as it might sound. To build a query expression, all you need to do is open a dialog box that contains a query builder interface (such as Select By Attributes), double-click an attribute, and then click an operator and a value.

The marketing information for your retail client shows online sales of groceries, home and garden supplies, and vacations. These values show the percentage of households (not the dollar amounts they spent). For example, the GROCERY attribute shows the percentage of households that used the Internet to purchase groceries within the past year. The related attributes GRCYL250 AND GRCYG250 show the percentage of households that spent less than $250 or more than $250, respectively, on groceries purchased online within the same time period.

Rather than target advertisements toward the entire neighborhood, your client wants to target specific postal codes. For groceries, your client wants to win over new customers and has determined that it wants to advertise to postal code areas where less than 5 percent of households purchase groceries online. For home and garden products, your client wants to advertise to postal code areas with an average household income of $75,000 to $150,000 and where households make online home and garden purchases of more than $250 (regardless of the percentage of population that spends this amount). Finally, for vacations, your client wants to target postal code areas that have an average income of $175,000 or more and where more than 35 percent of households spend more than $1,000 on online vacation purchases.

1 **In ArcMap, open the ex15b map document from your Chapter15 folder.** The map of neighborhood postal codes is displayed.

2 **On the menu bar, click Selection > Select By Attributes. On the Select By Attributes dialog box, set the Layer option to Cabbagetown Postal Code Areas.** The field names from the postal codes attribute table appear in the Fields list in the upper portion of the dialog box. When a field name is highlighted, you can display its attribute values in the Unique Values box, in the center of the dialog box. The buttons to the left of the Unique Values box are used to choose operators, such as <symbol>greater than or equal to (>=) or less than or equal to (<=).

Unique Values box

The first target you want to find is postal code areas with less than 5 percent of households purchasing groceries online.

3 In the Fields list, scroll to "GROCERY" and double-click it. Click the less than (<) button, click Get Unique Values, and then double-click 0.05 in the Unique Values box. Your query expression is displayed in the expression box in the lower portion of the dialog box. When you apply the query, ArcMap searches the attribute table for records that have a value in the GROCERY field that is less than 0.05. The corresponding features are selected on the map.

You can double-click an attribute value to add it to the expression without pressing the SPACEBAR, but if you type a value, you definitely need to add a space in front of it.

15b

16

4 **Click OK.** On the map, postal code areas that meet the desired criteria are selected, based on your query.

If you had just looked at the map at face value, you never would have known that these postal code areas have the selected attribute values. On the status bar, notice that 123 features are selected. (If you don't see it, switch to the List By Selection view to see how many are selected, and then switch back to the List By Drawing Order view.) Since this is one of the targets your client wants, you'll next create a layer from this selection.

15b

16

5 **Right-click the Cabbagetown Postal Code Areas layer, and then on the shortcut menu, click Selection > Create Layer From Selected Features.** A new layer is created, called Cabbagetown Postal Code Areas selection. It is symbolized by a random fill color.

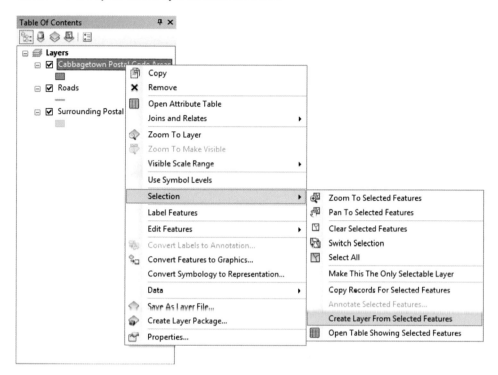

6 **Rename the new layer** Grocery target.

7 **On the menu bar, click Selection > Clear Selected Features.** You now have a new selection layer that you can provide to your client that shows postal code areas with less than 5 percent of households purchasing groceries online. Your client will now be able to increase advertisements to postal codes in this segment that generally do not make many online purchases.

Next, you'll work on the home and garden criteria your client wants. You'll target postal code areas whose households have an average income of $75,000 to $150,000 *and* in which at least one household spends more than $250 online each year for home and garden products.

8 **Turn off the Grocery target layer. On the menu bar, click Selection > Select By Attributes. Make sure Cabbagetown Postal Code Areas is chosen in the Layer list. Below the expression box, click the Clear button to clear the previous query.**

9 In the Fields box, scroll to "AvgHHIncom" and double-click it. Click the greater than or equal to (>=) button, press the SPACEBAR, and type 75000. Click the And button and double-click "AvgHHIncom" once more. Then click the less than or equal to (<=) button, press the SPACEBAR, and type 150000. Your query is displayed in the expression box. Click Apply.

> When building query expressions, incorrect syntax, such as unnecessary use of commas or periods, can cause errors. The Verify button can be used to check whether the query syntax is correct. For more information on building a query expression, in ArcGIS for Desktop Help, on the Contents tab, navigate to Mapping > Working with layers > Interacting with layer contents > Query expressions in ArcGIS > Building a query expression.

Select By Attributes

Layer: ◇ Cabbagetown Postal Code Areas ▾
☐ Only show selectable layers in this list

Method: Create a new selection ▾

```
"35_to_39"
"40_to_44"
"AvgHHIncom"
"GROCERY"
"GRCYL250"
```

[=] [<>] [Like]
[>] [>=] [And]
[<] [<=] [Or]
[_][%] [()] [Not]
[Is]   [Get Unique Values] Go To: [   ]

SELECT * FROM cabbagetown WHERE:

```
"AvgHHIncom" >= 75000 AND "AvgHHIncom" <= 150000
```

[Clear] [Verify] [Help] [Load...] [Save...]

[OK] [Apply] [Close]

When you apply the query, the corresponding features (93) that have an average household income of $75,000 to $150,000 are selected on the map. Next, you'll make the query slightly more complex to target the postal code areas that have households that spend more than $250 on home and garden products online (represented by the HMGRG250 attribute field).

10 On the Select By Attributes dialog box, click the And button. In the Fields box, double-click "HMGRG250", click the greater than (>) button, press the SPACEBAR, and then type 0. Your query should match the graphic, and then click OK.

ArcMap searches the HMGRG250 attribute field for values greater than zero (0), which means that all households that have spent more than $250 are chosen from the selection. On the map, fewer postal code areas (79) are selected because fewer of them satisfy both conditions.

15b

16

As with the previous grocery selection, you'll next create a new layer for your client.

11 Right-click the Cabbagetown Postal Code Areas layer, and on the shortcut menu, click **Selection > Create Layer From Selected Features**.

12 Rename the new layer Home and Garden target.

13 Switch to the List By Selection view. In the Selectable category, click the Click to clear layer selection button for the Cabbagetown Postal Code Areas layer. After that, don't forget to switch back to the List By Drawing Order view.

Finally, you will create one more query to target households that have an average income of $175,000 or more *and* that have a high proportion (35 percent or above) of online vacation purchases that exceed $1,000 (represented by the V_G1000 attribute field).

14 Turn off the Home and Garden target layer. On the menu bar, click Selection > Select By Attributes. Make sure Cabbagetown Postal Code Areas is chosen in the Layer drop-down list. Click Clear to clear the previous query.

15 In the Fields box, scroll to "AvgHHIncom" and double-click it. Click the greater than or equal to (>=) button, press the SPACEBAR, and type 175000. Click the And button and double-click "V_G1000". Then click the greater than or equal to (>=) button, press the SPACEBAR, and type 0.35. Your query is displayed in the expression box. Click OK.

There are far fewer areas (17, to be exact) to target, considering the combined criteria of a high level of income along with a high proportion of online purchases of vacations costing more than $1,000.

On your own

As with the previous segment selections, create a new layer from the current selection. Rename the new layer **Vacation target**. When you are finished, clear your selection.

Remember, these layer files are elements of the map document, pointing to cabbagetown.shp. If you want to make this a permanent shapefile or feature class, you can right-click the layer in the table of contents and click Data > Export Data to create it. Another option is to create a layer package (.lpk) that includes both the layer properties (symbolization, labeling, table properties) and the dataset referenced by the layer. It can be easily shared with ArcGIS Online through ArcMap. For more information, in ArcGIS for Desktop Help, on the Contents tab, navigate to Mapping > Working with layers > Managing Layers > Saving layers and layer packages.

16 If you want to save your work, rename your map document and save to your MyData folder for chapter 15.

17 If you are continuing to the next exercise, leave ArcMap open. Otherwise, exit the application and click No if prompted to save your changes.

In exercise 15c, you will make a report of the query results that your clients can take with them.

15b

16

Exercise 15c
Creating reports

Reports let you organize, format, and print the information contained in an attribute table. ArcMap has a built-in report generator that allows you to create simple reports that can be added to your map layout.

In exercise 15b, you selected postal code areas in the Cabbagetown neighborhood that meet certain criteria defined by your client. The client will target households in these postal code areas for customized marketing campaigns focused on three different products sold online (groceries, home and garden products, and vacation packages). The grocery target areas will receive advertisements designed to entice new customers who have rarely if ever purchased groceries online. The home and garden target areas will receive advertisements designed to appeal to existing online customers of such products. And the vacation target areas will get information about family and luxury resort vacation packages.

For this exercise, you will create a report so your client can create a customized mailing list. The report you make will be based on the Vacation target layer, where average household income is $175,000 or more, and 35 percent or more of households spend more than $1,000 annually on online vacation purchases.

1 **In ArcMap, open the ex15c map document from your Chapter15 folder.** The map displays the Vacation targets layer (with 17 postal code areas) at the top of the table of contents (see facing page).

15c

16

2 On the View menu, click Reports > Create Report. The Report
Wizard dialog box appears. The Layer/Table list specifies the layer or table
you will be working with. The Available Fields list lets you choose the attributes
to include in your report.

You're going to provide your client with information including the target postal codes' average household income, proportion of households that make online vacation travel purchases, and proportion of households that spend more than $1,000 annually on online vacation travel purchases.

3 Make sure the Layer/Table list shows Vacation target. In the Available Fields list, click POSTALCODE. Click the Add fields button (right-facing arrow) in the middle of the dialog box. The POSTALCODE field moves from Available Fields to Report Fields.

4 In the Available Fields list, double-click AvgHHIncom to add it to the Report Fields list. Do the same for the VACATRAV and V_G1000 fields.

5 In the upper-right portion of the dialog box, click the Report Viewer Contents Field arrow and click POSTALCODE. This means this field will be displayed when you preview the report.

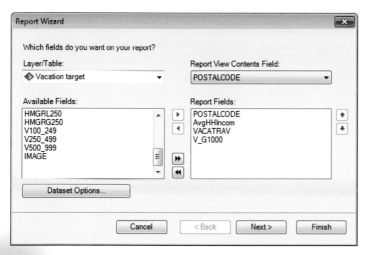

The Vacation target layer you created already has the postal code areas you want to report. If you didn't create this layer, you could still create a report that uses the selected postal code areas. To do this, click Dataset Options on this wizard panel and click the Selected Set option. This will output only the selected features to the report.

6 On the Report Wizard dialog box, click Next, and then click Next again. You will not be adding any grouping levels.

7 **In the next panel, in the Fields column, click V_G1000, and in the Sort column, click Descending from the first row in the list.** In the report, target postal code areas will appear in order from the highest to the lowest proportion of households that spent more than $1,000 annually on online vacation travel.

8 **Click Next. In the next panel, set the orientation to Landscape, and then click Next.**

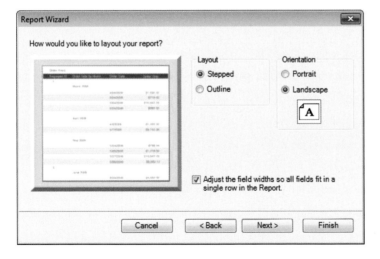

9 The Havelock (Default) style is selected. Scroll down the list and click Toronto, and then click Next. (These styles have more to do with colors than any real relation to the city.)

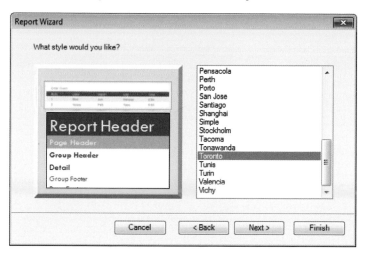

10 Replace the text in the title box with Vacation Target Report. Make sure the Preview the report option is selected and click Finish. The report is previewed. The size of your dialog box may be different, but you probably can't see the whole report. The list at the top of the dialog box lets you change the preview scale. (It doesn't change the printed size of the report.)

If you like the way the report looks, you could click Print to send it to your printer.

Most reports, including this one, need additional work. For example, the headings need to be better aligned. In steps 11–16, you will give the report a professional look by setting a few of its properties.

11 **In the upper-left corner of the Report Viewer window, click the Edit button.** The window changes to the Report Designer window (see facing page). The Report Designer is a template layout tool to help you edit the appearance of your reports. The upper-right box lists the report components you can customize. Below it are the element properties associated with the highlighted component and the current value of each property. When a report component is highlighted, the element is selected in the middle of the report template view. The template area represents how your report pages will look.

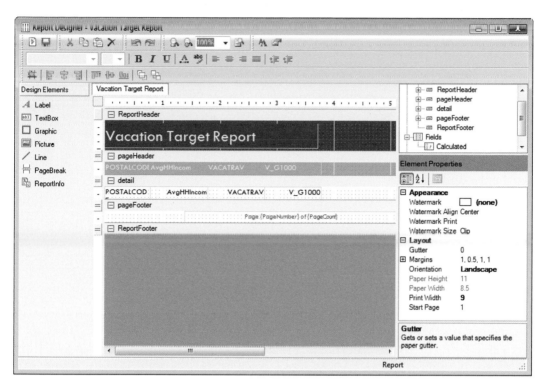

The headers are located in the pageHeader section of the template and need to be aligned properly, so you'll do that next.

12 **In the template area, click the AvgHHIncom pageHeader (not detail) component.** Items selected in the template area have their properties displayed under Element Properties. Scroll down the list to see the different categories: Appearance, Behavior, Data, Design, and Layout. You can always use this list to confirm that the correct item in the template has been selected (by viewing the Text section under Data).

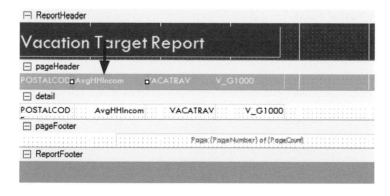

13 In Element Properties, under the Appearance category, change the Alignment property to Right. (Alternatively, click the Align Right button in the controls above the Design Elements list.)

14 Repeat the same right alignment for the VACATRAV and V_G1000 pageHeader components. Then in the upper-left portion of the window, click the Run Report button ▷ to preview the report.

You can also press F5 to run the report after editing your template.

Vacation Target Report

POSTALCODE	AvgHHIncom	VACATRAV	V_G1000
M4X1S3	191687	0.55	0.46
M4X1E7	197717	0.59	0.45
M4X1H7	181489	0.57	0.43
M4X1H7	181489	0.57	0.43
M4X1T6	182857	0.53	0.41
M4W3A2	189338	0.54	0.41
M4W3A2	189338	0.54	0.41
M5A2L9	179933	0.53	0.4
M4X1R7	201391	0.49	0.38
M4X1T7	243427	0.45	0.38
M5A2M1	195900	0.47	0.38
M4X1B3	205439	0.42	0.37
M5A2L7	238710	0.4	0.36
M4X1B4	224881	0.48	0.36
M4X1X7	199839	0.46	0.35
M4X1A3	282750	0.47	0.35
M4X1T8	234183	0.42	0.35

15c

16

The AvgHHIncom, VACATRAV, and V_G1000 headers are now aligned properly to the column values. The POSTALCODE header, however, is cut off a bit, so you'll fix that next.

15 In the upper-left corner of the Report Viewer window, click the Edit button.

16 In the pageHeader section of the template, click AvgHHIncom. Slide the left selection handle slightly to the right to show the entire POSTALCODE header.

On your own

Change some of the template element properties to see how it affects your report. You can edit the text of headers, change spacing between values, change text appearance, and more.

17 In the upper-left portion of the window, click the Run Report button to preview the report one more time.

18 If you want to print the report, in the upper portion of the Report Viewer window, click the Print button. On the Print dialog box, click Print to print the report.

19 If you want to save the report, on the Report Viewer window, click the Save Report output to file button. On the Save Report Layer dialog box, name your report and save to your MyData folder for chapter 15.

15c

16

20 **Close the Report Viewer window. If prompted to save your changes, click No.**

Adding reports to layouts

Within a map document, you can add a report to a layout by clicking the Add report to ArcMap Layout button on the Report Viewer window. If you have saved a report on disk, you can add it to any map layout by loading it from the Report Properties dialog box. For more information, in ArcGIS for Desktop Help, on the Contents tab, navigate to Mapping > Reports > Saving, exporting, loading, and running a report > Loading a report.

There is no need to save your map document. You haven't made any changes to it.

21 **If you are continuing to the next chapter, leave ArcMap open. Otherwise, exit the application and click No when prompted to save your changes.**

Summary

In this chapter, you learned how to identify and search for features by using tools on the Standard toolbar. You also used the Select By Attributes tool to generate queries and target specific features that satisfy them. Query results and selections don't change the original dataset, so you can always clear your query and start a new one. Finally, you learned how to create a report from ArcMap so you can easily share your query findings with others.

Terms

hyperlink

query expression

operator

Chapter 16
Joining and relating data

Joining data by attribute
Relating data
Joining data by location

Learning objectives

- *Examine tabular data and how it might be displayed spatially*
- *Join nonspatial data to spatial data to display new information on a map*
- *Select features on a map*
- *Modify the selectable layers on a map*
- *Export a selected set of features to a new feature class*
- *Determine how and why to use a join or a relate function*
- *Relate the tables of two datasets*
- *Spatially join two layers*
- *Symbolize join results to display new information*

LAYER ATTRIBUTE TABLES CONTAIN THE CHARACTERISTICS OF GEOGRAPHIC features, such as names (of cities, lakes, schools), codes (land use, zoning, feature type), and statistics (population, length, price). But a layer may not have all the attributes you want. If there is another data table that contains the desired attributes, you can join these attributes to the original attribute table, provided the two tables share a common field so the connection can be made—a Name field, for instance, or a field that contains a unique identification number.

15

16

Typically, an ***attribute join*** operation is used when you want to append nonspatial data (attributes that describe characteristics, but not geometry or spatial location) to a spatial dataset. In the following scenario, table 1 contains just a few basic spatial facts about world countries.

ObjectID *	Shape *	CNTRY_NAME	SQKM	SQMI
175	Polygon	Afghanistan	641358.44	247628.48
120	Polygon	Albania	28798	11118.91
106	Polygon	Algeria	2323510.25	897107.19
9	Polygon	American Samoa	84.2	32.51
107	Polygon	Andorra	336.24	129.82
217	Polygon	Angola	1252934.88	483758.22
26	Polygon	Anguilla	74.48	28.76
209	Polygon	Antarctica	12277658	4740404.5
27	Polygon	Antigua & Barbuda	255.13	98.51
18	Polygon	Argentina	2776912.75	1072166.13

An attribute table of a world countries layer.

Suppose you wanted to make a world map of average life expectancy. You would need some attributes like those found in table 2, which has socioeconomic data.

CNTRY_NAME	POP2007	BIRTH RATE	DEATH RATE	LIFE EXPECT
Afghanistan	31889923	46.21	19.96	43.77
Albania	3600523	15.16	5.33	77.6
Algeria	33333216	17.11	4.62	73.52
American Samoa	57663	21.83	3.24	76.25
Andorra	71822	8.45	6.45	83.52
Angola	12263596	44.51	24.81	37.63
Anguilla	13677	13.97	5.34	77.46
Antarctica	-99	-99	-99	-99
Antigua & Barbuda	69481	16.62	5.31	72.42
Argentina	40301927	16.53	7.55	76.32

A nonspatial table containing population, birth rate (number of births per 1,000 population), and mortality data for world countries.

Unlike the previous table, this one is a nonspatial table—lots of attributes, but no Shape field. In other words, it stores descriptions, but not spatial information. It's the sort of table you might find in a non-GIS database, or that you might even create yourself from facts based on your own research. By itself, this table can't be used to map data because its records have no relationship to features. You can append the attributes in the second table, however, to those of the first to make one big table that contains spatial information (see facing page). Once the new attributes are appended, the map can display new information.

ObjectID *	Shape *	CNTRY_NAME	SQKM	SQMI	POP2007	BIRTH RATE	DEATH RATE	LIFE EXPECT
175	Polygon	Afghanistan	641358.44	247628.48	31889923	46.21	19.96	41.77
120	Polygon	Albania	28798	11118.91	3600523	15.16	5.33	77.6
106	Polygon	Algeria	2323510.25	897107.19	33333216	17.11	4.62	73.52
9	Polygon	American Samoa	84.2	32.51	57663	21.83	3.24	76.25
107	Polygon	Andorra	336.24	129.82	71822	8.45	6.45	83.52
217	Polygon	Angola	1252934.88	483758.22	12263596	44.51	24.81	37.63
26	Polygon	Anguilla	74.48	28.76	13677	13.97	5.34	77.46
209	Polygon	Antarctica	12277658	4740404.5	-99	-99	-99	-99
27	Polygon	Antigua & Barbuda	255.13	98.51	69481	16.62	5.31	72.42
18	Polygon	Argentina	2776912.75	1072166.13	40301927	16.53	7.55	76.32

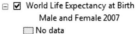

☐ ☑ World Life Expectancy at Birth
Male and Female 2007
☐ No data
■ 32 to 49 years
▨ 50 to 64 years
☐ 65 to 74 years
▨ Over 75 years

The output table, *top*, includes all
the attributes from both input tables.
Now countries can be symbolized,
left, based on one of the joined
attributes—in this case, average life
expectancy (see legend, *above*).
Data from Esri Data & Maps 2010,
courtesy of ArcWorld Supplement.

The process depends on record matches being made between the tables. For
records to be matched, the tables need to have a common attribute—that is, a
field of shared values. In the preceding example, the CNTRY_NAME field would
be a common attribute, since both tables have a field that contains country
names (the field name itself does not have to be the same, it just has to be
the same type—text fields are joined to text fields and numbers to numbers).
When the common attribute—also called a key—has been specified, records
can be matched—for example, the Afghanistan record in the nonspatial table
(along with all its attributes) can be matched to the Afghanistan record in the
layer attribute table. The same goes for all the rest of the countries in the world.
If a record in the **target table** (which contains the spatial information you
wish to maintain) doesn't have a match in the **join table** (which contains the
attributes that will be added to the spatial table), it will either be left out of the
output table or have null values where the joined attributes would be—you
choose which result you want when you set up the operation in ArcMap.

For a join operation to work, there can be only one matching attribute value in
each table. In the case of countries, in this example, there should never be more
than one feature that has the same country name, so the join will work correctly.

15

16

Sometimes, rather than join two tables, you'll want to *relate* the tables instead. A relate keeps the two tables separate but linked, so that record selections in one table lead to corresponding record selections in the other. Unlike a join operation, however, attributes from one table are not appended to the other.

Whether you join or relate tables depends on a relationship called cardinality between the tables. Cardinality refers to the numerical relationship (one-to-one, one-to-many, and so on) between a layer's attribute table and the table with the information you want to connect. You join tables when each record in the target table has no more than one matching record in the join table.

Common attribute

FID	Shape *	ObjectID	NAME	ST	POP2007
122	Point	836	Abilene	TX	115018
250	Point	2640	Akron	OH	211218
193	Point	1637	Albany (NY)	NY	94140
118	Point	813	Albuquerque	NM	503375
282	Point	3187	Alexandria	VA	139748
272	Point	3103	Allentown	PA	110279
119	Point	824	Amarillo	TX	183256
71	Point	559	Anaheim	CA	348374
1	Point	8	Anchorage	AK	283369
181	Point	1499	Ann Arbor	MI	118788
130	Point	878	Arlington (TX)	TX	373933
283	Point	3188	Arlington (VA)	VA	203123
256	Point	2754	Atlanta	GA	463421

iso3	Citycode	City	pm10
USA	8400173	Abilene	17.62378
USA	8400122	Akron	19.6925
USA	8400051	Albany (NY)	18.37535
USA	8400061	Albuquerque	19.65676
USA	8400188	Alexandria	22.52458
USA	8400065	Allentown	14.75936
USA	8400125	Amarillo	15.34999
USA	8400107	Anaheim	36.4622
USA	8400115	Anchorage	20.10121
USA	8400196	Ann Arbor	21.17848
USA	8400104	Arlington (TX)	23.16416
USA	8400138	Arlington (VA)	16.07828
USA	8400012	Atlanta	25.28407

US Cities data (target table)

City pollution data (join table)
Data courtesy of US Census and The World Bank (pm10 field).

FID	Shape *	ObjectID	NAME	ST	POP2007	iso3	Citycode	City	pm10
119	Point	836	Abilene	TX	115018	USA	8400173	Abilene	17.62378
247	Point	2640	Akron	OH	211218	USA	8400122	Akron	19.6925
190	Point	1637	Albany (NY)	NY	94140	USA	8400051	Albany (NY)	18.37535
115	Point	813	Albuquerque	NM	503375	USA	8400061	Albuquerque	19.65676
278	Point	3187	Alexandria	VA	139748	USA	8400188	Alexandria	22.52458
268	Point	3103	Allentown	PA	110279	USA	8400065	Allentown	14.75936
116	Point	824	Amarillo	TX	183256	USA	8400125	Amarillo	15.34999
69	Point	559	Anaheim	CA	348374	USA	8400107	Anaheim	36.4622
1	Point	8	Anchorage	AK	283369	USA	8400115	Anchorage	20.10121
178	Point	1499	Ann Arbor	MI	118788	USA	8400196	Ann Arbor	21.17848
127	Point	878	Arlington (TX)	TX	373933	USA	8400104	Arlington (TX)	23.16416
279	Point	3188	Arlington (VA)	VA	203123	USA	8400138	Arlington (VA)	16.07828
252	Point	2754	Atlanta	GA	463421	USA	8400012	Atlanta	25.28407

Output table Data courtesy of US Census and The World Bank (pm10 field).

Each record in the target table has just one match in the join table, so the information from both tables is included in the output table.

There is one record for each city in both the target table and the join table. Another way to express this relationship is that the table cardinality is one-to-one. (It's OK if one table has some cities that aren't found in the other—not every record needs to have a match, but each record can have *no more than one match*.) You can also join tables with a many-to-one cardinality—for example, if you have a layer of land-use polygons that defines land use using a three-letter code, you may want to join a nonspatial table of zoning code descriptions to it. This is a many-to-one cardinality (many land-use codes to one description for each). The resulting table would duplicate the description of each feature that shared the same zoning code.

You relate tables when each record in the target table may have more than one match in the relate table, as in the next example. When related tables are viewed, a selection highlighted in one will also select and highlight the matching features in the other.

Common attribute

ObjectID *	Shape *	CNTRY_NAME
175	Polygon	Afghanistan
120	Polygon	Albania
106	Polygon	Algeria
9	Polygon	American Samoa
107	Polygon	Andorra
217	Polygon	Angola
8	Polygon	Anguilla
209	Polygon	Antarctica
27	Polygon	Antigua & Barbuda
18	Polygon	Argentina

CNTRY_NAME	CITY_NAME
Afghanistan	Baghlan
Afghanistan	Mahmud-E Eraqi
Afghanistan	Charikar
Afghanistan	Bamian
Afghanistan	Mehtar Lam
Afghanistan	Kabul
Afghanistan	Mayda Shahr
Afghanistan	Baraki Barak
Albania	Bajram Curri
Albania	Shkoder

World countries data (target table)   Cities data (related table)

The table cardinality is one-to-many. Each record in the target table, *left*, has many matches in the related table, *right* (because there are many cities in a country).

If you simply joined the tables, you would lose information. For example, only one of the matching records for Afghanistan in the cities table could be joined to the target layer attribute table. In this case then, it is better to relate the tables rather than join them so information won't be lost.

You can also relate tables that have a one-to-one cardinality (if you don't want to create a permanent join) or a many-to-many cardinality—for example, relating a buildings layer to a table of building tenants, in which some tenants might be chain stores located in multiple buildings.

For more information about cardinality rules for joining and relating tables, in ArcGIS for Desktop Help, on the Contents tab, navigate to Geodata > Data types > Tables > Joining and relating tables by attributes > About joining and relating tables, and any of the others listed.

15

16

Sometimes, rather than joining data based on a common attribute, you'll want to join data based on location, or spatial relationship. This operation is called a *spatial join*. It allows you to define a spatial relationship between two layers and combine their attributes in a new output layer. Depending on the types of data you want to join, you can choose how the spatial relationship is calculated. For instance, you may choose to generate a Count field that summarizes the number of features in the joined layer that intersect each feature in the target layer. Or you may generate a Distance measurement

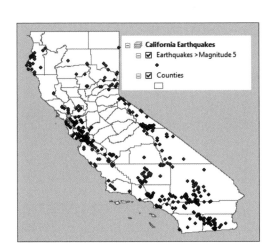

FID	Shape *	NAME
0	Polygon	Butte
1	Polygon	Alpine
2	Polygon	Tehama
3	Polygon	San Bernardino
4	Polygon	San Diego
5	Polygon	San Francisco
6	Polygon	Kings
7	Polygon	Amador
8	Polygon	San Luis Obispo
9	Polygon	Ventura
10	Polygon	Yuba

Counties table (target features)

Earthquake table (join features)

FID	Shape *	MAG	MMI	LOCATION	YEAR
0	Point	5.9	7	Near San Diego, California	1862
1	Point	5	6	Near Calexico, California	1918
2	Point	6.95	10	In the Imperial Valley, near El Centro, California	1940
3	Point	5.04		In the California - Baja California border region	1979
4	Point	5.5		In the California - Baja California border region	1940
5	Point	5.5	9	In the Imperial Valley, near El Centro, California	1940
6	Point	5		In the California - Baja California border region	1940
7	Point	5.5		In the California - Baja California border region	1940
8	Point	5.5	6	In the Imperial Valley, near Holtville, California	1917
9	Point	5.7	6	Near San Diego, California	1894
10	Point	5.5	8	North of Calexico, Imperial Valley, California	1915

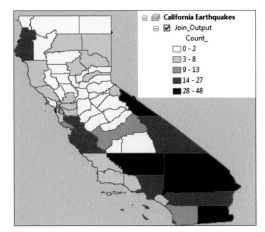

FID	Shape	NAME	Count_	Avg_MAG
0	Polygon	Butte	4	5.5375
1	Polygon	Alpine	3	5.166667
2	Polygon	Tehama	1	5
3	Polygon	San Bernardino	48	5.397292
4	Polygon	San Diego	13	5.759231
5	Polygon	San Francisco	1	5.1
6	Polygon	Kings	0	0
7	Polygon	Amador	0	0
8	Polygon	San Luis Obispo	7	6.004286
9	Polygon	Ventura	5	5.44
10	Polygon	Yuba	0	0

Output table

Above, a layer of California counties is spatially joined to a layer of earthquake points. *Below,* the output is a new feature class that includes a Count field for the total number of earthquakes in each county and an average of the earthquake attributes that fall within each county. Here, you see the average magnitude, but you can choose different ways to calculate the attributes—a sum, for instance. The output layer is symbolized to show the number of earthquakes in each county. (Data includes earthquakes above magnitude 5, since 1769.) Data from Esri Data & Maps 2010, courtesy of National Atlas of the United States, US Geological Survey, ArcUSA, US Census, Esri (Pop2010 field), and ArcWorld.

between features in two layers. For example, you could define a relationship between city points and earthquake points based on distance, generating an output that shows the nearest city to each earthquake.

Data

The exercises in this chapter use the following datasets, stored in a geodatabase called Wisconsin.gdb:

- FoodDeserts—polygons showing locations of census tracts that qualify as food deserts (result of join operation performed in exercise 16a) *(Source: US Department of Agriculture's Economic Research Service)*
- USstates—polygons representing all 50 states, including demographic attributes *(Source: Tele Atlas, US Census, Esri)*
- WisconsinCities—points representing cities (population greater than 10,000) in the state of Wisconsin *(Source: US Census)*
- WisconsinCounties—polygons representing Wisconsin county boundaries *(Source: Tele Atlas, US Census, Esri)*
- WisconsinTracts—polygons representing Wisconsin census tracts, including demography *(Source: Tele Atlas, US Census, Esri)*

The chapter data also includes:

- FoodDesert_Data.xls—a nonspatial table compiling research results from a food desert study *(Source: US Department of Agriculture's Economic Research Service)*
- A layer file symbolizing Wisconsin cities

15

16

Exercise 16a
Joining data by attribute

Once tables are joined, you can use the appended attributes of the nonspatial table to symbolize, label, query, and analyze the features in a layer.

A table join is preserved only within a map document—the tables remain separate on disk—and can be undone at any time. You can permanently append attributes by exporting a joined layer as a new feature class.

For this exercise, you are a free-lance GIS analyst and mapmaker. The Economic Research Service (ERS), an agency of the US Department of Agriculture (USDA), has conducted a study to identify food deserts, defined as "low-income neighborhoods with high concentrations of people who are far from a grocery store." The Wisconsin state Legislature has asked you to create a series of maps that will present the data effectively. The maps you create will be used by a task force that is charged with making nutritious food more readily available in food deserts.

The raw dataset is in tabular form. To visualize the data on a map, you need to use GIS to join the tabular data to geographic information. That way, you aren't required to peruse a list of data to find the information you need—rather, you can access a location, and then quickly visualize the associated information. That's just what ERS did. To share important information that pertains to location, ERS used ArcGIS software to create an interactive web map, called the Food Desert Locator, to serve both the public and government agencies. The Healthy Food Financing Initiative, conceived under the President Barack Obama administration, was designed to make nutritious food more readily available in food deserts.

In this exercise, you will examine the Food Desert Locator. Then you will create the same type of map by joining a portion of the ERS raw data to a layer of Wisconsin census tracts.

1 To examine the results of the Food Desert Locator study, open a web browser, browse to http://www.ers.usda.gov/data/fooddesert, and click Enter Locator (see facing page). Use the navigation tools to pan and zoom around the map if you wish. (If the site is unavailable or you don't have an Internet connection, skip to step 4.)

Do you live in or near a food desert? Do you know anyone who is affected by limited access to groceries, especially fresh produce?

2 On the Locator, click any pink square. Scroll through the pop-up box and notice the information that is provided. You will see this data later, in another form.

Is there any information provided for the tracts that are not designated as food deserts?

Number of people with low access	1008
Percentage of people with low access	68.4
Number of low-income people with low access	480
Percentage of total population that is low-income and has low access	32.6
Number of children age	234

15

16a

3 Click the About the Locator tab. Read the sections titled Overview, Objectives, and "How is a food desert defined?"

How does a census tract qualify as a "low-income" community?

How does a census tract qualify as a "low-access" community?

You will learn how to create a similar map yourself, using the ERS dataset provided on this website, and limit the study area to the state of Wisconsin. The food desert data has already been downloaded for you.

4 **Start ArcMap and open the ex16a map document from your C:\EsriPress\GTKArcGIS\Chapter16 folder.** This map document contains US states and Wisconsin census tracts. Notice the blue lakes—they are not actual features but symbolized by means of a blue background color applied to the data frame.

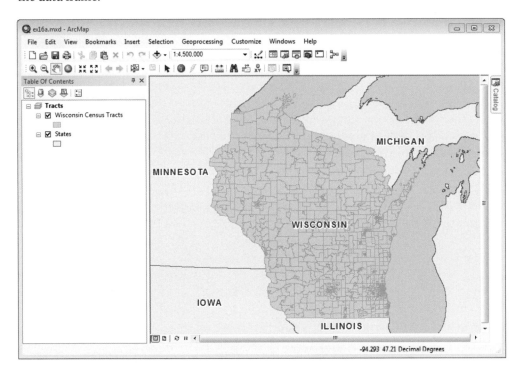

5 **Open the attribute table of Wisconsin census tracts and examine the information stored there. Notice the FIPS_TRACT field** (see facing page). To join nonspatial data to spatial data, the two datasets must have a common field. Recall that the ERS data was collected for census tracts, each of which has a unique identification number (identified by the Federal Information Processing Standard, or FIPS).

Next, you will examine the food desert dataset that has been downloaded from the ERS Food Desert Locator website. The raw dataset is stored in a Microsoft Excel file.

6 Close the attribute table.

Steps 7–10 require the use of Microsoft Excel. As an alternative, you can preview the tables' sheets using the Item Description command in the ArcMap Catalog window.

7 Start Excel and open FoodDesert_Data.xls from your C:\EsriPress\ GTKArcGIS\Chapter16\Data folder.

	A	B	C	D	E	F	G	H	I	J	
1	State	CountyName	TRACT_FIPS	URBAN	TOTALPOP	PERCENT_LOWA_Pop	LOWA_POP	PERCENT_LOWI	LOWI	PERCENT_HUNV	HU
2	AL	Autauga	1001020200	Yes	1892	43.9	830	17.5	331	3.1	
3	AL	Autauga	1001021100	No	2982	59.2	1766	27.6	822	6.5	
4	AL	Baldwin	1003010600	Yes	3724	44.2	1645	18.3	681	4.6	
5	AL	Barbour	1005950200	No	4342	11.9	515	5.4	236	2.3	
6	AL	Barbour	1005950400	No	4068	28.2	1147	16.1	655	5.8	
7	AL	Barbour	1005950500	Yes	4549	25.9	1177	11.1	507	2.7	
8	AL	Bullock	1011952100	No	1705	84.1	1434	53.8	917	14.6	
9	AL	Bullock	1011952400	No	1818	41.2	748	21.6	392	6	
10	AL	Butler	1013953000	No	1693	42.7	724	24.5	415	1.1	
11	AL	Calhoun	1015000400	Yes	2777	50.5	1402	29	807	6.5	
12	AL	Chambers	1017954400	Yes	1455	59.3	863	28.7	417	10	
13	AL	Choctaw	1023956700	No	3883	73.8	2865	41.8	1621	13	
14	AL	Clarke	1025957800	No	6076	15.2	922	11.1	674	2.2	
15	AL	Clarke	1025958000	No	3160	32.2	1019	18	569	3.1	
16	AL	Colbert	1033020100	Yes	3779	17.4	658	4.8	181	1	
17	AL	Conecuh	1035960400	No	3580	35.5	1270	19.5	698	3.6	
18	AL	Conecuh	1035960600	No	4925	17.6	868	7	345	1.8	
19	AL	Covington	1039962900	No	1666	40.2	670	16.5	275	0.9	
20	AL	Crenshaw	1041963400	No	1855	80	1485	41.3	766	11.4	

ReadMe / Variable_Lookup / **Data**

15

16a

8 Scroll down through the rows of data. Notice that the spreadsheet includes data for all 48 continental states (Hawaii and Alaska weren't included in the study).

9 Scroll back to the top of the spreadsheet. Now scroll across and examine the column headers. To decipher the abbreviations, click the Variable_Lookup tab at the bottom of the spreadsheet and read the variable descriptions.

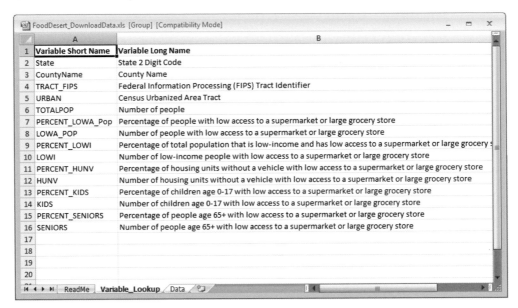

10 Click the ReadMe tab to get more information about the dataset. When you're finished examining the spreadsheet, close Excel.

11 In ArcMap, on the Standard toolbar, click the Add Data button. Navigate to your \Chapter16\Data folder, double-click FoodDesert_ Data.xls, and click Data$. Then click Add.

15

16a

Excel worksheets added to ArcMap have a $ symbol appended to their names.

The Data table is added to ArcMap. When a table is added to ArcMap, the button at the top of the table of contents is switched from List By Drawing Order to List By Source. That's because a table isn't drawn in ArcMap, but it is part of the map document.

12 **In the table of contents, right-click the Data table and click Open.**
The TRACT_FIPS attribute identifies census tracts and is the common attribute that will be used to join the tables. Notice the size of the table—there are 6,500 records! This table includes every census tract in the United States, but only the records that have a match in the Wisconsin census tracts attribute table will be brought over in the join operation. In other words, ArcMap will find all the food desert information that pertains to the state of Wisconsin and append it to the Wisconsin tracts data. Then you will be able to display it on the map.

CountyName	TRACT_FIPS	URBAN	TOTALPOP	PERCENT_LOWA_Pop	LOWA_POP	PERCENT_LOWI	LOWI	P
Autauga	1001020200	Yes	1892	43.9	830	17.5	331	
Autauga	1001021100	No	2982	59.2	1766	27.6	822	
Baldwin	1003010600	Yes	3724	11.2	1646	18.3	681	
Barbour	1005950200	No	4342	11.9	515	5.4	236	
Barbour	1005950400	No	4068	28.2	1147	16.1	655	
Barbour	1005950900	Yes	4549	25.9	1177	11.1	507	
Bullock	1011952100	No	1705	84.1	1434	53.8	917	
Bullock	1011952400	No	1818	41.2	748	21.6	392	
Butler	1013953000	No	1693	42.7	724	24.5	415	
Calhoun	1015000400	Yes	2777	50.5	1402	29	807	
Chambers	1017954400	Yes	1455	59.3	863	28.7	417	
Choctaw	1023956700	No	3883	73.8	2865	41.8	1621	

1 ▶ ▶I (0 out of 6500 Selected)

Data$

15

16a

13 Scroll down to find the data records for Wisconsin (WI in the State column). Notice that there are often several tracts per county. This will be important in exercise 16b.

14 Right-click TRACT_FIPS, and on the shortcut menu, click Freeze/ Unfreeze Column. Scroll all the way to the right. The TRACT_FIPS column is moved to the leftmost position in the table and does not move when you scroll.

Table □ ×

Data$

TRACT_FIPS	LOWI	PERCENT_HUNV	HUNV	PERCENT_KIDS	KIDS	PERCENT_SENIORS	SENIORS
55081950800	447	3.9	53	7.7	316	3.5	144
55083990500	366	1.7	18	8.5	204	10.2	245
55083990600	290	1.5	17	7.9	225	5.7	161
55085970100	250	0.5	10	3.3	177	2.8	151
55087010100	368	6.6	75	2.9	94	4.5	148
55093960100	634	4.6	50	19.1	557	13.2	387
55093960800	325	1	12	11.3	412	6.2	224
55097960400	543	2.3	39	6	414	5.8	405
55099970600	511	2.1	22	12.7	329	9.3	240
55099970700	1021	6.2	69	25	688	15	413
55101000200	1404	7.6	174	16.8	1053	4.6	288
55101000300	1904	15.3	167	38.2	1389	5.3	192

I◄ ◄   0 ► ►I ▣ ▣ (0 out of 6451 Selected)

Data$

What does the PERCENT_HUNV field measure?

(HInt: If you've forgotten, revisit the Excel table's Variable_Lookup sheet—you will have to open it in read-only mode since it's in use by ArcMap—or the Food Desert Locator website.)

Both the FoodDesert_Data table and the Wisconsin census tracts layer attribute table contain one record for each census tract. The record relationship between the tables is therefore one-to-one: for each record in the Wisconsin census tracts table, there is no more than one matching record in the
FoodDesert_Data table, thus making a good choice for a join operation.

15

16a

15 Close the table. In the ArcMap table of contents, right-click WIsconsIn Census Tracts and click Joins and Relates > Join.

You can also execute a join operation using the Add Join geoprocessing tool in ArcToolbox (Data Management Tools > Joins toolset). For large join operations, the geoprocessing tool is recommended.

16 In the upper portion of the Join Data dialog box, in the first drop-down list, make sure Join attributes from a table is selected. In the first entry box, click the drop-down arrow and click FIPS_TRACT. This specifies FIPS_TRACT as the common attribute in the Wisconsin Census Tracts table.

17 In the second entry box, make sure the Data$ table is selected. In box 3, make sure the TRACT_FIPS field is selected. This specifies TRACT_FIPS as the common attribute in the Food Desert table. Notice that ArcMap automatically populates the second and third entry boxes, but you should always check to be sure the correct fields are selected.

15

16a

18 **In the Join Options section, you want to keep all the records, so maintain the default setting.** Note that choosing to "keep all records" doesn't mean you are bringing all the food desert records over to the census tracts table; rather, it means all Wisconsin census tract records will remain, even if there's not a match in the food deserts table. In fact, in this case, there won't always be a match, because the food deserts table does not have a record for every census tract, but only for tracts that qualify as food deserts. In the resulting table, census tracts that do not qualify will have null values in the newly joined fields.

19 **Make sure your dialog box matches the graphic, and then click OK.** ArcMap joins the attributes of the FoodDesert_Data table to the Wisconsin Census Tracts table, matching the records by their FIPS tract numbers.

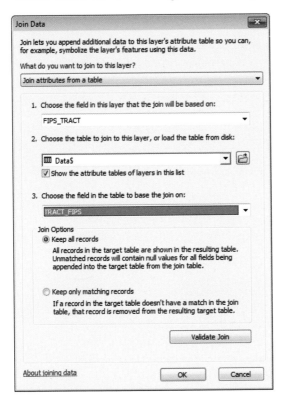

20 **Open the attribute table of the Wisconsin Census Tracts layer. Scroll to the right until you see the TRACT_FIPS field.** To the right of TRACT_FIPS are the fields from the FoodDesert_Data table, which are shown using field aliases (see table on facing page). Field aliases are a more user-friendly description of the content of the field. Any alias can be changed in the field properties.

Table

Wisconsin Census Tracts

SQMI	OBJECTID *	Shape_Length	Shape_Area	TRACT_FIPS	State	CountyName	URBAN	TOTALPOP
4.89	1	0.203001	0.001397	<Null>	<Null>	<Null>	<Null>	<Null>
96.02	2	0.950889	0.027287	<Null>	<Null>	<Null>	<Null>	<Null>
0.14	3	0.027575	0.000041	<Null>	<Null>	<Null>	<Null>	<Null>
2.18	4	0.13566	0.000624	<Null>	<Null>	<Null>	<Null>	<Null>
0.6	5	0.094002	0.000173	<Null>	<Null>	<Null>	<Null>	<Null>
0.35	6	0.041807	0.000099	<Null>	<Null>	<Null>	<Null>	<Null>
0.12	7	0.030908	0.000033	<Null>	<Null>	<Null>	<Null>	<Null>
76.44	8	0.917819	0.02237	<Null>	<Null>	<Null>	<Null>	<Null>
0.5	9	0.054798	0.000144	<Null>	<Null>	<Null>	<Null>	<Null>
2.35	10	0.126024	0.00068	<Null>	<Null>	<Null>	<Null>	<Null>
2.64	11	0.193319	0.000775	<Null>	<Null>	<Null>	<Null>	<Null>
0.53	12	0.065392	0.000151	<Null>	<Null>	<Null>	<Null>	<Null>

I◄ ◄   1 ► ►I   (0 out of 1323 Selected)

Wisconsin Census Tracts

21 Turn off the field aliases by clicking the Table Options button 🔲, and then click **Show Field Aliases** to clear the selection. Now all field names show a prefix (in this case, Data$) that identifies the source table that the attribute comes from. This is a good way to distinguish joined fields from original fields if you're not sure.

22 Resize the **Data$.TRACT_FIPS** field to see its entire name.

Table

Wisconsin Census Tracts

WisconsinTra	Data$.TRACT_FIPS	Data$.St	Data$.CountyN	Data$.U	Data$.TOTA
0.001397	<Null>	<Null>	<Null>	<Null>	<Null>
0.027287	<Null>	<Null>	<Null>	<Null>	<Null>
0.000041	<Null>	<Null>	<Null>	<Null>	<Null>
0.000624	<Null>	<Null>	<Null>	<Null>	<Null>
0.000173	<Null>	<Null>	<Null>	<Null>	<Null>
0.000099	<Null>	<Null>	<Null>	<Null>	<Null>
0.000033	<Null>	<Null>	<Null>	<Null>	<Null>
0.02237	<Null>	<Null>	<Null>	<Null>	<Null>
0.000144	<Null>	<Null>	<Null>	<Null>	<Null>
0.00068	<Null>	<Null>	<Null>	<Null>	<Null>

23 Use the knowledge you gained in chapter 15 to perform a Select By Attributes query 🔲 using this SQL statement:

"Data$.TRACT_FIPS" IS NOT NULL

15

16a

24 Close the Select By Attributes dialog box. Show only the selected records in the attribute table.

	Wisconsi	WisconsinTra	WisconsinTr	Wiscon	WisconsinTra	Wisconsin	WisconsinTra	Wisconsin	WisconsinT
▸ Polygon	55	009	001200	55009001200	2696	9296.6	2670	9206
Polygon	55	079	013000	55079013000	1764	7056	1769	707
Polygon	55	123	960500	55123960500	3222	19.1	3562	21
Polygon	55	123	960400	55123960400	4871	263.2	5014	270
Polygon	55	025	011505	55025011505	4406	4362.4	4593	4547
Polygon	55	025	001402	55025001402	5341	1431.9	5362	1437
Polygon	55	071	000700	55071000700	5581	1671	5568	1667
Polygon	55	059	000900	55059000900	4523	8534	4616	8709
Polygon	55	139	000500	55139000500	5825	6131.6	6118	644
Polygon	55	001	950200	55001950200	3815	15.5	4026	16
Polygon	55	101	001201	55101001201	4051	1695	4079	1706
Polygon	55	079	014900	55079014900	1654	11814.3	1627	11621

I◀ ◀   1 ▸ ▸I   ▤ ▥   (110 out of 1323 Selected)

Wisconsin Census Tracts

How many census tracts in the state of Wisconsin qualify as food deserts?

25 Move the table out of the way so you can see the map display. The selected records are also selected on the map. The tracts outlined in cyan are food deserts. Because you have joined the food desert data to spatial data, you can present this information graphically on a map.

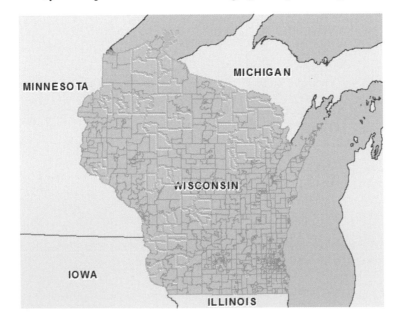

26 Close the table. Tracts will remain selected on the map.

Joins exist only within a map document—they aren't permanent. However, at this point, you can create a new feature class of food desert polygons that includes their census attributes as well as their ERS attributes. To do this, you must export the selected features to a new, permanent feature class, so you'll do that next.

27 In the table of contents, right-click Wisconsin Census Tracts and click Data > Export Data. Notice that the first option is set to export only the selected features of the current layer. Maintain the first two default options.

28 For the output location, browse to your MyData folder for chapter 16. Click the Save as type arrow and click File and Personal Geodatabase feature classes. Double-click MyWisconsin.gdb, and for Name, type FoodDeserts. Click Save.

29 When your Export Data dialog box matches the graphic at right, click OK. When the operation is complete, click Yes to add the new layer to the map.

15

16a

30 On the menu bar, click Selection > Clear Selected Features. Change the table of contents to the Drawing Order view, and change the name of the new layer to Wisconsin Food Deserts.

On your own

Symbolize the new layer using a 10% Ordered Stipple pattern, Gray 40% outline, and 35% transparency.

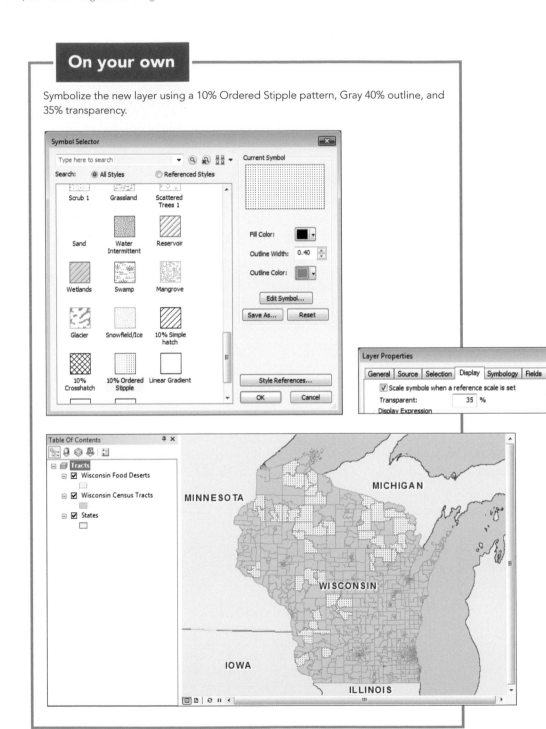

31 If you want to save your work, rename your map document and save to your MyData folder for chapter 16.

32 If you are continuing to the next exercise, leave ArcMap open. Otherwise, exit the application and click No if prompted to save your changes.

15

16a

Exercise 16b
Relating data

Tables should be associated by a relate instead of a join when records in the layer attribute table (the target) may have many matches in the relate table (a one-to-many or a many-to-many relationship). When tables are related, you can highlight records in either table and see the matching records highlighted in the other. Whereas joins are performed to create new layers for symbolizing, labeling, querying, and analysis, relates are used for data exploration. Although the result is not a new dataset, it is valuable nonetheless in providing an organized view of the relationships between existing datasets.

The food desert data was calculated for census tracts in exercise 16a, although it's county governments, in cooperation with federal programs, that are likely to be the ones tackling the problem. In this exercise, you will overlay food desert census tracts on Wisconsin counties to get a quick visual of how many food desert tracts there are in each county. Then you will relate the food desert table to a county layer to make it easy to pull up all the food desert details associated with each county. Because some counties contain multiple census tracts that qualify as food deserts, the relationship is one-to-many (one county, many census tracts). In this case, a join isn't helpful—you need to relate the tables instead.

1
2
3
4
5
6
7
8
9
10
11
12
13
14
15
16b
17
18
19
20

1 In ArcMap, open the ex16b map document from your Chapter16 folder.

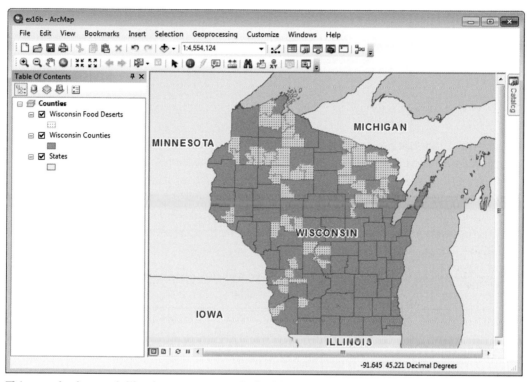

This map looks much like the map you worked with in exercise 16a, but instead of census tracts, you see polygons representing Wisconsin counties. Before you relate the tables, you will first examine the relationship between counties and food deserts.

2 Turn the Wisconsin Food Deserts layer off and on to get an idea of where the food desert tracts are in relation to county boundaries.

You can see that some counties have no food deserts, and others have multiple food deserts. But even with a carefully chosen symbology scheme, it can be hard to tell which food desert tracts fall within which counties. So next, you will use the Select Features tool to get a better understanding of the data before relating tables.

3 Change the table of contents view to List By Selection. The map document was saved with a selection scheme that allows only the counties to be selectable.

4 On the Tools toolbar, click the Select Features tool ![icon] . (Just click the button, not the arrow next to it.) The default tool option for Select Features is to select by rectangle. This means you can either draw a rectangle to select features that intersect the rectangle or simply click to select a feature.

5 On the map, click any county that looks as if it has multiple food desert census tracts.

6 Click around the map to select other counties, one at a time. You may also try drawing a small rectangle to select a county. If you accidentally select several counties at once, just click outside the state of Wisconsin to clear the selection and start over.

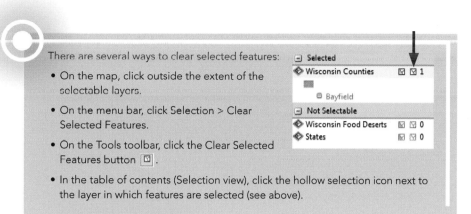

There are several ways to clear selected features:

- On the map, click outside the extent of the selectable layers.

- On the menu bar, click Selection > Clear Selected Features.

- On the Tools toolbar, click the Clear Selected Features button.

- In the table of contents (Selection view), click the hollow selection icon next to the layer in which features are selected (see above).

On your own

Make Wisconsin Food Deserts selectable by clicking the first icon to the right of the layer in the table of contents. With the Select Features (Select by Rectangle) tool active, draw small rectangles to select individual counties and the food desert tracts within them for a quick visual analysis.

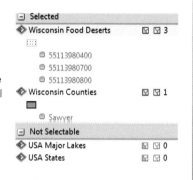

Notice in the table of contents that the NAME attribute is displayed for Wisconsin Counties, and the FIPS_TRACT attribute is displayed for Wisconsin Food Deserts. You can change the Display field on the Display tab of the Layer Properties dialog box.

A more accurate method of selecting food deserts within a specific county would be to run an attribute query using the Select By Attributes command, which you learned in chapter 15.

7 Clear any selections and switch the table of contents to Drawing Order view.

8 On the Tools toolbar, click the Select Elements tool ▶ to activate it.

15

16b

9 Right-click Wisconsin Counties and click Properties. On the Layer
Properties dialog box, click the Display tab. Verify that the Display
Expression Field option is set to NAME. Select the "Show
MapTips using the display expression" check box. When your
dialog box matches the graphic, click OK.

10 Next, pause your mouse pointer over any county on the map. Notice
that a MapTip appears, quickly showing a county's name. Using
MapTips is another quick way to get an idea of which counties contain food
deserts.

Can you name a county whose entire extent qualifies as a food desert?

Where population is higher, there may be many, smaller census tracts. To see some food deserts that aren't easily visible at the full extent, you need to zoom in, which you'll do next.

⓫ Go to the Southeast bookmark.

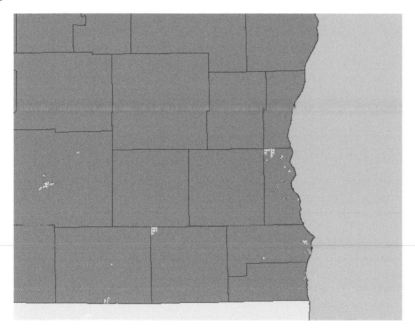

To get a better sense of where food deserts are in relation to large cities, you will next add a cities layer that includes cities with a population of at least 10,000.

15

16b

12 **Add Wisconsin Cities.lyr to the map (located in the Data folder for chapter 16).** The cities are symbolized by graduated symbols to show population ranges.

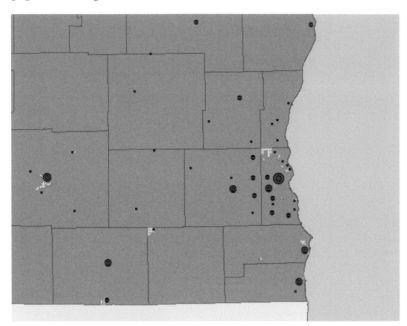

On your own

Turn on MapTips for Wisconsin Cities.

What is the name of the largest city in the area?

You've done some good visual analysis. Now, you are ready to relate tables so you can easily get all the detailed information you need.

13 **Zoom to the full extent of the map** 🔘 **.**

14 In the table of contents, right-click Wisconsin Counties and click Joins and Relates > Relate. In the upper portion of the dialog box, read the description of the Relate command. If you were to examine the attribute tables of Wisconsin Counties and Wisconsin Food Deserts, you would see a common field: CNTY_FIPS. In this case, the field in each table has the same name, but this is not always necessary, as long as both fields contain values that match. The common attribute is the county code, in accordance with FIPS.

15 On the Relate dialog box, for option 1, click the arrow and click CNTY_FIPS. For option 2, click Wisconsin Food Deserts. For option 3, click CNTY_FIPS. For option 4, maintain the default name. When your dialog box matches the graphic, click OK.

16 Open the Wisconsin Counties attribute table. Be sure the Show all records button at the bottom of the table is activated. Notice there are 72 records—one for each county. No records or fields are added in a relate operation.

15

16b

17 In the table, select the third record, for Vernon County, by clicking the gray tab to the left of the OBJECTID column.

18 At the top of the attribute table, click the Related Tables arrow and click the only related table, Relate1: FoodDeserts. The related table appears, showing the three food desert tracts that are located within the boundaries of Vernon County. This allows county and state officials to quickly view the details of the food deserts under their jurisdiction.

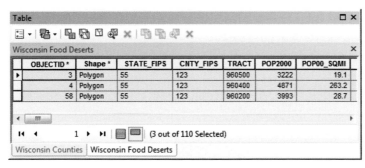

19 Again, click the Related Tables arrow and click Relate1: Wisconsin Counties. Only the selected record is shown. In this way, you can switch back and forth between related tables, showing either all the records or only those that are currently selected.

You can also click the tabs at the bottom of the table to switch between the related tables.

20 If necessary, move the table so you can see the ArcMap display. Click the Identify tool and click Vernon County, which is selected. Make sure you click a part of the county that is not a food desert—you want to select the county, not the food desert.

21 In the Identify window, notice the plus sign (+) symbol next to Vernon, in the upper panel. Expand the contents of Vernon. Now expand the contents of Wisconsin Food Deserts. Conveniently, related features are visible in the Identify window.

15

16b

22 Close the Identify window. At the bottom of the table, click the Show all records button to return the table to its original view.

On your own

Find and select Milwaukee County and view the information pertaining to its food deserts.

How many food deserts does Milwaukee County have?

Relationship classes

If your data is stored in a geodatabase, and you are using an ArcGIS for Desktop Standard or Advanced license, you can permanently relate data in the database, not just in a map. To do this, you would create a relationship class between datasets. (Relationship classes are read-only with a Basic license.) A relationship class is its own entity in the geodatabase, with user-defined rules, properties, and optionally, its own attribute table. You can set up a relationship class so that modifications trickle down to related features (for instance, if a utility pole is deleted, an attached transformer would be deleted as well). To learn more, in ArcGIS for Desktop Help, on the Contents tab, navigate to Geodata > Data types > Relationships and related objects > Benefits of relationship classes.

23 Close the table.

24 If you want to save your work, rename your map document and save to your MyData folder for chapter 16.

25 If you are continuing to the next exercise, leave ArcMap open. Otherwise, exit the application and click No if prompted to save your changes.

Exercise 16c
Joining data by location

In exercises 16a and 16b, you associated datasets according to common attributes; in this exercise, you will associate datasets according to common location. When you perform a spatial join operation, the output is a new feature class with new attributes based on the parameters selected on the tool dialog box.

For the final stage of your research project, you will create a map showing which counties have the highest number of food desert tracts. These counties should receive top priority for state resources dedicated to reducing food deserts.

1 **In ArcMap, open the ex16c map document from your Chapter16 folder.** You see the map you have been working with, with Wisconsin counties and food deserts, and a background States layer.

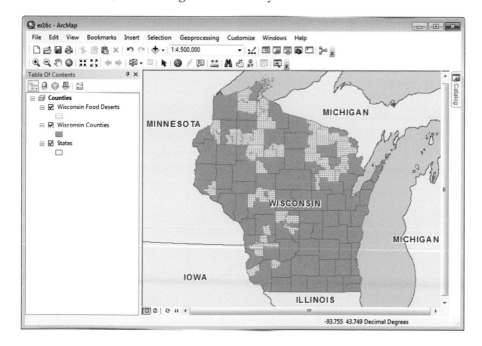

You will join the Wisconsin Food Deserts layer to the Wisconsin Counties layer.

Instead of performing a table join between a spatial layer and a nonspatial table, as you did in exercise 16a, you will create a spatial join to combine two spatial layers. Specifically, you want to create a spatial join based on a containment relationship between two polygon layers (food desert tracts contained within counties). Although the Join dialog box can create spatial joins, there is another option that provides more options for handling data relationships, the Spatial Join geoprocessing tool.

The Spatial Join geoprocessing tool is in ArcToolbox. You will learn more about ArcToolbox and use many more of its tools in the upcoming chapters on analysis (chapters 18, 19, and 20).

2 On the Standard toolbar, click the ArcToolbox button 🗔 . In ArcToolbox, expand the Analysis Tools > Overlay toolset.

3 Right-click the Spatial Join tool and click Open.

4 On the Spatial Join dialog box, click the Target Features arrow and click Wisconsin Counties. For Join Features, click Wisconsin Food Deserts. Set the Output Feature Class location to \Chapter16\ MyData\MyWisconsin.gdb. Call the feature class CountyStats.

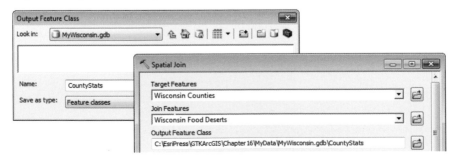

5 For Join Operation, maintain the default of JOIN ONE TO_ONE.
You may wonder why you are using the one-to-one option here. This isn't about
cardinality—yes, you have many food deserts that you want to join to each
county. However, you want only one row returned for each target feature. If you
chose one-to-many, you would get a separate row for each county match.

6 Maintain the default for Keep All Target Features. This means even
counties without food deserts will remain in the output feature class.

> ## On your own
>
> Click the Show Help button to read more about each tool parameter.

7 In the Field Map of Join Features panel, scroll down until you see
the TOTALPOP (Double) field. Right-click it and click Merge Rule >
Sum. The output will now show the sum of all the populations that live within
the food deserts in each county. Note that the default merge rule is "First." This
means that when there are multiple join features that share space with a target
feature, only the first attribute value will appear in the output table.

> The TOTALPOP field is in the lower third of the list. Expand the dialog
> box for easier scrolling, or rotate the scroll wheel on your mouse to see it.

Next, you will change the merge rule for all the attributes related to the food
desert study.

8 So you can see the average percentage of those living in food deserts who have low access to grocery stores, change the PERCENT_LOWA_Pop merge rule to Mean. Then, scrolling down as necessary, change the merge rule for the next nine fields, summing the numerical figures and averaging (mean) the percentages (those with field names that begin with PERCENT_).

Next, it's important to consider the correct match option to get the desired result. For instance, if you were to choose the INTERSECT option, the results would include all food deserts adjacent to, as well as within, county boundaries. You want to join just the food deserts that are contained within the boundaries of each county feature.

9 Change the Match Option to CONTAINS.

The Count field—which tallies all join features within each target feature—will be added automatically to the target table.

10 Click OK to run the Spatial Join tool. When the process is completed, the output layer is added to the map with a randomly chosen fill color.

11 Close ArcToolbox. Then move the Wisconsin Food Deserts layer to the top of the drawing order in the table of contents.

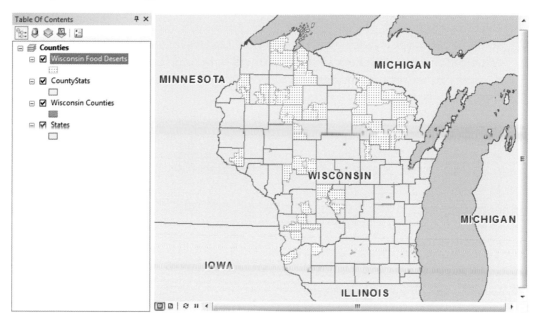

12 Open the attribute table of CountyStats. Notice the Join_Count field. Sort it by ascending order.

15

16c

13 **Scroll through the rows and note how many food deserts are in each county.** There are a lot of attributes now. Remember, the food deserts layer was the result of a join to census tracts, so you now have attributes from counties, census tracts data, and the food desert research all within one table.

Which county has the most food desert tracts?
How many does it have?

14 **Scroll to the right. Notice that the counties that have a Join_Count value of 0 have null values for census tracts data and food desert data. If you select one of these rows, you can see on the map that the county has no food deserts.**

15 **Scroll farther to the right. The food desert attributes you are interested in begin with TOTALPOP (total number of people who live in food deserts in each county). Sort the TOTALPOP field by descending order, and then scroll to the top of the table.**

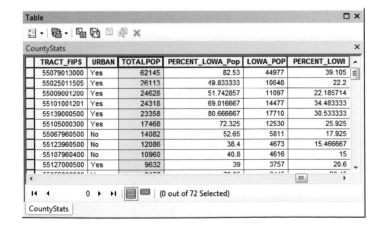

16 Examine the other food desert attributes, sorting fields as you like.

Which county's food desert population has the highest percentage of children?

Next, you want to symbolize the map to show the number of people in each county who have low access to grocery stores.

17 Close the table and clear any selections. Open the layer properties of CountyStats and click the Symbology tab.

18 Under Show, click Quantities > Graduated colors. In the Fields frame, for Value, scroll down and click LOWA_POP.

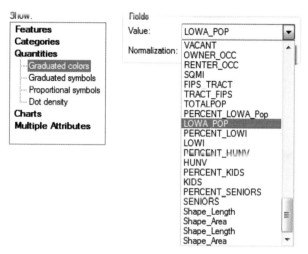

19 Choose any color ramp you like, and then, in the Classification frame, click Classify.

15

16c

20 On the Classification dialog box, in the Data Exclusion frame, click Exclusion. On the Data Exclusion Properties dialog box, on the Query tab, build the following query: "LOWA_POP" IS NULL

21 Next, on the dialog box, click the Legend tab. Select the Show symbol for excluded data check box and set the symbol color to white or another color that is easily distinguishable from the color ramp you chose in step 19. For Label, type No food deserts.

22 Click OK to close the Data Exclusion Properties dialog box, and then OK again to close the Classification dialog box.

23 Back on the Symbology tab, above the center panel, click the Label tab and click Format Labels.

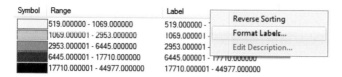

24 On the Number Format dialog box, in the Rounding frame, change the number of decimal places to 1 and click OK.

25 When your layer properties dialog box matches the graphic (with a color ramp of your choice), click OK.

15

16c

26 **Turn off Wisconsin Food Deserts. Rename CountyStats** Low Access Population **and rename the legend** Total by County.

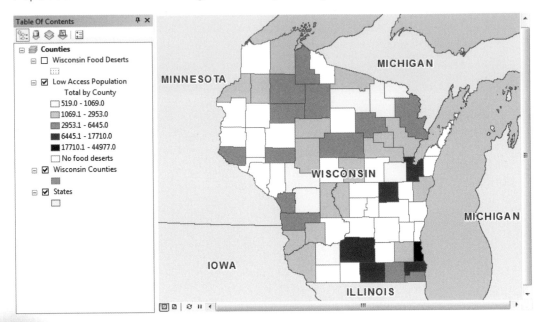

To map other attributes, such as the number of seniors in food deserts or the percentage of the food desert population that doesn't have a vehicle, just open the layer properties and on the Symbology tab, change the field value. You don't have to re-create the exclusion query.

On your own

Save Low Access Population as a layer file, and then create additional graduated-color layer files (with unique legend and layer file names) that capture other food desert statistics.

27 **If you want to save your work, rename your map document and save to your MyData folder for chapter 16.**

28 If you're not continuing to the next chapter, close ArcMap. Otherwise, exit the application and click No if prompted to save your changes. Creating a spatial join has allowed you to create a map that effectively tells an important story. County, state, and federal resources can be better prioritized using this tool.

Summary

In this chapter, you learned about how and why you might join or relate tabular data to spatial data in ArcMap. You also learned how to join two spatial datasets to calculate the feature attributes of one layer that are spatially coincident to features in another layer. Joining and relating data are common tasks for many GIS professionals. Additional software functions that were reviewed include MapTips, identifying features, selecting features, layer symbology, and legend formatting.

Terms

attribute join

target table

join table

relate

spatial join

15

16c

Analyzing

feature relationships

Chapter 17
Selecting features by location

Using location queries
Combining attribute and location queries

Learning objectives

- *Create a selected set of features according to location-based queries*
- *Create a new map layer from selected features*
- *Select features by combining attribute and location queries*
- *Create a new layer from a selected set of features*

IN CHAPTER 15, YOU WERE INTRODUCED TO THE SELECT BY ATTRIBUTES TOOL, so you could select features based on specific criteria. In other chapters, you manually selected features by highlighting rows in an attribute table. In this chapter, you'll use the Select By Location tool to create a *location query* (also called a *spatial query*) that selects features according to their spatial relationship to other features, whether in another layer or within the same layer.

The four types of spatial relationships that can be used are distance, containment, intersection, and adjacency.

   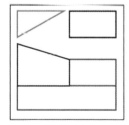

Left to right: points within a given distance of the red point are selected, points contained by the red polygon are selected, lines that intersect the red line are selected, and polygons adjacent to the red polygon are selected.

17

18

19

20

ArcGIS defines 15 spatial relationships, each a variation of one of the four basic types. For example, there is containment (where the contained feature may touch the boundary of the containing feature) and complete containment (where the boundaries cannot touch).

Data

The exercises in this chapter use the following datasets:

- BayArea.shp—polygons of the counties in the Bay Area region of California *(Source: County of Sonoma GIS Central)*
- Cities.shp—points representing cities, towns, and unincorporated communities near the Russian River in Northern California *(Source: National Atlas of the United States)*
- Guerneville.jpg—orthophotograph of downtown Guerneville and adjacent area *(Source: County of Sonoma GIS Central)*
- GuernevilleParcels.shp—polygons showing parcel locations within the limits of the Guerneville community *(Source: County of Sonoma GIS Central and its Permit and Resource Management Department)*
- RussianRiver.shp—line feature representing the Russian River *(Source: County of Sonoma GIS Central)*
- RussianRiverFloodZone.shp—polygon layer identifying areas that could potentially be at risk of flooding from the Russian River and its tributaries *(Source: County of Sonoma GIS Central)*

Exercise 17a
Using location queries

To select features by location, you choose two layers and specify a spatial relationship. Features in the first layer are then spatially compared to features in the second layer and selected if they satisfy that relationship.

If the spatial relationship is one of distance, you also choose the distance value and measurement units you want to apply.

In this example, cities are selected if they lie within one mile of rivers. ArcMap measures the distances from cities to rivers and selects cities that satisfy that relationship. Data from Esri Data & Maps 2010, courtesy of ArcUSA, US Census, Esri (Pop2010 field), National Atlas of the United States, and US Geological Survey.

In the case of Pennsylvania, cities that lie within a mile of a river can be easily shown (see map on next page).

17a

18

19

20

ObjectID *	Shape *	NAME	CLASS	ST
2654	Point	New Castle	City	PA
2656	Point	Sharon	City	PA
2672	Point	Scott	Census Designated Place	PA
2677	Point	Pittsburgh	City	PA
2973	Point	Butler	City	PA
2974	Point	Oil City	City	PA
2975	Point	Warren	City	PA
2980	Point	Johnstown	City	PA

The Pennsylvania cities selected on the map, *left*, and in the table, *right*, lie within one mile of a river. Data from Esri Data & Maps 2010, courtesy of ArcUSA, US Census, National Atlas of the United States, and US Geological Survey.

For more information, in ArcGIS for Desktop Help, on the Contents tab, navigate to Mapping > Working with layers > Interacting with layer contents > Using Select By Location.

In this exercise, you will be working with parcels within a small town bordering a major river in Sonoma County, California. For this exercise, you are an emergency response analyst who has been asked to identify parcels in the town of Guerneville that may be at risk of flooding. Once the at-risk parcels are identified, they will be related to phone data to generate an emergency call list in accordance with the county's disaster plan.

1 **In ArcMap, open the ex17a map document from your C:\EsriPress\ GTKArcGIS\Chapter17 folder.** The map is centered on Sonoma County (see facing page). You can see a blue line feature representing the Russian River, selected Sonoma County cities and towns represented with red points (only towns near the river are included in this dataset), and adjacent county polygons. Two layers, Guerneville Parcels and Guerneville.jpg, are turned off at this scale—they are visible only when you zoom in closer.

17a

18

19

20

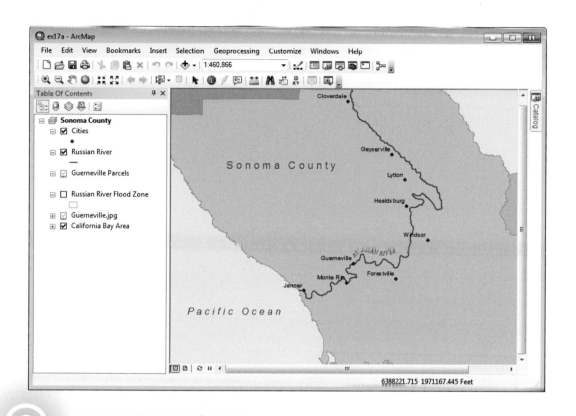

You can set a scale dependency for a layer's symbology on the Layer Properties dialog box, on the General tab.

On your own

Perhaps you want to have only the city of Guerneville displayed at this scale. Create a definition query to limit the display of cities so only Guerneville is shown. You learned how to do this in chapter 7. (*Hint:* Explore the layer properties.)

2 **On the menu bar, click Bookmarks > Guerneville 1.**

17a

18

19

20

You are zoomed in on the city limits of Guerneville, and the Guerneville Parcels layer is turned on.

Depending on the size of your ArcMap window, the dynamic label for the Russian River may be in a different spot.

3 Turn on the Russian River Flood Zone layer.

This layer shows areas at risk of flooding from the Russian River. This is a dataset generated by Sonoma County, based on Federal Emergency Management Agency (FEMA) 100- and 500-year flood zones along the Russian River. By selecting parcels that fall within this flood zone, the county can generate a call list to warn residents of possible flooding. You can get a general idea of which parcels fall within the flood zone by looking at the map, but next you will create a new selected set of parcels so you can generate a parcel call list.

4 Zoom to the Guerneville2 bookmark. Turn off the Cities and Russian River Flood Zone layers.

The orthophotograph is turned on at this scale. You will keep your map display focused here, on the center of town, but remember, your selections may include parcels that are not in view. First, you will select all parcels that intersect the flood zone layer. This means that any parcel that has a boundary within or touching the boundary of the flood zone layer will be selected.

17a

18

19

20

5 **On the menu bar, click Selection > Select By Location.**

6 **On the Select By Location dialog box, maintain the default Selection method option to select features from. In the Target layer(s) list, select Guerneville Parcels. For Source layer, click the arrow and click Russian River Flood Zone. Maintain the default spatial selection method for target layer feature(s) that intersect the source layer feature. Make sure your dialog box matches the graphic, and then click OK to make the selection.** Selected parcels are outlined in cyan.

The target layer is the layer to which the selection is applied, based on its relationship to the source layer. Notice that you can choose any layers in the map for the source and target layers, regardless of whether they are turned on or not.

7 Turn on Russian River Flood Zone again so you can see that the parcels that intersect the flood zone are selected, even if it's just a corner or a boundary of the parcel that intersects the flood zone.

You can change the selection color in the layer properties on the Selection tab.

8 Open the attribute table of Guerneville Parcels and dock it or place it so it does not obscure the map display.

How many parcels are selected?

Perhaps you are also interested in locating parcels that are within a certain distance of the flood zone.

17a

18

19

20

9 Open the Select By Location tool dialog box. Again, select Guerneville Parcels as the Target layer and Russian River Flood Zone as the Source layer. For the spatial selection method, again select "intersect the source layer feature."

10 Select the Apply a search distance check box and set the search distance to 400 feet.

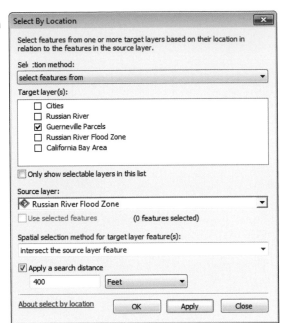

The search distance settings will be recalled the next time you use the tool, so be sure to clear the check box the next time you use the tool if you don't want to apply a search distance to the selection.

11 Click Apply and move the Select By Location dialog box out of the way.

Now how many parcels are selected?

You want to refine your selection to only those parcels that are mostly within the flood zone. This will weed out the parcels that have only a small part of land in the zone. All residents of parcels within the flood zone will be contacted, but you want to determine how many parcels are likely to require emergency assistance. By calculating the number of parcels that have boundaries mostly within the flood zone, you can safely estimate and request emergency resources for this area.

12 On the Select By Location dialog box, maintain the previously chosen target and source. For the spatial selection method, click the arrow, scroll to the bottom of the list, and click "have their centroid in the source layer feature."

13 Clear the Apply a search distance check box. Click Apply. Notice how the selection is reduced slightly.

Now how many parcels are selected?

On your own

Select target features that are within the source layer feature and examine the results. When you are finished, close the Select By Location tool dialog box.

14 If you want to save your work, rename your map document and save to your MyData folder for chapter 17.

15 If you are continuing to the next exercise, leave ArcMap open. Otherwise, exit the application and click No if prompted to save your changes.

Exercise 17b
Combining attribute and location queries

Location and attribute queries can be used together to solve a problem. For example, if you're doing emergency planning, you might want to select cities that have a population of 100,000 or greater (an *attribute query*) and that lie within 10 miles of a major fault line (a location query). In this exercise, you will select parcels that are zoned residential (an attribute query) and that intersect the flood zone (a location query).

1 In ArcMap, open the ex17b map document from your Chapter17 folder. This map shows the results of exercise 17a, through step 10. The Guerneville parcels that intersect the flood zone have been selected.

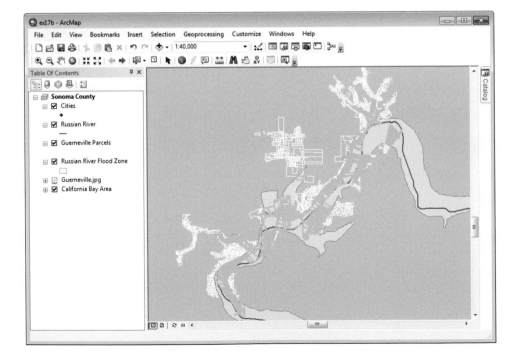

If you accidentally cleared the selection, just repeat the "intersect" selection described in step 6 of exercise 17a.

In this exercise, you want to further fine-tune the selection so it includes only parcels that are zoned residential. In this way, you will combine an attribute query and a location query. You completed the location query in exercise 17a, so next, you will do the attribute query.

2 On the menu bar, click Selection > Select By Attributes.

3 On the Select By Attributes dialog box, for Layer, click the arrow and click Guerneville Parcels. For Method, change the default setting to Select from current selection.

4 In the fields list, double-click "BASEZONING" to add it to the expression box. Click the equals sign (=) to add it to the expression.

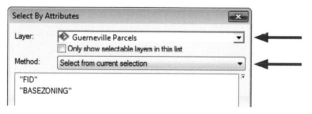

5 Click Get Unique Values. The values for the BASEZONING field are listed.

17b

18

19

20

6 In the list of values, double-click 'Residential' to add it to the expression. Click OK to make the selection.

The selection now indicates parcels that are within the flood zone and zoned as residences.

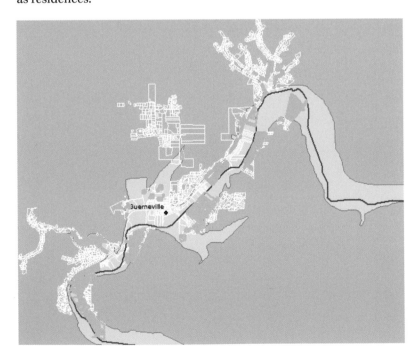

17b

18

19

20

7 Open the attribute table of Guerneville Parcels.

How many parcels are currently selected?

Next, you will save the selection as a new layer on the map.

8 In the table of contents, right-click Guerneville Parcels and click **Selection > Create Layer From Selected Features**. A new layer called Guerneville Parcels selection is added to the map.

> To save the selected features permanently, you could export the selection to a new feature class (right-click the layer and click Data > Export Data).

9 Close the attribute table, clear the current selection, and turn off Guerneville Parcels.

10 In the table of contents, click the default polygon symbol for the new layer. In the Symbol Selector, use a hollow fill (no color) and a red outline, and then click OK.

11 Zoom to the Guerneville **2** bookmark and turn off the flood zone layer.

17b

18

19

20

As expected, most of the downtown area is not included in the selection. Thanks to your work, a flood watch call list for at-risk residences can be created.

12 If you want to save your work, rename your map document and save to your MyData folder for chapter 17.

13 If you are continuing to the next chapter, leave ArcMap open. Otherwise, exit the application and click No if prompted to save your changes.

Summary

In this chapter, you learned how to perform location queries using the Select By Location tool and how to combine location and attribute queries by applying the Select By Attributes tool to the results of a location query. These are common GIS tasks that will facilitate many different projects and workflows. The results of a selection aren't permanent, so you can always clear your selection and rebuild the query until you get the correct results. You can save a selected set of features to a new layer in the map document.

Terms

location query

attribute query

17b

18

19

20

Chapter 18
Preparing data for analysis

Dissolving features
Clipping layers
Exporting data
Running tools in a model

Learning objectives

- *Understand the basic steps of GIS analysis*
- *Perform common geoprocessing functions*
- *Select and export features*
- *Use ModelBuilder to create a geoprocessing model*

GIS ANALYSIS ALLOWS YOU TO INVESTIGATE AND UNDERSTAND SPATIAL relationships between and among real-world features. Beginning with multiple datasets, you can perform various types of geoprocessing and mathematical operations to get new information. GIS analysis is widely used in many industries, from conservation to emergency response, to answer questions and make educated decisions. To use GIS properly, it is important to understand the problem at hand, and then follow a defined process to obtain the solution. This chapter and chapter 19 will introduce some basic geoprocessing tools and forms of geographic analysis.

When you perform an operation on spatial data that creates a new dataset (often a slightly modified version of the original), you are *geoprocessing* data. Geoprocessing tasks are accomplished with geoprocessing tools. In ArcGIS, these tools are stored in ArcToolbox, which is accessible from ArcMap and ArcCatalog.

17

18

19

20

Typical analysis workflow includes the following steps:

1. *Frame the problem as a question.* Begin a GIS analysis by determining what type of information you need. This is most often accomplished by forming a relevant question: How many schools are within 500 feet of a busy intersection? How much land is covered by forest within a particular boundary? Which neighborhoods have the highest incidence of auto theft? Be as specific as possible about the question you want to answer. This will help you decide how to approach the analysis, which geoprocessing tools to use, which data you need, and how to present the results.

2. *Choose the appropriate data.* Choose good, quality data to use in your analysis. Data can come from any number of sources—databases within your organization, commercial data providers, government organizations, the Internet, ArcGIS Online, and more. The data you choose and where you get it depends on your needs and your budget. Most critical is that the data be of good quality and accurate, because the output produced is only as good as the input data you use for geoprocessing.

3. *Choose an appropriate analysis method.* Decide on an analysis method based on your proposed questions and how the results of the analysis will be used. Depending on your situation, you may have to compare different analysis results. For example, if you are doing a quick study of schools close to busy intersections in a city, you may map the individual schools and road network to look for patterns. If the information will be used to determine the location of a new school, you might want a more precise measure of the locations of schools, demographic data about families in the neighborhood, and proximity to intersections.

4. *Organize and process the data.* Once you select an analysis method, you'll need to process your data in a way that makes sense for the problem you're trying to answer. When you are mapping feature locations, you may need to assign geographic coordinates, such as an address or latitude and longitude, to your data and assign category values to the data. When you are mapping quantities, such as the number of tree types in a forest or park, you may need to choose a classification scheme and decide on how many classes to use to represent your data. When you are trying to find out what is inside an area of interest, you may need to measure an area or combine different layers of information.

   Datasets are not always in exactly the condition you need for a project. For example, features may be either too detailed or too generalized for the map scale you are working with. Another problem may be not having all the attributes you want. And sometimes the problem is even one of too much data—you may have thousands of features that clutter your area of interest or extend well beyond it. You'll also need to figure out which ArcMap geoprocessing tools to use. ArcToolbox contains hundreds of tools to help you modify datasets. Simplify data by dissolving a group of features with a common attribute value into a single feature. Trim a dataset to your area of interest by using features in one layer to clip features in another. Almost all ArcToolbox tools can be used in ModelBuilder to create simple or complex models. ModelBuilder can be used not only to automate multiple processes, but also to create a nice workflow diagram to visualize the analysis.

5. *Review, display and share the results.* The final step in GIS analysis is to look at the results of your analysis and take action based on those results. Results can be displayed as a digital map, printed as a paper map, combined with tables or charts, and shared with others. Although a lot of emphasis in GIS is on making maps, the software is flexible enough to allow you to display your results in the format that best suits your needs.

Data

The exercises in this chapter use the following datasets, stored in a geodatabase called OregonForest.gdb:

- cirlook—point feature class of fire lookout towers in the state of Oregon *(Source: Oregon Department of Forestry)*
- fwz100k—polygon feature class of Fire Weather Zones in the state of Oregon *(Source: Oregon Department of Forestry)*
- forestland—polygon feature class of forested land in the state of Oregon *(Source: Oregon Department of Forestry)*
- populated_jurisdictions—polygon feature class of inhabited areas in the state of Oregon *(Source: Oregon Department of Forestry)*

17

18

19

20

Exercise 18a
Dissolving features

In ArcMap, a *dissolve* creates a new layer in which all features in an input layer that have the same value for a specified attribute are dissolved into a single feature. In the example shown here in the maps, Wisconsin census tracts are dissolved into counties to form a new layer of counties made up of census tracts.

Wisconsin census tracts.

Census tracts dissolved into counties. Data from Esri Data & Maps 2010, courtesy of Tele Atlas, US Census, Esri (Pop2010 field), and ArcUSA.

17

18a

19

20

Some 1,323 census tracts are dissolved into counties to form a new layer of 72 counties. The new layer's attribute table has the standard geometry (Shape) and feature identifier (OBJECTID) attributes, along with the attribute used in the dissolve (CNTY_FIPS). You can include other attributes as well. For each census tract, the input table contains attributes that store population counts and other demographic data. In the output table, these values can be summed for each tract.

Input census tracts table

Output counties table

The attributes of the input census tracts table, *left*, are summed in the output counties table, *right*. Data from Esri Data & Maps 2010, courtesy of Tele Atlas, US Census, Esri (Pop2010 field), and ArcUSA.

For this exercise, you work for a forestry consulting company that plans to conduct research on forest wildfires and their proximity to urban areas in Oregon. As the population grows, towns and cities expand outward—residential homes, infrastructure, and buildings are built in or near forests. This has caused an increased risk of human-caused fires and a greater risk of local and widespread forest fires, damage to property, and injuries, or even death.

Fire weather zones (FWZs) are used to forecast fire weather in the state of Oregon. There are 32 FWZs in the state. Each FWZ belongs to one of four county warning areas (CWAs), each of which has a weather forecast office that has the responsibility of forecasting fires and providing warnings to the communities it covers. Each FWZ carries its own fire risk based on the number of communities and the different types of forest and terrain.

17

18a

19

20

1
2
3
4
5
6
7
8
9
10
11
12
13
14
15
16
17
18a
19
20

The forests of Oregon, shown in green, cover approximately 30,472,000 acres (about 123,315 square kilometers). Populated communities and their fire risk assessments (low, medium, or high) are shown in yellow, orange, and red. Data courtesy of Oregon Department of Forestry.

As your consulting company's GIS analyst, your job in this chapter and chapter 19 is to calculate the risk area of a number of communities in low-, medium-, and high-risk situations in one CWA. You will also use active fire lookouts used by fire agencies in Oregon to create visibility buffers. Your analysis will help your company provide a proper risk assessment to the Oregon state government and supporting agencies.

You have a polygon layer of FWZs provided by the forest service. Its attributes include the names of each CWA and a number that uniquely identifies each zone. In this exercise, you'll dissolve the FWZs into the four CWAs.

1 **Start ArcMap and open the ex18a map document from your C:\EsriPress\GTKArcGIS\Chapter18 folder.** The map (see facing page) shows the Forest polygon layer and the Fire Weather Zones polygon layer symbolized by a dark-red outline and hollow fill—the hollow fill allows the zones to be overlaid on top of the forest layer without obscuring it.

2 **Open the attribute table of the Fire Weather Zones layer.** The Fire Weather Zones layer has 32 zones, each represented by one record in the table. The FIREWXZONE attribute (where WX stands for "weather") contains a unique identifier for each FWZ (ranging from 601 to 638).

	OBJECTID *	Shape *	AREA	PERIMETER	FWZ100K_	FWZ100K_ID	CTYWARNARE	FIREWXZONE	
▶	1	Polygon	37760025700T.2	1000571.00051	2	394	PQR	601	
	2	Polygon	56140588582.3	1518881.22349	3	151	PQR	602	
	3	Polygon	73046316781.7	1310096.12297	4	125	PDT	635	
	4	Polygon	08686461697.7	1974894 30688	5	127	PDT	633	
	5	Polygon	171986611376	2325791.04396	6	132	PDT	631	
	6	Polygon	29145395679.2	964356.864051	7	132	PQR	609	
	7	Polygon	88381221756.8	3343966.54675	8	149	PQR	604	
	8	Polygon	36299443649.2	1285236.13579	9	153	PQR	607	
	9	Polygon	28535980707.1	1287446.39496	10	393	PQR	605	
	10	Polygon	15614262027.8	767286.013133	11	126	PDT	634	

1 ▶ ▶I ▦ ▭ (0 out of 32 Selected)

Fire Weather Zones

The CNTYWARNARE attribute contains an abbreviation indicating the four CWAs: Boise (BOI), Medford (MFR), Pendlton (PDT), and Portland (PQR).

Each FWZ belongs to one of these four CWAs. (Each CWA covers the extent of several counties.) The AREA attribute shows how large each zone is (in square feet). You want to dissolve the boundaries between FWZs so you end up with four polygons representing the four CWAs—each one will comprise the representative FWZs.

On your own

Click several records in the table. Determine which CWA the records belong to, and notice their locations on the map. When you are finished, clear your selection.

3 **Close the table. On the menu bar, click Geoprocessing > Search For Tools.** The Search window appears. It may be tabbed to the side of the main ArcMap window, or docked or floating. This is the same Search window you used in chapter 4; however, because you started it from the Geoprocessing menu, it is automatically set to search for tools rather than maps or data.

The Search window displays many system toolboxes. Each toolbox has tools for related geoprocessing tasks. You can access tools directly from the ArcToolbox window, but if you don't know exactly where a tool is, you should use the Search window instead. In this case, you know you want to use the Dissolve tool, but you don't know which toolbox it's in. So next, you'll use the quick way to search for it with the Search window.

④ In the Search window, type dissolve and press ENTER (or click the Search button, which looks like a magnifying glass).

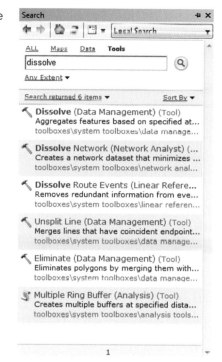

⑤ In the Search results, click the first return for Dissolve (Data Management). Be sure to click the first link for the tool to open the tool dialog box. Then close or hide the Search window.

17

18a

19

20

Although each tool has its own functionality and settings, they all use a similar dialog box with some common elements. For example, each tool has a Show Help button in the lower-right corner that describes the tool and its settings.

Tools that have lots of settings, such as the Dissolve tool, have a scroll bar on the right. As an alternative to scrolling, you can resize the dialog box by dragging on a side or a corner. Once you resize a dialog box, ArcMap remembers the new size and applies it to the next tool dialog box you open. (So if the dimensions of your Dissolve dialog box are different from the graphic, you may have previously resized this or some other tool. That's fine—the dialog box can be whatever size you like.)

The Dissolve tool requires that you select an input layer of features to dissolve, a location where the newly created layer will be saved on disk, and an attribute on which to base the dissolve of the input features. You'll set these selections next.

6 **In the Input Features list, click Fire Weather Zones.** The list of attributes that dissolves can be based on appears in the Dissolve_Field(s) box.

You can save the output as a shapefile. However, you'll save it as a geodatabase feature class, since the rest of the Oregon forest data is in this format. To leave the original OregonForest geodatabase intact, you'll save the output to a duplicate geodatabase called MyOregonForest that has been created for you.

7 **Click the Browse button next to the Output Feature Class box. On the Output Feature Class dialog box, browse to your \GTKArcGIS\ Chapter18\MyData folder. Double-click MyOregonForest.gdb.**

8 **In the Name box, type** CountyWarningAreas, **and then click Save.** The output feature class information is updated.

9 **In the Dissolve_Field(s) box, select the CTYWARNARE check box.** Each set of polygons with the same CWA name (meaning they are in the same CWA) will be dissolved into a single feature. The four possible areas created from the dissolve are BOI for Boise, MFR for Medford, PDT for Pendlton, and PQR for Portland.

ArcMap can also summarize the attribute values of dissolved features by a variety of statistics. To find out the area of each CWA, you will next summarize the AREA attribute by the Sum statistic type. This will total the area (in square feet) of all FWZs dissolved into a CWA. (You can also get the mean, the range, the standard deviation, and other measures of numeric attributes.)

10 On the Dissolve dialog box, scroll down to the Statistics Field(s) area. Click the Statistics Field(s) arrow and click AREA to add it to the list.

The field name is displayed in the Field column. The red circle with a white X indicates that you still need to specify the type of statistic you want, which you'll do next.

17

18a

19

20

11 **Click the first cell in the Statistic Type column. Then click the arrow and click SUM.**

12 **Click OK to run the tool.** ArcToolbox tools are executed in the background, which means you can continue working in ArcMap as a tool runs. You'll see a progress bar at the bottom of the main window that displays the name of the currently running tool. When the tool process is completed, a pop-up notification appears in the notification area, at the far right of the taskbar. (It will disappear in a few seconds.)

Before the pop-up notification disappears, you can click the tool name (a link) to open the Results window. This window displays all the details of running an ArcToolbox tool, including what layers were involved, which fields were calculated, and the start, finish, and elapsed time. You can also access this information from the menu bar (Geoprocessing > Results).

17

18a

19

20

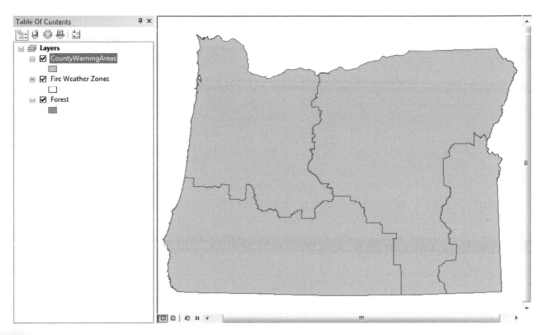

The new CountyWarningAreas layer is added to the map.

> If you don't see the new layer in the table of contents, on the Standard toolbar, click the Add Data button, browse to your MyOregonForest geodatabase, and add the layer. Then on the menu bar, click Geoprocessing > Geoprocessing Options. On the Geoprocessing Options dialog box, in the lower portion, select the "Add results of geoprocessing operations to the display" check box. Click OK. Any new geoprocessing layers will automatically be added to ArcMap.

When a new geoprocessing layer is created and added to ArcMap, it is displayed in a random color. (The color of the layers you create in the exercises may not match the graphics in the book.)

13 Open the attribute table of the CountyWarningAreas layer. If necessary, resize the table to show all fields. The original 32 FWZs have been aggregated into four CWA polygons.

Table — CountyWarningAreas

OBJECTID *	Shape *	CTYWARNARE	SUM_AREA	Shape_Length	Shape_Area
1	Polygon	BOI	279878427941	3102491.987324	279878427921.677
2	Polygon	MFR	736221086179.01	4944792.093709	736221086177.203
3	Polygon	PDT	1045625881886.1	6154950.839309	1045625881891.75
4	Polygon	PQR	641846600985.2	5536239.311169	641846600966.71

1 ▸ ▸| (0 out of 4 Selected)

CountyWarningAreas

17

18a

19

20

CNTWARNARE is the attribute you based the dissolve on. SUM_AREA is the statistical field you requested. It contains the sum of all FWZ values for area (in square feet).

Shape_Length and Shape_Area are measurement attributes that are automatically maintained by ArcMap for geodatabase feature classes. (Shape_Length measures feature perimeters for a polygon feature class.) Depending on the accuracy of the input attribute, there can be a slight discrepancy between the total calculated area and the area statistics.

14 **Close the table.**

15 **Right-click the CountyWarningAreas layer and click Label Features.**
Each CWA is labeled with the appropriate label.

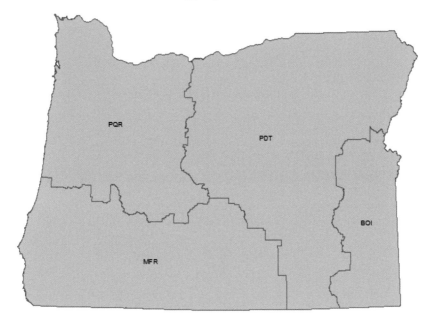

17

18a

19

20

On your own

In the table of contents, drag the Fire Weather Zones layer to the top. Then symbolize CountyWarningAreas based on the CTYWARNARE field. Turn the Fire Weather Zones layer off and on. Doing so will help you visualize how the FWZs were dissolved into CWAs.

16 If you like, rename your map document and save to your MyData folder for chapter 18.

17 If you are continuing to the next exercise, leave ArcMap open. Otherwise, exit the application and click No if prompted to save your changes.

17

18a

19

20

Exercise 18b
Clipping layers

A *clip* operation is a type of data extraction function. It trims features in one layer using the boundaries of polygon features from another layer, called the clip layer. It's like having a pair of scissors, or a cookie cutter, to trim data you don't need for your project, leaving you with a clip of what you want. For example, you might have a feature class of all the streets in your county, yet your study area encompasses just one ZIP Code within that county. If you have a polygon feature that represents that ZIP Code, you can use it to clip just the streets you need. Clipping can be a matter of convenience—because smaller datasets are easier to process—but it can also be important for analysis. To find out the total length of roadway within a ZIP Code, for instance, you must exclude streets and street segments that lie outside it.

The clip layer must be a polygon, although the dataset being clipped can be points, lines, or polygons. In the following example, a layer of interstate highways is clipped at the boundaries of a river basin in the state of Florida.

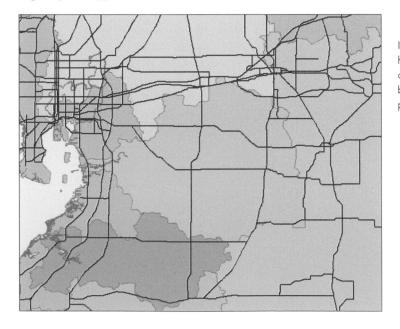

Input layer of US interstate highways and a clip layer of the Alafia River Basin boundary in Florida (center polygon).

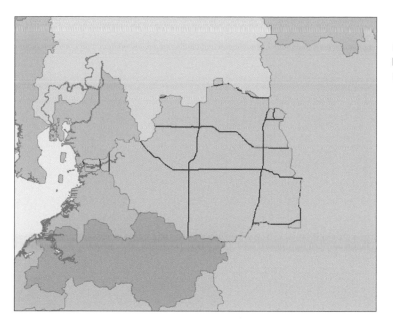

Highways clipped at polygon boundary of the Alafia River Basin boundary.

For this exercise, your forestry consulting company has just determined that, of the four CWAs, Portland (PQR) is the area of interest for this study. Your company will direct its attention to analyzing this area in depth.

Not every square foot of the CWAs is covered by forest or has a degree of fire danger. However, certain communities are built close to forestland, and these are the areas of concern. To aid in early smoke and fire detection, the state of Oregon strategically placed fire lookout towers throughout the forests. You have layers of forests, communities, and lookout towers that cover all four CWAs, but now you'd like to work with datasets that cover the area of Portland only.

In this exercise, you'll clip several layers to the Portland CWA, or PQR, boundary.

17
18b
19
20

1 **In ArcMap, open the ex18b map document from your Chapter18 folder.** The map displays the four CWAs. Below it is the Forest layer, which contains 3,771 forest polygons, and the Fire Weather Zones layer, which is turned off.

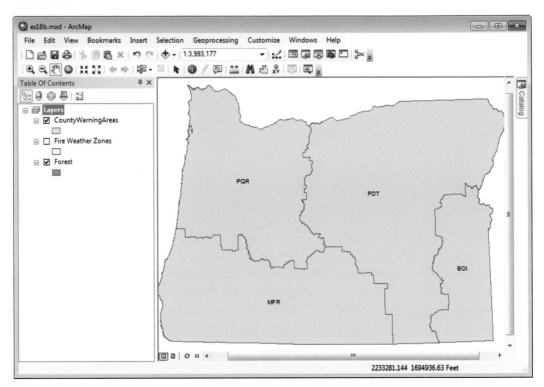

To clip the forest to the PQR boundary, you must first select the area boundary, which you'll do next. You'll then turn the selection into a new layer so you can look at the Portland area apart from the others.

2 **In the table of contents, click the List By Selection button. Click the toggle key next to Fire Weather Zones and Forest to make them not selectable.**

3 **On the Tools toolbar, click the Select Features tool** **.**

4 On the map, click PQR to select it. It is outlined in cyan.

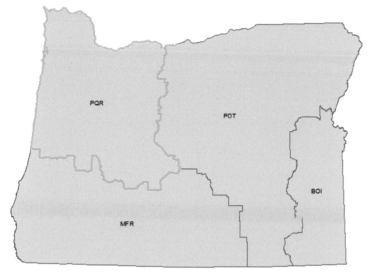

5 In the table of contents, right-click the CountyWarningAreas layer and click Create Layer From Selected Features. A layer called CountyWarningAreas selection is added to the table of contents. This layer references the same geodatabase feature class as the CountyWarningAreas layer. Remember, creating a selection layer doesn't create a new dataset. You can, however, export a selection layer, or a set of selected features, as a new dataset if you wish. You will do this in exercise 18c.

6 Click any white area of the map to clear the feature selection.

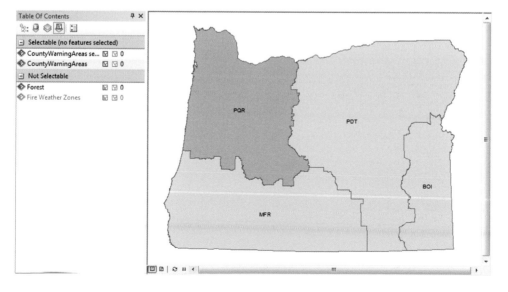

17

18b

19

20

7 **In the table of contents, click the List By Drawing Order button. Turn off the CountyWarningAreas layer and drag the CountyWarningAreas selection layer to the bottom of the table of contents.** The forest cover for the state is now visible again, and it covers the CountyWarningAreas selection layer (shown here in a light purple color, although it may be a different color on your map).

Commonly used geoprocessing tools are listed on the Geoprocessing menu (such as the Dissolve tool you searched for previously.) Next, you'll use the Clip tool—accessed from the Geoprocessing menu rather than ArcToolbox—to clip the forest to the boundary of the Portland CWA.

8 **On the menu bar, click Geoprocessing > Clip.** The Clip tool dialog box appears.

17

18b

19

20

Using this dialog box, you'll select the layer with the input features to be clipped, the layer to clip with, and an output layer. You will store the output in your MyOregonForest geodatabase.

9 In the Input Features list, click Forest. In the Clip Features list, click CountyWarningAreas selection.

10 Click the Browse button next to the Output Feature Class list. On the Output Feature Class dialog box, browse to the contents of MyOregonForest. gdb in your MyData folder for chapter 18. In the Name box, type PortlandForest, and then click Save. The output feature class information is updated on the Clip dialog box.

11 Click OK to run the Clip operation. When the Clip tool processing is completed, a pop-up notification will appear in the notification area. Click the Clip link in the pop-up notification to view the Clip tool's results report.

12 Review the information in the Results window, and then close it.

13 **In the table of contents, turn off the Forest layer.** The PortlandForest layer (assigned a random color), which consists of forest polygon features clipped to the boundary of the Portland CWA, is added to the map.

14 **Open the attribute table of the PortlandForest layer.** The clip operation has reduced the number of Forest layer features from 3,771 to 1,514. The table has attributes that include forest codes for some of the features. You will be using this layer again in exercises 18c and 18d.

OBJECTID *	Shape *	CODE	Shape_Length	Shape_Area
1	Polygon	501M	509411.103168	2468599970.94606
2	Polygon	501H	304781.195281	549909126.80713
3	Polygon	510M	180386.352783	673950646.986624
4	Polygon	501L	44960.027439	94429687.786234
5	Polygon	501H	230553.040124	377859218.570934
6	Polygon	510H	1589180.959444	5713951169.67372
7	Polygon	512L	993982.270403	4069452768.4625
8	Polygon	501M	166980.029765	825944965.355526
9	Polygon		22338.154755	25580647.577657
10	Polygon	501L	52732.424023	94172637.45044

(0 out of 1514 Selected)

PortlandForest

15 **Close the table.**

On your own

In the table of contents, turn on the Fire Weather Zones layer and drag it to the top. You can see how you have gone from a complete Fire Weather Zones layer to a dissolve to create CWAs and now to a clip of the Portland CWA.

16 **If you want, rename your map document and save to your MyData folder for chapter 18.**

17 **If you are continuing to the next exercise, leave ArcMap open. Otherwise, exit the application and click No if prompted to save your changes.**

Exercise 18c
Exporting data

Operations such as dissolving and clipping automatically create new datasets. Recall that another way to make a new dataset from an existing one is to make a selection on a layer and export the selected features, as you learned in previous chapters. This is a very common task when creating maps and preparing your data for analysis.

In this exercise, in addition to the Forest layer, you have a layer of fire lookouts covering all four CWAs. You could clip this layer just as you did the forest, but this time you'll use another method to change the dataset's boundaries. Because lookout towers are points contained within polygons, you can use Select By Location to select the points inside the Portland CWA. Then you will export the selected set of points to create a new dataset that includes Portland CWA lookout towers only. You will also learn how to import symbology from another layer.

1 In ArcMap, open the ex18c map document from your Chapter18 folder. The map displays a layer of lookout tower locations and the Portland CWA selected from exercise 18b (CountyWarningAreas selection). The Lookouts layer is symbolized by blue and white lookout tower symbols—the colors in the book graphic may be different from your map if you experimented with changing colors in previous exercises.

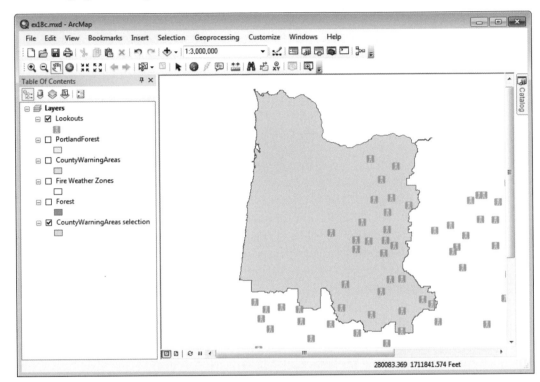

2 On the menu bar, click Selection > Select By Location. If necessary, click the Selection method arrow and click "select features from."

3 For Target layer(s), select the Lookouts check box.

4 For Source layer, click CountyWarningAreas selection.

5 For the Spatial selection method, click "are completely within the source layer feature," and then click Apply.

How many lookout towers in the Portland CWA are selected?
(Hint: The count is displayed on the far left of the status bar after the Select operation is completed.)

Select By Location

Select features from one or more target layers based on their location in relation to the features in the source layer.

Selection method:

select features from

Target layer(s):

☑ Lookouts
☐ PortlandForest
☐ CountyWarningAreas
☐ Fire Weather Zones
☐ Forest
☐ CountyWarningAreas selection

☐ Only show selectable layers in this list

Source layer:

CountyWarningAreas selection

☐ Use selected features   (0 features selected)

Spatial selection method for target layer feature(s):

are completely within the source layer feature

☐ Apply a search distance

100000.000000   Feet

About select by location   OK   Apply   Close

6 Close the Select By Location dialog box. In the table of contents, right-click the Lookouts layer and click Data > Export Data. The Export list is correctly set to Selected features. The option to use the same coordinate system as the layer's source data is also correct.

7 Click the Browse button next to the Output feature class box. On the Saving Data dialog box, make sure the Save as type list is set to File and Personal Geodatabase feature classes and browse to the contents of MyOregonForest.gdb in

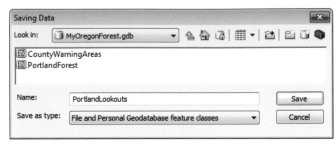

your MyData folder for chapter 18. In the Name box, replace Export_Output with PortlandLookouts, and then click Save.

8 On the Export Data dialog box, click OK.

ArcMap exports the selected lookout towers to a new feature class and prompts you to add the data to the map.

9 **Click Yes.** You don't see the point features of the new PortlandLookouts layer because they are covered by the highlighted selection on the Lookouts layer.

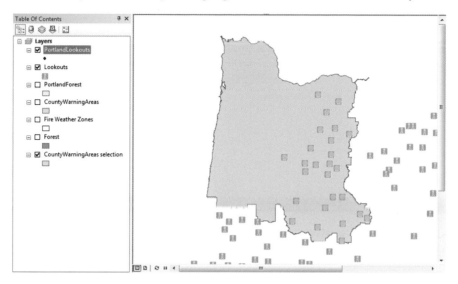

10 **In the table of contents, turn off the Lookouts layer.** The new layer contains lookout towers within the Portland CWA only. The default symbol, however, is not as informative as the one in the Lookouts layer.

17

18c

19

20

11 In the table of contents, double-click the PortlandLookouts layer. On the Layer Properties dialog box, click the Symbology tab, and in the upper-right corner of the "Draw all features using the same symbol" section, click Import.

The Import Symbology dialog box appears.

You want to import symbology from another layer in the map, so the first option is set correctly. The Layer list is also set correctly to Lookouts (to the only other point layer on the map).

12 Click OK. Notice that the symbol is updated in the layer properties.

13 On the Layer Properties dialog box, click OK to apply the change.

On your own

Turn on the PortlandForest layer. Use the Import method to change the symbology to match the original Forest layer.

14 If you want, rename your map document and save to your MyData folder for chapter 18.

15 If you are continuing to the next exercise, leave ArcMap open. Otherwise, exit the application and click No if prompted to save your changes.

You're almost done preparing the data for your analysis. You still have to prepare one more layer in exercise 18d. You'll do some more clipping to create a layer that shows the populated jurisdictions (or communities) that are in the Portland CWA. However, this time you'll learn how to run tools in ModelBuilder.

17

18c

19

20

Exercise 18d
Running tools in a model

ModelBuilder is a graphic user interface for diagramming solutions to spatial analysis problems. The diagrams you make are called *models*, and they have many advantages. The ModelBuilder interface allows you to organize the inputs, tools, and outputs used to perform a GIS analysis. ModelBuilder is not a free-standing software application like ArcMap or ArcCatalog, but it's more than just a geoprocessing tool—it's a design environment for creating workflow diagrams. The ModelBuilder window has its own menu and toolbar. Many of the controls have the same icons and functions as ArcMap controls.

Individual geoprocessing tools in ArcToolbox, such as Dissolve, Clip, Buffer, and Union, are useful for completing particular tasks, but you can also string them together to create a complex tool that performs an entire workflow with just one click. ArcGIS will carry out the individual procedures in order, according to your plan, or model. A complete model is an interconnected set of processes, with each process consisting of input data, a tool, and output data, which can be fed into another process.

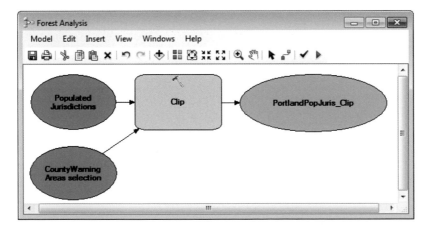

A model using the Clip tool in ModelBuilder.

The elements of a process are distinguishable by their symbology. By default, input data is represented by a blue oval. Tools are represented by yellow rectangles. Output data is represented by a green oval. The representative shapes, colors, and other properties can be changed, if you wish.

> Right-click an element in your model to access the Display Properties dialog box. This allows you to change properties such as background and border color, change the shape and font of selected elements, and change the arrow style and size of selected connectors. You can even change the look of ToolTips that appear when the mouse is paused over an element.

Once you have diagrammed a model, you can run it with a single click. ModelBuilder runs each of the processes in turn, passing the output of one process to the next and from that process to the next after that, until a final dataset is produced.

Various combinations of input parameters can be set to output different kinds of results. This is the real power of modeling. For example, a model that calculates traffic noise in proximity to residential neighborhoods can have several main inputs, such as the decibel measure and the distance from noise origin to homes. The model can then be run quickly and repeatedly using various inputs and parameter settings to achieve different results. Without ModelBuilder, you would need to open each tool dialog box in ArcToolbox, set the tool parameters, and run each tool. Then you'd have to reopen the tool dialog boxes each time to set different parameters. Each iteration would take you just as long (or almost as long) as the first time. ModelBuilder can save huge amounts of time if you perform repetitive spatial analyses. It also lets you effectively evaluate the changing conditions and what-if scenarios that arise in decision-making scenarios.

Furthermore, models are a good way to evaluate the soundness of your methodology, to document it for others, and to develop and refine it over time. Sharing models with colleagues helps to standardize decision-making processes. If you worked for a government agency, for example, you could use models to exchange knowledge across the spectrum and establish uniform practices.

For more information about spatial modeling and ModelBuilder, in ArcGIS for Desktop Help, on the Contents tab, navigate to Geoprocessing > ModelBuilder > What is Model Builder?

Another great resource is a companion book to this series, *Getting to Know ArcGIS ModelBuilder* (Esri Press 2011). It is recommended for students with limited exposure to GIS, the intermediate GIS user doing independent study, and the seasoned GIS professional seeking more advanced skills. The book

17

18d

19

20

illustrates how components are combined to form complex models and shows how ModelBuilder is used in more of a programming role to perform advanced tasks using Python scripts.

Like tools, models are stored in a toolbox. In earlier chapters, you used some of the tools and toolboxes that come with ArcMap. In this exercise, you will learn how to create a new toolbox and put a model inside it. A new toolbox can be created within ArcToolbox, within an ordinary file folder, or within a geodatabase. Because models must be stored in toolboxes, these are also the three locations in which models can be created.

In this exercise, you will continue preparing your data for analysis and create a toolbox to store your model. Then you'll create a model and add one process to it—clipping populated jurisdictions to the Portland CWA. Populated jurisdictions are areas that include residential and commercial areas. The data preparation you did in exercises 18a, 18b, and 18c is assumed to be complete, although you could include those operations in your model as well. Typically, ModelBuilder is used for multiple tasks, but here, you will create a single-operation model as an introduction to the application.

1 **In ArcMap, open the ex18d map document from your Chapter18 folder.** The map contains layers for lookout towers, forest, and the Portland CWA. It should look familiar, as it has the same data and symbology as the map document in exercise 18c. A new layer called Populated Jurisdictions has been added to the map for you. It contains communities and populated areas for the entire state.

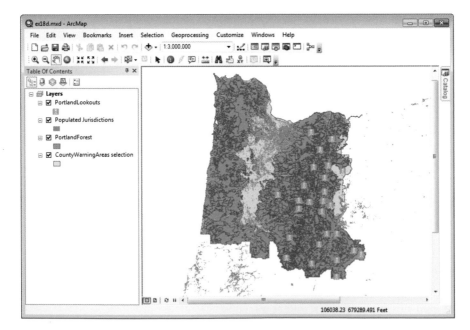

2 **Display the Catalog window. In the Catalog window, navigate to your MyData folder for chapter 10. Right-click MyData and click New > Toolbox.** A new toolbox is added with its default name highlighted.

3 **In the Catalog window, change the new toolbox name to** Oregon Forest Analysis **and press ENTER. In the Catalog tree, expand your MyData folder if necessary.** You have created a toolbox in an ordinary file folder. If you have activated file extensions, such as in Windows Explorer, it appears with a .tbx extension.

Depending on your ArcGIS setup, you may have file extensions hidden. To change how file extensions are viewed in the Catalog window, open ArcCatalog, and on the menu bar, click Customize > ArcCatalog Options and clear the Hide file extensions check box.

Next, you will create a model within the toolbox.

4 **In the Catalog window, right-click the Oregon Forest Analysis toolbox and click New > Model.**

17

18d

19

20

This opens the Model window. Next, you'll change the name in the title bar to something more meaningful.

5 **On the ModelBuilder menu bar, click Model > Model Properties. On the General tab, in the Name box, replace Model with** ForestAnalysis **(the model name cannot contain any spaces). In the Label box, replace Model with** Forest Analysis **(the model label can contain spaces), and then click OK.** The title bar is updated with the new name. Next, you will save and close the model so you can work with it in ArcMap.

6 **On the ModelBuilder toolbar, click the Save button** 💾 **.**

7 **Close the ModelBuilder window.**

8 **In the Catalog window, if necessary, click the plus sign (+) next to the Oregon Forest Analysis toolbox.**

You may access your model from the Catalog window, or you can add it to ArcToolbox. Next, you will add it to ArcToolbox, but it won't be a permanent change—it will affect the appearance of ArcToolbox in the current session only.

9 **On the Standard toolbar, click the ArcToolbox button. (Depending on how you previously set up your ArcToolbox window, it may be floating, docked, or tabbed to the side of the main window.)**

You can still drag items between the Catalog and ArcToolbox windows, even when they are tabbed. With your mouse pointer, pause over the tab, and the window will expand into view.

10 In the Catalog window, click the Oregon Forest Analysis toolbox and drag it to the bottom of the ArcToolbox window. (Alternatively, you can right-click the white background in ArcToolbox and click Add Toolbox, navigate to the Oregon Forest Analysis toolbox, and click Open.) The Oregon Forest Analysis toolbox is automatically added and alphabetized with the other toolboxes.

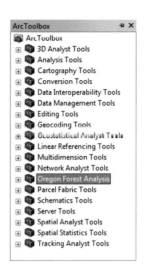

To save a customization to ArcToolbox, right-click the background and click Save Settings. You can save to a file or change the default ArcToolbox settings.

11 Close or hide the Catalog window, since it will no longer be needed.

12 In the ArcToolbox window, click the plus sign (+) next to Oregon Forest Analysis to expand its contents. Right-click the Forest Analysis model and click Edit. When the ModelBuilder window appears, move it so it doesn't overlap either the ArcToolbox window or the ArcMap table of contents. (It may be easier to work with both windows if your ArcToolbox window is floating.)

Next, you will create a process to clip populated jurisdictions to the Portland CWA. You will drag the Clip tool from ArcToolbox and then the Populated Jurisdictions layer from the ArcMap table of contents to the model window.

13 In the ArcToolbox window, expand Analysis Tools and then the Extract toolset.

17

18d

19

20

14 Click the Clip tool and drag it to the center of the model window.

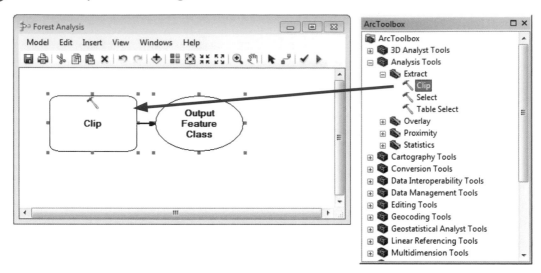

The Clip tool, represented by a rectangle, is added to the model. An arrow connects the tool to its output data, represented by an oval. Blue selection handles indicate that the elements are currently selected. The fact that they are not colored in means they are incomplete and not ready to run—they are waiting for input data to be attached, which you'll provide next.

15 In the ArcMap table of contents, press and drag the Populated Jurisdictions layer to the model window and drop it directly on top of the Clip tool rectangle. A shortcut menu appears—click Input Features.

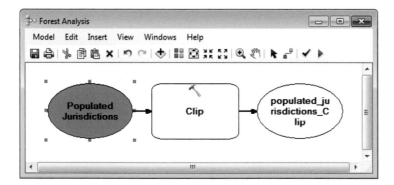

The Populated Jurisdictions layer, represented by a blue oval, is automatically connected to the tool it was dropped onto. Since clipping requires two layers, you will set the Portland CWA as a model parameter as well.

17

18d

19

20

16 In the ArcMap table of contents, press and drag the CountyWarningAreas selection layer to the model window and drop it directly on top of the Clip tool rectangle. A shortcut menu appears—click **Clip Features**. The second input is placed directly on top of the first, so the first is no longer visible.

17 Put your mouse pointer on the blue oval and drag the CountyWarningAreas selection element to a position below the Populated Jurisdictions element so you can see both inputs.

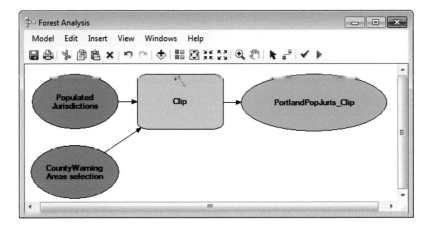

Next, you will have to change the output data settings. By default, the output feature class is set to your user Default.gdb. You will change the output location to the geodatabase in your MyData folder for chapter 18.

Each map document has a default geodatabase, which is the home location of the spatial content of your map. This location is used for adding datasets and for saving resultant datasets created by various editing and geoprocessing operations. To learn more, in ArcGIS for Desktop Help, on the Contents tab, navigate to Mapping > Working with ArcMap > Working with the Catalog window > Working with the Home folder and default geodatabase.

17

18d

19

20

18 Double-click the output element in the model. Click the Browse button next to the Output Feature Class box and browse to the contents of MyOregonForest.gdb in your MyData folder for chapter 18. In the Name box, type PortlandPopJurs, and then click Save. The output feature class information is updated on the output feature class dialog box.

19 Click OK to close the dialog box.

All the elements are colored in, which means the process is ready to run. When you run the process, a new feature class will be created and saved on disk. However, the output data is not automatically added as a layer to ArcMap. If you want the data added to the map, you will have to set a property on the output data element, which you'll do next.

20 In the model, right-click the PortlandPopJurs output oval and click Add To Display so it will be turned on in the table of contents.

21 On the ModelBuilder toolbar, click the Run button ▶ to run the model (or on the ModelBuilder menu bar, click Model > Run). As a model runs, each of its tools turns red (see facing page). Your model has only one tool, so if you weren't watching closely (or if the progress report covered the model), you might not have seen it. When the tool is finished running, it turns yellow again. Drop shadows behind the tool and output data elements indicate that the process has been run. Once a process has been run, you cannot run it again until you either change a parameter or click Run Entire Model or Validate Entire Model on the Model menu.

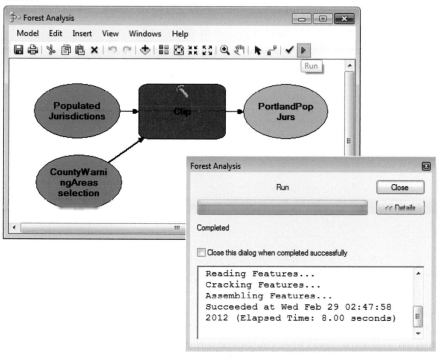

22 Close the progress report. Notice that after a process runs, drop shadows appear behind the model elements. The PortlandPopJurs layer is added to ArcMap. (You may need to move the model window out of the way to see it.) It is symbolized in a random color.

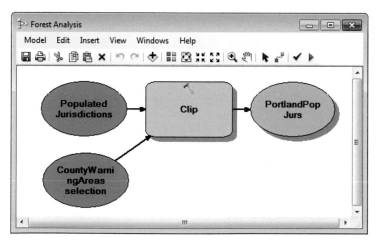

17

18d

19

20

23 If necessary, clear the check box next to the Populated Jurisdictions layer to turn it off. Open the PortlandPopJurs layer properties and change the symbology to match the orange color of the Populated Jurisdictions layer.

24 On the ModelBuilder toolbar, click the Save button to save the model.

25 Close the ModelBuilder and ArcToolbox windows. The model and all its settings are stored in the Oregon Forest Analysis.tbx file you created at the beginning of the exercise.

26 If you want, rename your map document and save to your MyData folder for chapter 18.

27 If you are continuing to the next chapter, leave ArcMap open. Otherwise, exit the application and click No if prompted to save your changes.

In chapter 19, you will be analyzing spatial data. This will involve comparing spatial relationships among features and trying to draw conclusions from analysis. You will learn how to use more geoprocessing tools and see how your analysis is coming together.

1

2

3

4

5

6

7

8

9

10

11

12

13

14

15

16

17

18d

19

20

Summary

In this chapter, you learned some basic steps on how to approach GIS analysis. Many geoprocessing tools are commonly accessed from the ArcMap Geoprocessing menu, from a search using the Search window, or by using ArcToolbox directly. You learned how to use the Dissolve tool to aggregate fire weather zones into county warning areas. You also used the Clip tool twice—once on its own and once in a model. Learning how to use ModelBuilder efficiently will facilitate and automate many projects and workflows, from the simple to the complex. You will build a larger model and connect more geoprocessing tools together in chapter 20.

Terms

geoprocessing

dissolve

clip

model

17

18

19

20

Chapter 19
Analyzing spatial data

Buffering features
Overlaying data
Calculating attribute values
Creating graphs

Learning objectives

- *Use geoprocessing functions such as buffering and overlay*
- *Use ModelBuilder to run multiple processes*
- *Calculate attribute values*
- *Create graphs*

ECONOMIC DEVELOPMENT ISSUES, HEALTH-CARE REFORM, ENVIRONMENTAL impacts, industrial growth, and a large number of other concerns demand sophisticated problem-solving strategies. Most of the problems you solve with GIS involve comparing spatial relationships between features—in one layer or in different layers—and drawing conclusions. Problem solving in GIS involves spatial analysis and using a wide range of geoprocessing and analysis tools to visualize, generalize, and theorize about geographic data. Techniques of spatial analysis can include everything from measuring the distance between points to modeling the behavior of ecosystems to using a multivariable model to compute flow characteristics.

Spatial analysis today is highly visual, and so are many of the geoprocessing tools in ArcToolbox. Through the aid of ArcMap and ModelBuilder, you can create visual analysis models by piecing together inputs, processes, and outputs. Models facilitate data preparation and spatial analysis. In this chapter, you'll continue your work with the Oregon forest fire analysis and create layers that are extremely useful in spatial analysis: buffers and overlays.

17

18

19

20

A *buffer* is an area drawn at a uniform distance around a feature—whether a point, a line, or a polygon. A buffer represents a critical zone, such as a floodplain, a protected species habitat, or a municipal services area, wherein features lying inside the buffer can have a different status from features that lie outside of it. You can specify buffer distances manually or use attribute values to determine varying sizes of buffers, and even create multiple buffer rings. Buffers created on a single layer can often overlap. For example, cellular phone towers need to be built within a certain proximity of each other so that their service radii can cover enough of an urban area to keep calls from being dropped. Overlap between individual buffers is acceptable, although it may be more beneficial to dissolve the overlaps to form one large feature.

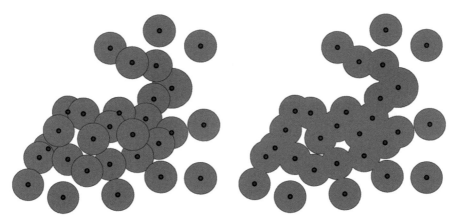

Buffers are created around point features, with some buffers overlapping.

Dissolved buffers create large polygons. Nonoverlapping buffers remain the same.

Map *overlays* (union and intersect) identify overlaps between features in two layers and create a dataset in which the lines of overlap define new features. In a *union* overlay, nonoverlapping areas are included in the output dataset. A union dataset, then, has three types of features: those found only in the first input layer (with layer 1 attribute values), those found only in the second input layer (with layer 2 attribute values), and those created by areas of overlap between the two layers (with both layer 1 and layer 2 attribute values). In an *intersect* overlay, only the overlapping geometry is preserved, and features have attributes from both input layers.

17

18

19

20

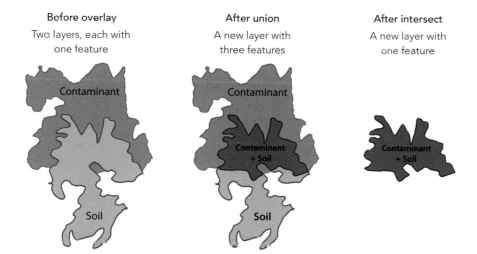

Before overlay	After union	After intersect
Two layers, each with one feature	A new layer with three features	A new layer with one feature

Left, an area of soil and a spilled contaminant zone overlap each other. Overlay analysis—whether union, *center*, or intersect, *right*—creates new features where input layers overlap (contaminated soil). In this case, both types of analysis identify areas where contaminated soil must be treated or cleaned up.

Layers overlaid on other layers can sometimes cause unexpected results, called slivers. Slivers are small polygons that are created because of uneven or unequal boundaries. In certain cases, boundaries may not match up perfectly, so attention must be paid to removing these extraneous features, if required. For example, a shared parcel lot boundary measured using two different methods (say, GPS versus field surveying) may not line up exactly. To combat inaccuracies, ArcGIS supports topology to enforce data integrity rules (such as, no gaps allowed between polygons). In geoprocessing, and GIS in general, higher quality data input results in higher quality data output.

Data

The following datasets, stored in a geodatabase called OregonForest.gdb, are used in this chapter:

- CountyWarningAreas—polygon feature class of county warning areas composed of fire weather zones in the state of Oregon, derived from exercise 18a *(Source: Oregon Department of Forestry)*
- PortlandForest—polygon feature class of forested land in the state of Oregon, derived from exercise 18b *(Source: Oregon Department of Forestry)*
- PortlandLookouts—point feature class of fire lookout towers in the state of Oregon, derived from exercise 18c *(Source: Oregon Department of Forestry)*
- PortlandPopJurs—polygon feature class of inhabited areas in the state of Oregon, derived from exercise 18d *(Source: Oregon Department of Forestry)*

Exercise 19a
Buffering features

Buffers are based on the concept of proximity. Buffers are created as polygons in a new layer. Buffers can be drawn at a constant distance around every feature in a layer (for example, 100 meters) or at a distance that varies according to attribute values. For example, buffers representing the size of earthquakes might vary according to an attribute that describes the strength of magnitude. Buffers can also be concentric rings representing multiple distances, such as the areas within 100, 500, and 1,000 meters of a well.

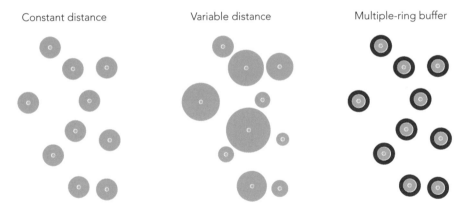

Constant distance   Variable distance   Multiple-ring buffer

Buffers can be determined by distance in different ways.

Buffers can also be created for lines and areas. If features are close together, their buffers may overlap. You can preserve the overlaps or remove them by dissolving them.

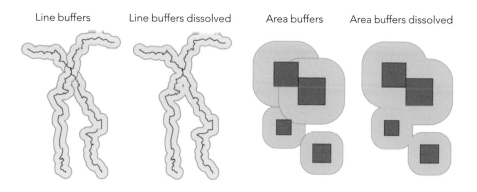

Line buffers   Line buffers dissolved   Area buffers   Area buffers dissolved

Buffers for lines and areas can overlap or be dissolved.

You'll continue your analysis of the fire safety of the Portland County Warning Area (PQR CWA). In chapter 18, you clipped forestland, populated jurisdictions, and lookout towers to the extent of the study area. Your goal is to determine which of the populated jurisdictions are in proximity to forestland and to lookout towers, because this will indicate which areas have been deemed to have a higher probability of risk due to a forest fire. In this exercise, you'll buffer 1,000 feet around populated jurisdictions because you want to find the areas that are close to forestland, not just adjacent to it. You'll also buffer the lookout towers to show the range of visibility of each tower. The visible-range value of each tower was precalculated and included as an attribute (RADIUS2). Lookout towers were built in areas with a high probability of fire and with the least resources to handle a sudden fire outbreak (that is, with no fire stations nearby).

1 **Start ArcMap. Open the ex19a map document from your C:\EsriPress\GTKArcGIS\Chapter19 folder.**

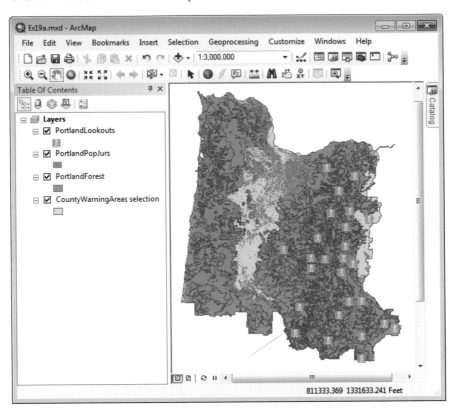

The map shows the lookout towers, populated jurisdictions, forestland, and the PQR CWA (CountyWarningAreas selection). You'll begin by buffering the populated jurisdictions using the Buffer geoprocessing tool.

2 **On the menu bar, click Geoprocessing > Buffer.** This menu (see facing page) provides access to commonly used geoprocessing tools. You can also access the tools through ArcToolbox (perform a search to locate them quickly.) The Buffer tool dialog box appears.

17

18

19a

20

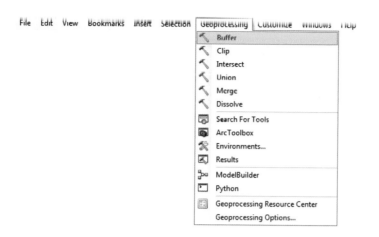

Next, you'll select the layer with the input features to be buffered, designate an output feature class, enter a buffer distance, and select a dissolve option.

3 **On the Buffer tool dialog box, click the Input Features arrow and click PortlandPopJurs. (Alternatively, drag the PortlandPopJurs layer from the ArcMap table of contents to this location.)**

> Only layers from an active data frame can be dragged to an ArcToolbox tool. If you drag a layer from an inactive data frame to a geoprocessing tool, warnings or errors are produced inside the tool, and you will not be able to proceed.

4 **Click the Browse button next to the Output Feature Class box. On the Output Feature Class dialog box, navigate to \GTKArcGIS\ Chapter19\MyData and double-click MyOregonForest.gdb. In the Name box, type** PopJursBuffer, **and then click Save.** The output feature class information is updated with the new location and name.

17

18

19a

20

5 For the buffer distance, make sure the Linear unit option is selected. In the Linear unit box, type 1000. Maintain Feet as the measurement unit. Scroll down if necessary, and make sure Dissolve Type is set to NONE.

6 Click OK. (This process may take a few minutes because of the large number of communities.) The new PopJursBuffer layer is added to the map. It looks similar to the original populated jurisdictions layer, except that it has a 1,000 ft buffer surrounding the features. (If you want, zoom in to view the buffer. Just be sure to return to your previous extent to continue the exercise.)

7 Change the color of both the buffer fill and outline to Anemone Violet (see facing page).

17

18

19a

20

8 **Open the attributes of the PopJursBuffer layer.** In addition to its standard attributes, the output table has attributes from the input table. (Because you did not dissolve the buffers, the output and input tables have the same number of records—there is one buffer for each populated jurisdiction. This correspondence of record to record makes it possible to copy attributes from one table to the other.) An attribute called BUFF_DIST is added to the table, which shows the buffer distance you specified. (You may have to scroll across the table to see it.)

Table   □ ×

PopJursBuffer   ×

COMMUNITY	COMID	RISK	HAZARD	PROTECTION	VALUE	OVERALL	FED_TRIBAL	BUFF_DIST	
▶ ADAIR RFPD (RFPD)	1	L	M	L	H	M	Y	1000.002	
ALBANY FIRE DEPT (RFPD)	4	L	L	L	L	L	Y	1000.002	
ALSEA RFPD (RFPD)	5	M	M	M	H	M	Y	1000.002	
AMITY FIRE DIST (RFPD)	6	L	L	L	L	L	Y	1000.002	
AUMSVILLE RFPD (RFPD)	8	L	L	L	H	L	Y	1000.002	
AURORA RFPD (RFPD)	9	L	L	L	H	L	Y	1000.002	
Adair Village (City)	11	L	L	L	L	L	Y	1000.002	
Albany (City)	14	L	L	L	L	L	Y	1000.002	
Amity (City)	15	L	L	L	L	L	N	1000.002	

I◀ ◀   1   ▶ ▶I   (0 out of 305 Selected)

PopJursBuffer

9 **Close the PopJursBuffer attribute table.**

Next, you'll buffer the lookout towers layer. The buffer sizes for this layer will vary according to the visibility range of each tower.

10 Open the attribute table of the PortlandLookouts layer.

The attribute table contains an attribute called RADIUS2, which is the visibility range value (in feet) of each tower. Next, you'll specify this attribute as the buffer distance value.

11 Close the PortlandLookouts attribute table.

12 On the menu bar, click Geoprocessing > Buffer.

13 On the Buffer tool dialog box (see facing page), click the Input Features arrow and click PortlandLookouts. (Alternatively, drag the PortlandLookouts layer from the ArcMap table of contents to this location.)

14 Save the output feature class as LookoutsBuffer to MyOregonForest.gdb.

15 For the buffer distance, click the Field option. In the Field list, click RADIUS2.

16 Click the Dissolve Type arrow and click ALL. Then click OK. Wherever buffer polygons overlap each other, the overlapping boundaries will be dissolved to make a single feature.

When the operation is completed, the new LookoutsBuffer layer is added to the map.

17 If necessary, change the color of the buffers so they are visible against the background. In the table of contents, drag the LookoutsBuffer layer below the PopJursBuffer layer.

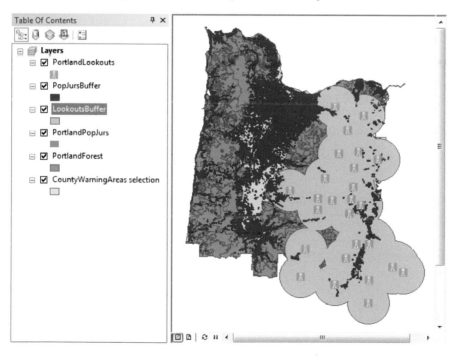

18 Open the attribute table of the LookoutsBuffer layer. The table contains just a single record. Buffers created with the dissolve ALL option form a single feature, called a *multipart polygon* (a polygon with discontinuous boundaries).

The attributes are the standard four for a geodatabase polygon feature class: OBJECTID, Shape, Shape_Length, and Shape_Area. Because of the dissolve, none of the PopJursBuffer layer attributes are passed to the output table.

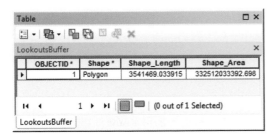

19 Close the attribute table.

20 If you want to save your work, rename your map document and save to your MyData folder for chapter 19.

21 If you are continuing to the next exercise, leave ArcMap open. Otherwise, exit the application and click No if prompted to save your changes.

The two buffers, LookoutsBuffer and PopJursBuffer, define the areas where a probability of fire near urban areas exists. In exercise 19b, you'll use overlays to define the areas in which the communities will need to be isolated from forest areas.

17

18

19a

20

Exercise 19b
Overlaying data

A union overlay combines the features of two input layers to create a new dataset that includes all combinations of polygons. In the following example, one polygon layer represents a land parcel and the other an oil spill. Part of the parcel lies outside the spill, and part of the spill lies outside the parcel; in the middle, the two layers overlap. When a union function is applied to the two layers, the two original polygons become three. The area of overlap becomes a new feature, and the nonoverlapping areas become a multipart polygon (individual polygons that are separated but still belong to the same group).

The parcel and oil spill layers are unioned to create three new polygons.

What kind of attributes does a union layer have? In addition to its own standard attributes, it includes all the attributes of both input layers. This doesn't mean, however, that every record has a value for every attribute. In this example, feature 1 (OBJECTID 1) in the output table has no Spill_Type value because it is outside the oil spill, and feature 2 has no Owner or Landuse values because it is outside the parcel. Only feature 3, which spatially coincides with both input layers, has a value for every attribute.

17

18

19b

20

Input tables

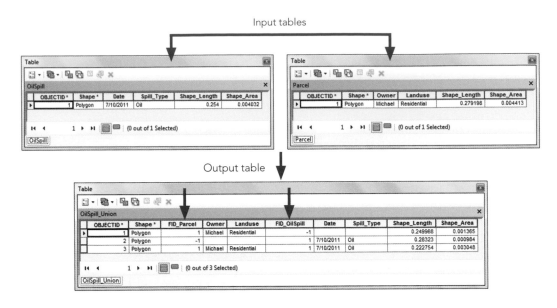

Output table

The output table contains the attributes of both input layers. Output features get the attribute values of input features with which they are spatially coincident.

What happens to the unique identifier attribute of each layer? One of the standard attributes of the union layer, assuming it is a geodatabase feature class, is the OBJECTID field, which assigns a unique identifier (ID) number to each feature. In other data formats, the identifier attribute is called FID (known as feature ID) or OID (known as object ID). In this case, both the input layers also have OBJECTID fields. ArcMap will not delete these fields—they may be useful to you—but at the same time, there shouldn't be a table with three OBJECTID fields.

To resolve the conflict, the OBJECTID fields from the input layers are renamed in the union layer. The new name consists of the prefix "FID_" followed by the input layer name. Thus, the parcel layer's identifier attribute is renamed "FID_Parcel," and the oil spill layer's is renamed "FID_OilSpill."

In an intersect overlay, the process is a little simpler, because only the area of overlap is preserved. An intersect of the oil spill and parcel layers would consist of a single feature, as shown in the graphic on the facing page.

17

18

19b

20

Oil spill and parcel
overlap each other

Intersect overlay of
oil spill and parcel

Two polygons

One polygon

An intersect layer consists of a single feature.

As with a union layer, an intersect layer includes its own standard attributes, plus all the attributes of both input layers. In an intersect, each attribute is populated for every record. A union overlay requires that both input layers be polygon layers. In an intersect, the input layers can be either two polygon layers or a polygon layer and a line layer. In the latter case, shown in the graphic below, the output is a line layer.

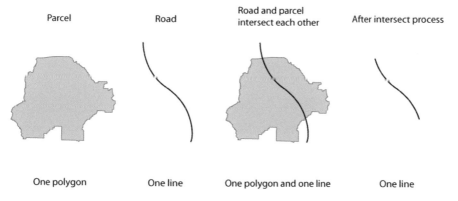

Parcel

Road

Road and parcel
intersect each other

After intersect process

One polygon

One line

One polygon and one line

One line

In this exercise, you'll union the populated jurisdiction buffer layer and the forest layer to create a single output dataset that contains all the attributes of both inputs. Then you'll union this layer with the lookout towers buffer layer. Because new features will be created wherever forest, populated jurisdictions buffer, and lookouts buffer polygons overlap, each output feature will lie either entirely inside or entirely outside a buffer. Once you have performed these union operations, you'll see how the output provides the correct attributes to determine which of the populated jurisdiction buffers are in proximity to both forestland and lookout towers. This will provide an output that shows which areas have a high probability of a forest fire within the visibility of lookout towers.

17

18

19b

20

1 **In ArcMap, open the ex19b map document from your Chapter19 folder.** The map shows the buffer for populated jurisdictions and the Portland forest layer. The other layers are turned off.

2 **On the menu bar, click Geoprocessing > Union.**

The Union tool and other analysis tools can also be accessed from ArcToolbox. On the Standard toolbar, click the ArcToolbox button. In the ArcToolbox window, expand Analysis Tools and then the Overlay toolset. Then double-click the Union tool.

Next, you'll specify the layers to overlay and designate an output feature class.

3 **On the Union tool dialog box, click the Input Features arrow and click PopJursBuffer. (Alternatively, drag the layer from ArcMap to the Union dialog box.)**

4 **Click the Input Features arrow again and click PortlandForest.** The selected layers are added to the list of layers that will be processed in a union.

⑤ Save the output feature class as AreaAtRisk **to MyOregonForest.gdb.**
By default, the output layer attribute table contains its standard attributes, plus the attributes of both input layers. Optionally, you can omit the input layer identifier attributes. Alternatively, you can omit all input attributes, except the identifiers. This creates a smaller output table and is convenient if you don't need to work with attributes. (If you later decide that you need the input attributes, you can join or relate back to the input layer tables using the common identifier attribute.)

For now, you'll include all the layer attributes.

⑥ For the JoinAttributes parameter, make sure ALL is selected. Click OK to run the tool.

When the operation is completed, the AreaAtRisk layer is added to the map. The new layer, which consists of all the areas from the two input layers, defines the zone where communities are in proximity to forestland and have a probability of either causing a fire or being at risk of damage from a fire. The layer is symbolized uniformly with a single color.

7 **Open the attribute table of the AreaAtRisk layer.** The table contains all the attributes included in both unioned layers. Along with the four standard polygon feature class attributes (OBJECTID, Shape, Shape_Length, and Shape_Area), the table also contains the renamed identifier attributes from the input layers: FID_PopJursBuffer and FID_PortlandForest.

FED_TRIBAL	BUFF_DIST	FID_PortlandForest	AREA	PERIMETER	FORSTLND_	FORSTLND_I	CODE	SYM
Y	1000.002	-1	0	0	0	0		
Y	1000.002	-1	0	0	0	0		
Y	1000.002	-1	0	0	0	0		
Y	1000.002	-1	0	0	0	0		
	0	1	25167	497990.375	2	12	501M	
	0	2	55360	303881.969	3	13	501H	
	0	3	67535	178216.953	4	15	510M	
	0	4	94601	45036.75	5	18	501L	
	0	5	37850	230620.766	6	27	501H	

I◄ ◄   1 ► ►I   (0 out of 8845 Selected)

AreaAtRisk

In the exercise introduction, you saw that not every attribute may be populated for every record in a union attribute table. Identifier attributes, however, are completely populated for every record. Every record in the AreaAtRisk attribute table has not only an OBJECTID value, but also an FID_PortlandForest value and an FID_PopJursBuffer value—regardless of which input feature is spatially coincident with the output feature.

Adjust or move the columns to examine or compare the attributes in the table.

For example, the 10th record in the AreaAtRisk table has an OBJECTID value of 10. This value is the feature's new identifier. The record's FID_PopJursBuffer value is 13. This means that the feature coincides spatially with the feature that has the OBJECTID of 13 in the PopJursBuffer table. The record's FID_PortlandForest value is –1. The –1 value means the feature does not coincide spatially with any feature in the PortlandForest layer. If you scroll down the attribute table to OBJECTID 246, you'll notice that the record has a FID_PopJursBuffer value of –1. This means the feature does not coincide spatially with any feature in the PopJursBuffer layer. So, at this point in the analysis, the areas at risk are all features that have an FID_PopJursBuffer value other than –1.

17

18

19b

20

8 Later in this exercise, you will see how to use this information to your advantage. Close the table. Turn on the LookoutsBuffer layer, and then drag it above the AreaAtRisk layer. It will give you a better view of how this layer will look after the union overlay.

To find the areas at risk near lookout towers, you will next union the AreaAtRisk layer with LookoutsBuffer. After the overlay, you will be able to select the polygons that represent communities at risk in the PQR CWA that are within visible range of a lookout tower.

9 On the menu bar, click Geoprocessing > Union.

10 On the Union tool dialog box (see facing page), click the Input Features arrow and click AreaAtRisk to add it to the list of input layers. Do the same for LookoutsBuffer.

11 Save the output feature class as Final to MyOregonForest.gdb. Leave the Join Attributes (optional) list set to ALL (see facing page). You will include all input attributes in the Union layer. (You will need several of the attributes to calculate values and use for symbology in exercise 19c.)

12 **Click OK to run the tool.** When the operation is completed, the Final layer, which has more than 9,000 features, is added to the map.

At this scale, it's hard to tell if you're looking at the result of a spatial analysis or just a jumble of lines. In a moment, you'll zoom in, but first you'll look at the attribute table of the Final layer.

13 **Open the attribute table of the Final layer.** The FID_AreaAtRisk attribute is the renamed OBJECTID from the AreaAtRisk layer. If a record has a

17
18
19b
20

value other than –1 in this field, it means the output feature coincides spatially with a buffer feature—in other words, it's an area at risk of fire. If a record has the value of –1, it means the output feature does not coincide with a buffer and is therefore not at risk. Similarly, if a record in the FID_PortlandForest attribute has a value other than –1, it means the output feature coincides spatially with forested areas. If not, it means the output feature does not coincide with the forest area and so is not considered at risk.

	FID_AreaAtRisk	FID_PopJursBuffer	COMMUNITY	COMID	RISK	HAZARD	P
▶	1	1	ADAIR RFPD (RFPD)	1	L	M	L
	2	2	ALBANY FIRE DEPT (RFPD)	4	L	L	L
	3	4	AMITY FIRE DIST (RFPD)	6	L	L	L
	4	5	AUMSVILLE RFPD (RFPD)	8	L	L	L
	5	6	AURORA RFPD (RFPD)	9	L	L	L
	6	8	Albany (City)	14	L	L	L
	7	9	Amity (City)	15	L	L	L
	8	10	Astoria (City)	19	H	L	L
	9	11	Aumsville (City)	21	L	L	L

(0 out of 9018 Selected)

Final

Since you're also attempting to correlate areas at risk to areas that are visible from lookout towers, you'll have to also consider the FID_LookoutsBuffer attribute. If a record has a value of 1 in this field, it means it coincides with areas at risk. If it has a value other than 1, it does not coincide.

The communities at risk in the Portland forest within visibility of a lookout tower, therefore, consist of all polygons in the Final layer that have a value other than –1 in the FID_AreaAtRisk field, a value other than –1 in the FID_PortlandForest layer, and a value of 1 in the FID_LookoutsBuffer field. However, at this point, it may still be difficult to see exactly which areas are at risk. So next, you'll create a definition query to isolate these areas specifically.

⑭ Close the Final attribute table. In the table of contents, turn off all layers except Final.

⑮ Double-click the Final layer. Click the Definition Query tab, and then click Query Builder (see facing page). In the Query Builder window, use the inputs to build the query:

"FID_AreaAtRisk" <> -1 AND "FID_PortlandForest" <> -1 AND "FID_LookoutsBuffer" = 1

Then click OK.

16 Move the Layer Properties window aside to see the main ArcMap window and click Apply.

The applied query isolates the forested areas and populated communities within visibility of a lookout tower. However, you'll need to symbolize the layer to see exactly where the communities at risk are. So, you'll do that next.

17 On the Layer Properties dialog box, click the Symbology tab. Under Show, click Categories, and then click Unique values. In the Value Fields list, click FID_PopJursBuffer, and then click Add Values.

18 On the Add Values dialog box, in the Select the value(s) to add list, click −1, and then click OK.

19 Double-click the −1 value color chip (it is assigned a random color from the color ramp) and change the fill color to Macaw Green and the outline color to Tarragon Green.

17

18

19b

20

20 Similarly, double-click the <all other values> color chip and change the fill color to Fire Red and the outline color to No Color.

21 Finally, click the Label value for <all other values> and replace it with Communities at risk. **Click** tthe <Heading> label and type Not at risk. **For the second item in the list, change the Label value −1 to** Forested area. **Make sure your settings match the graphic at right, and then click OK.**

22 On the menu bar, click Bookmarks > Close-up.

To see which areas are at risk in visible proximity of lookout towers, you will next turn on labels.

23 In the table of contents, double-click the Final layer. On the Layer Properties dialog box, click the Labels tab. Select the Label features in this layer check box. In the Text String frame, click the Label Field arrow and click FID_PopJursBuffer. Then click OK.

On the map, red polygons labeled with values not equal to −1 represent community areas at risk (exactly how you symbolized them).

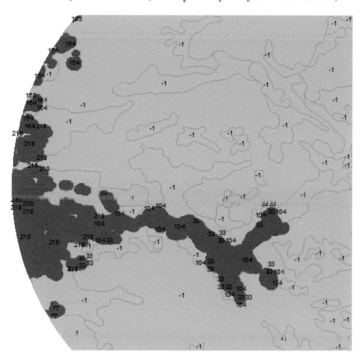

24 If you want to save your work, rename your map document and save to your MyData folder for chapter 19.

25 If you are continuing to the next exercise, leave ArcMap open. Otherwise, exit the application and click No if prompted to save your changes.

In exercise 19c, you will recalculate the final layer's area attribute to get an accurate total of how much area could potentially be affected by forest fire.

17

18

19b

20

Exercise 19c
Calculating attribute values

You can write an expression to calculate attribute values for all records in a table or for just selected ones. For numeric attributes, the expression can include constants, functions, or values from other fields in the table. For text attributes, the expression can include character strings that you type or text values from other fields.

In this exercise, you'll adjust the definition query to display communities at risk only, and then view the risk ratings. This will isolate these areas for easier calculation. Then you'll calculate values to determine how much area is actually affected.

1 **In ArcMap, open the ex19c document your Chapter19 folder.** The map contains the Final layer you created in exercise 19b (see facing page). The other layers were removed from the document so you can concentrate on just the Final layer.

17
18
19c
20

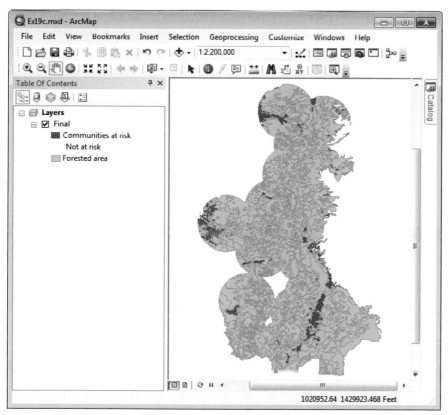

In exercise 19b, you unioned the populated jurisdictions buffer and the Portland forest to create a layer that contains communities in proximity to forestland. You then unioned this layer to the lookout towers buffer to show which areas are within the visibility of lookout towers and used this definition query to represent it:

> "FID_AreaAtRisk" <> -1 AND "FID_PortlandForest" <> -1 AND
> "FID_LookoutsBuffer" = 1

You'll adjust the existing definition query to display only the communities at risk that are within the visibility of a lookout tower (temporarily removing the forested area).

2 In the table of contents, double-click the Final layer. On the Layer Properties dialog box, click the Definition Query tab.

3 Click Query Builder to open the Query Builder dialog box.

17

18

19c

20

4 In the expression box, click the white space so that the cursor appears at the end of the expression. Click the And button, and then in the Fields box, double-click "FID_PopJursBuffer" to add it to the expression box. Click the not equal to (<>) button, and then click Get Unique Values. In the Unique Values list, double-click −1. Click OK.

The query is displayed in the Definition Query box.

Definition Query:

> "FID_AreaAtRisk" <> -1 AND "FID_PortlandForest" <> -1 AND "FID_LookoutsBuffer" = 1 AND "FID_PopJursBuffer" <> -1

Query Builder...

5 On the Layer Properties dialog box, click OK. On the map, only features that satisfy the query are displayed. These are the communities at risk in the Portland forest within visibility of a lookout tower. The layer attribute table will show only the records that correspond to these features.

17

18

19c

20

6 **Open the attribute table of the Final layer.** The table shows 649 records, rather than the original 1,598. However, take a look at the COMMUNITY field of the first three records in the attribute table. These three records are for a single community. And it's not the only case of duplicate communities. How did this happen?

During the unions, the populated jurisdictions buffer layer (PopJursBuffer) was overlapped by the forest layer (PortlandForest), which produced small area slivers. To recombine the communities, you'll next perform a dissolve.

7 **Close the Final attribute table.**

8 **On the menu bar, click Geoprocessing > Dissolve.**

9 **On the Dissolve tool dialog box, in the Input Features list, click Final. Save the output as FinalDissolve to MyOregonForest.gdb. In the Dissolve Field(s) box, select the check boxes for COMMUNITY, RISK, HAZARD, PROTECTION, VALUE, and OVERALL, and then click OK.**

17

18

19c

20

The FinalDissolve layer is added to the map in a random color.

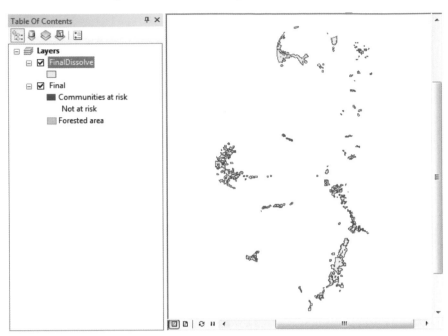

Each set of polygons with the same community name was dissolved into a single feature. Next, you'll quickly symbolize the layer to see the overall risk rating.

10 Open the FinalDissolve Layer Properties dialog box. On the Symbology tab, click Import.

11 On the Import Symbology dialog box, click the Layer Browse button and browse to \GTKArcGIS\Chapter19\Data, click FinalDissolve.lyr, and then click Add.

⑫ On the Import Symbology dialog box, click OK. On the Import Symbology Matching Dialog box, make sure the Value Field list is set to OVERALL and click OK. The layer is categorized based on the OVERALL risk rating attribute.

⑬ Click OK.

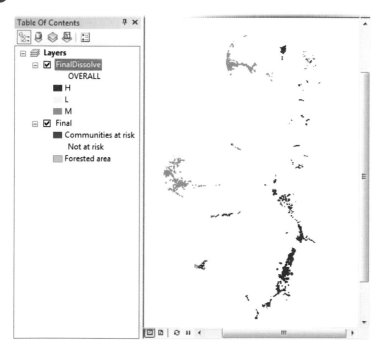

14 Open the attribute table of the FinalDissolve layer.

There are 49 communities at risk on the map. In addition, the other attributes chosen as dissolve fields—RISK, HAZARD, PROTECTION, VALUE, and OVERALL—have risk rating values of high (H), medium (M), or low (L). These were included by the agency to show each populated jurisdiction's risk rating, hazard rating, protection capability risk rating, value at risk rating, and overall risk rating. Throughout the geoprocessing tasks, ArcMap has automatically updated the Shape_Area attribute, which shows the area of each community. The Shape_Area attribute value is shown in square feet, but next you'll create a new field and recalculate it in acres. (***Hint:*** 1 acre is equal to 43,560 square feet.)

15 Above the table title bar, click the Table Options button and click Add Field.

16 On the Add Field dialog box, in the Name box, type ACRES. In the Type list, click Double and click OK.

The Double data type provides double-precision floating-point numbers, which means they are precise to many decimal places. For more about data types stored in ArcGIS, in ArcGIS for Desktop Help, on the Contents tab, navigate to Geodata > Data types > Tables > Designing a table > ArcGIS field data types.

17 Scroll to the far right of the attribute table. Right-click the field name ACRES and click Calculate Geometry. If necessary, click Yes on the message warning that you cannot undo the calculation. The Calculate Geometry dialog box appears.

18 Make sure Area is selected in the Property list. In the Units list, click Acres US [ac] and click OK. If necessary, click Yes on the message warning that you cannot undo the calculation.

This new field now contains values in acres.

HAZARD	PROTECTION	VALUE	OVERALL	Shape_Length	Shape_Area	ACRES
H	L	H	H	153635.101138	276404112.663791	6345.365304
H	H	H	H	355161.363752	450868039.969955	10350.505968
H	M	H	H	63178.752541	143167654.178085	3286.677093
H	M	H	H	30185.742407	49781249.076521	1142.820227
L	L	L	M	117784.060367	140150025.854501	3217.401879
H	L	L	H	60553.278745	97761218.050043	2244.288752
M	L	H	M	177514.083936	154551497.203756	3548.014169
H	H	H	H	225817.003084	244792304.223109	5619.65804
H	M	H	H	68354.281376	131174425.656638	3011.350451

(0 out of 49 Selected)

17
18
19c
20

19 **Right-click the ACRES field name and click Statistics.** The sum of the values is 323,332.127756. Thus, the total area of communities potentially at risk of a forest fire within visibility of a lookout tower is just over 323,332 acres—remember that this also includes a 1,000 ft buffer that puts these communities in contact with the Portland forest.

20 **Close the statistics window and the FinalDissolve attribute table.**

21 **If you want to save your work, rename your map document and save to your MyData folder for chapter 19.**

22 **If you are continuing to the next exercise, leave ArcMap open. Otherwise, exit the application and click No if prompted to save your changes.**

Your agency will use this information to establish future fire safety procedures. A more detailed risk analysis would consider additional factors, such as the location of existing roads, the slope of the land, and the density of surrounding vegetation.

Exercise 19d
Creating graphs

The ArcGIS Graph Wizard lets you create many different kinds of graphs, including column, pie, area, and scatterplot graphs. Using the wizard, you can set properties for such elements as titles, axes, and graph markers (for instance, the bars in a bar graph or the symbols in a scatterplot). Graphs can be saved with a map document or as files with a .grf extension that can be added to any map document.

In exercise 19c, you dissolved the Final layer to show all 49 communities within visibility of a lookout tower. You also calculated a new ACRES field in the attribute table. In this exercise, you'll present the ACRES values in a graph and add the graph to a map layout.

1 In ArcMap, open the ex19d map document from your Chapter 19 folder. The map shows the at-risk areas you created in exercises 19b and 19c.

2 **On the View menu, click Graphs > Create Graph.** The Create Graph
Wizard appears. The panel on the left has all the settings to create the graph. A
preview of it will appear on the right.

3 **Click the Graph type arrow and click Horizontal Bar.**

4 Make sure Layer/Table is set to FinalDissolve. Click the Value field arrow and click ACRES.

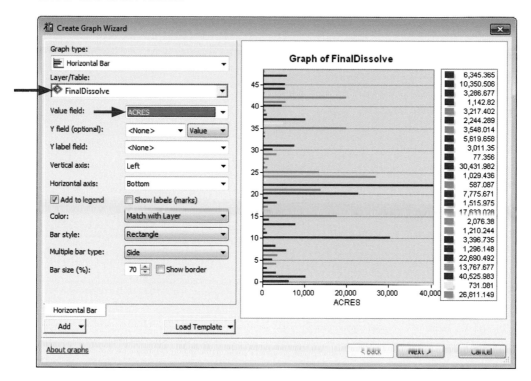

Next, you'll label the y-axis. As you make changes, the graph preview will be updated.

5 Click the Y label field arrow and click COMMUNITY. The community values are displayed on the y-axis, although it looks a bit messy.

17

18

19d

20

6 Clear the **Add to legend** check box. Make sure the wizard matches the graphic, and then click **Next.**

On the second panel of the wizard, you'll change the default graph title, add a footer, and remove the axis titles.

7 In the **General graph properties frame, in the Title box, replace Graph of FinalDissolve** with Overall Fire Risk Area. **In the Footer box, type** Includes 1000 ft buffer.

8 In the Axis properties frame (in the lower portion of the wizard), make sure the Left tab is selected. In the Title box, replace COMMUNITY with Community. Click the Bottom tab, and in the Title box, replace ACRES with Acres. Make sure the Visible check box is selected.

Axis properties

Left	Right	Bottom	Top

Title: Acres

Visible   ☑

Logarithmic   ☐

9 **Click Finish.** The graph is displayed in a window that floats on the application window. The graph makes it easy to compare the area values. If you resize the graph window, values displayed along the left axis may change.

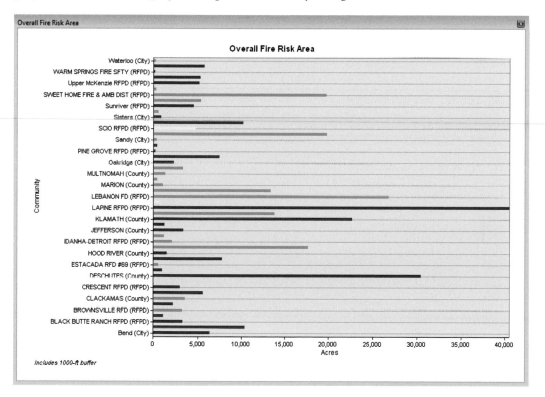

La Pine Rural Fire Protection District (RFPD) has the greatest area at risk. You'll add the graph to the map layout next.

17

18

19d

20

10 **Right-click the graph title bar. On the shortcut menu, click Add to Layout. Close the graph window.** ArcMap switches to layout view.

The graph is displayed in the middle of the layout page, marked with cyan selection handles. (Other layout elements, such as the map title and axis labels, have been added for you.)

11 **On the Tools toolbar, make sure the Select Elements tool is selected. Drag and resize the graph to fill the lower part of the layout, as shown in the graphic. Click outside the layout page to clear the graph selection.**

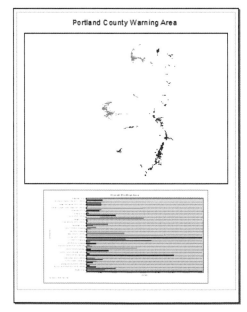

Next, you'll add a few layers to complete the look of the map.

12 Add the CountyWarningAreas layer from \Data\OregonForest.gdb. Change its order in the table of contents so that it is at the bottom of the layers list. Give it an appropriate color and label it.

13 Add the PortlandLookouts layer and give it an appropriate symbol.

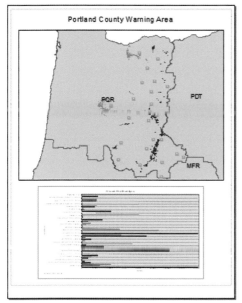

14 If you want to save your work, rename your map document and save to your MyData folder for chapter 19.

15 If you are continuing to the next chapter, leave ArcMap open. Otherwise, exit the application and click No if prompted to save your changes.

1
2
3
4
5
6
7
8
9
10
11
12
13
14
15
16
17
18
19d
20

Summary

In this chapter, you built on the foundation gained in chapters 17 and 18. You became familiar with some common geoprocessing tools and got practice running tools. You used common analysis tools including Dissolve, Clip, Buffer, and Union. You now have the skills to go back on your own and try re-creating this chapter's workflow using ModelBuilder to automate some of the geoprocessing tasks. It keeps track of the processes you run, as well as their results and their interdependencies. ModelBuilder gives you a convenient way to build a large project from its component parts and run its processes separately or together. You can change your model at any time to incorporate new data, add new conditions, or try different assumptions (best case, worst case, what if . . .). You will use more of ModelBuilder in chapter 20 to perform raster analysis.

Terms

buffer

overlay

union

intersect

multipart polygon

Chapter 20
Analyzing raster data

Creating raster surfaces
Combining raster surfaces

Learning objectives

- *Understand the fundamentals of raster models*
- *Explore tools available with the ArcGIS Spatial Analyst extension*
- *Continue using ModelBuilder to run a series of geoprocessing tools*
- *Create density and distance raster surfaces using ArcGIS Spatial Analyst*
- *Modify geoprocessing environment settings*
- *Convert polygons to a raster layer*
- *Reclassify rasters so they can be logically combined*
- *Overlay rasters to calculate a multiple-criteria suitability surface*

IN CHAPTERS 18 AND 19, YOU LEARNED HOW TO USE GEOPROCESSING TOOLS such as Dissolve, Buffer, and Clip. As you have seen so far, the exercises in this book work mainly with vector data—shapefiles and geodatabase feature classes. However, the raster data model is just as important to learn. As a student learning GIS, it is useful to know that both vector and raster data models have inherent advantages and disadvantages. This chapter provides an introduction to fundamental concepts of the raster data model and to raster analysis using the ArcGIS Spatial Analyst extension to ArcMap.

As discussed in chapter 1, the first thing you should understand about a raster is that it is composed of pixels, instead of discrete x,y coordinates, that define geographic entities. The pixels form a mesh of grid **cells**, which are used to record and define geographic phenomena on the surface of the earth. Each raster cell represents a portion of the earth, such as a square meter or square mile, and usually has an attribute value associated with it, such as soil type or vegetation class. A surface can be **discrete data**, meaning it shows distinct and discernible regions on a map, such as soil types or land cover (see facing page), or it can be **continuous data**, meaning there are smooth transitions between variations in the range of data. Elevation data is one of the most popular uses of a continuous raster. Typical raster datasets are very detailed and store a large amount of information, because information about each and every cell must be recorded.

This raster of continuous data represents elevation. Data courtesy of Lake Tahoe Data Clearinghouse, US Geological Survey, Office of Geographic Information (MassGIS), and the Commonwealth of Massachusetts Information Technology Division.

1
2
3
4
5
6
7
8
9
10
11
12
13
14
15
16
17
18
19
20

This raster of discrete data represents land cover. Data courtesy of Lake Tahoe Data Clearinghouse, US Geological Survey, Office of Geographic Information (MassGIS), and the Commonwealth of Massachusetts Information Technology Division.

Map algebra—a language that combines GIS layers—is fundamental to raster analysis. Raster cell values can be evaluated using functions and mathematical, logical, or Boolean operators. An output raster is the result of a cell-by-cell function performed on two input rasters.

A 3 x 3 grid added to another 3 x 3 grid produces a result grid, in which the cell values represent the total of the inputs.

As of the latest release, ArcGIS Spatial Analyst contains 170 geoprocessing tools to perform comprehensive, raster-based spatial analysis. These tools allow you to perform basic mathematical and logical operations, as well as raster dataset creation and processing. The most common way to access ArcGIS Spatial Analyst functionality is to run individual tools from ArcToolbox and use the graphic user interface of ModelBuilder to create models and automate workflows or, for advanced GIS users, to work in the Python window.

ArcGIS Spatial Analyst performs raster analysis using **Esri grids**. In this context, a grid is a raster data storage format native to Esri and consists of two types: integer and floating point. Integer grids typically represent discrete data, and floating-point grids typically represent continuous data. Rasters can exist in a variety of formats such as TIFF, MrSID (Multi-Resolution Seamless Image Database), and DEM (digital elevation model) but need to be converted to the grid format before this data can be used in analysis. When calculations are performed, the input grid type will usually be carried through to the output. Although every cell in a grid has a value assigned to it, cells without an actual value can be designated as **NoData** cells. However, NoData and zero (0) values are not the same thing. Zero is still a valid value, but NoData cells cannot be used in calculations. So, raster calculations ignore NoData cells.

Examples of the types of spatial analysis you can perform with ArcGIS Spatial Analyst include terrain analysis, surface and density modeling, suitability modeling, and image classification.

The exercises in this chapter require ArcGIS Spatial Analyst, an extension of ArcGIS software. If you are using the trial software, you have access to this extension. Make sure your ArcGIS Spatial Analyst extension has been authorized before proceeding.

How to authorize extensions

If you did not authorize ArcGIS Spatial Analyst when you installed the trial software, follow these instructions:

1. On the Start menu, go to All Programs > ArcGIS > ArcGIS Administrator.

2. In the ArcGIS Administrator window, click Desktop in the left panel and Authorize Now in the right panel.

3. In the Software Authorization Wizard, click the second option ("I have already authorized the software and need to authorize additional extensions"), and then click Next.

4. Click the appropriate authorization method (the first option, using the Internet, is preferable).

5. On the next two panels, enter the authorization information requested.

6. On the Authorize Software Extensions panel, click the Spatial Analyst text box and enter the authorization code from the back of the book. Then click Next.

7. Select any additional extensions you would like to evaluate, and then click Next.

8. When authorization is complete, click Finish to close the wizard.

As this is the final chapter in the analysis section, it is assumed you have already completed chapters 18 and 19, and have gained a general familiarity with ModelBuilder and geoprocessing tools. As such, the exercises in this chapter are written to challenge your recall, with streamlined instructions.

Data

The exercises in this chapter use the following datasets, stored in the New_Clinic geodatabase:

- Blockgroups—selected US Census block group polygons for Santa Rosa, California. Other layers were clipped to the extent of this layer. *(Source: Tele Atlas, US Census, Esri)*
- CityBusStops—points representing bus stops in Santa Rosa, California. *(Source: County of Sonoma Department of Health Services)*
- ClientList—point locations of current and past clients of a local clinic serving low-income families with young children. *(Source: County of Sonoma Department of Health Services)*
- streets—street lines symbolized by type. *(Source: County of Sonoma Department of Health Services)*

17

18

19

20

Exercise 20a
Creating raster surfaces

In this exercise, you will create three raster surfaces, each representing a suitability criterion for an analysis study. In exercise 20b, you will combine the raster surfaces to create a final suitability surface.

For this exercise, you are a GIS consultant for the state of California. A state-funded medical clinic serving low-income families with young children in the town of Santa Rosa, in the North Bay area of California, needs to relocate to a larger location capable of serving more clients. You have been asked to create a suitability map based on location and accessibility by existing and potential clients. Once your study is complete, a real estate search will be conducted in the high-suitability areas.

The new location should meet the following criteria: it should be located near a high concentration of existing clients, be within easy walking distance of a bus stop, and be where there is a high percentage of potential clients (specifically, families with children under age five).

1 **In ArcMap, open the ex20a map document from your C:\EsriPress\ GTKArcGIS\Chapter20 folder.** The map is centered on the town of Santa Rosa (see facing page). The map includes the point locations of past and current local clients of the existing clinic, bus stops, streets, and census block groups for Santa Rosa.

2 Turn on the ArcGIS Spatial Analyst extension (on the menu bar, Customize > Extensions). When you are done, close the Extensions dialog box.

17
18
19
20a

3 Open the geoprocessing environment settings (on the menu bar, click Geoprocessing > Environments). In the Raster Analysis section, change Cell Size to As Specified Below and enter the value of 100.

4 Set the mask to Study Area Block Groups and click OK.

These *environment settings* apply to your current ArcMap session and all the tools in your current model. (However, the environment can be overridden for specific tools if needed—you will see how this works later in the exercise.)

A *mask* is a means of identifying the areas that will be included in the analysis.

Some geoprocessing tools will give you different outputs depending on your environment settings. Sometimes the best way to determine the correct environment settings is through trial and error. If the geoprocessing output isn't what you expected, you might try modifying the environment settings. The raster cell size of 100 will be your new default, but most raster tools allow you to choose a different size if you wish.

5 Open the Catalog window embedded in ArcMap and create a new, empty toolbox in your \Chapter20\MyData folder (right-click MyData and click New > Toolbox). Maintain the default name.

6 Open ModelBuilder ⊞ and ArcToolbox ⊡ .

7 In ArcToolbox, navigate to the Spatial Analyst Tools > Density toolset. Click the Point Density tool and drag it to the ModelBuilder window.

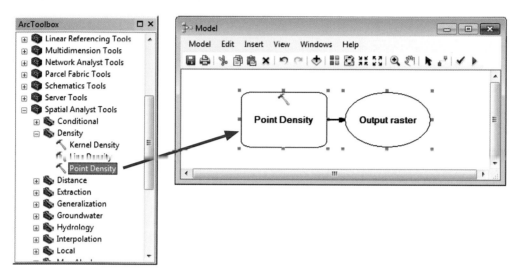

8 In the ModelBuilder window, double-click the empty Point Density tool. The tool dialog box appears. You can ignore the error message.

9 On the Point Density dialog box, for Input point features, click the arrow and click Clients. For Output raster, browse to your \Chapter20\MyData folder and name the raster density. The output cell size should be 100. In the Neighborhood Settings frame, change the neighborhood radius to 2000 map units, and then click OK.

17

18

19

20a

The Point Density tool is ready to run in ModelBuilder. Next, you will run it to see the results of the client density calculation.

10 Run the model (on the model menu bar, click Model > Run). When the process is completed, right-click the density element in the model and click Add To Display. Be sure to keep your model open for the extent of this exercise.

11 In the table of contents, drag the raster layer above the block groups layer. Turn off the bus stops layer, and turn the clients layer on and off to get a better understanding of the analysis results. The Point Density tool creates a raster surface that shows where points are clustered. The darker areas have a higher concentration of current clients, and the lighter areas have a lower concentration.

12 Turn the bus stops layer back on and return to the ModelBuilder window. From ArcToolbox, drag the Euclidean Distance tool (Spatial Analyst Tools > Distance toolset) to your model below the Point Density tool.

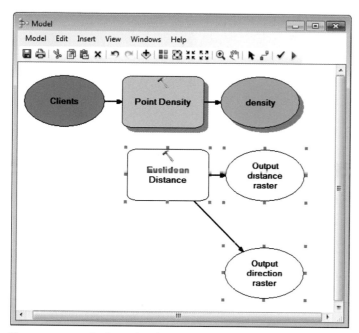

You want to calculate the distance from bus stops. You aren't interested in areas that are farther than a half-mile walk from a bus stop (and the most desirable areas will be much closer than that), so you'll choose a maximum output distance of a half mile (2,640 feet).

13 In ModelBuilder, double-click the empty Euclidean Distance tool. On the Euclidean Distance dialog box, for Input raster or feature source data, click the arrow and click City Bus Stops. For Output distance raster, browse to your \Chapter20\ MyData folder and name the raster distance. For Maximum distance, enter 2640 (this distance is calculated in map units, which is feet). Make sure Output cell size is 100, and leave Output direction raster (optional) blank.

17

18

19

20a

Raster file names cannot exceed 13 characters and cannot include special characters.

On your own

Show the Help panel for the Output direction raster parameter. Can you think of a scenario where it would be important to calculate the direction from the source layer?

14 Click OK to apply the parameters of the Euclidean Distance tool. In ModelBuilder, the Euclidean Distance tool is ready to run. (The optional direction output will remain empty.)

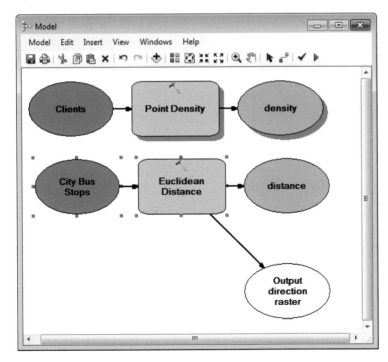

Trial and error has shown that the default processing extent of the Euclidean Distance tool results in an undesirable output with some edges cut off at the input points. Because you want to see a buffer around all bus stop points, even those on the outer edge of the input layer, you will change an environment setting for this tool only (again, trial and error has shown that changing the environment for the entire model would cause subsequent tools—ones you haven't added yet—to fail). Changing the setting for only one tool will provide more desirable results, so you'll do that next.

15 In the model, right-click Euclidean Distance and click Properties. Select the Processing Extent check box, and then click Values. On the Environment Settings dialog box, expand Processing Extent. Change the extent to Same as layer Study Area Block Groups. Click OK, and then OK again to close both windows.

17

18

19

20a

16 This is a good time to save your model. On the model menu bar, click Model > Save As. Save the model to the toolbox you created (Toolbox.tbx), with the default or a custom name, in your \Chapter20\MyData folder.

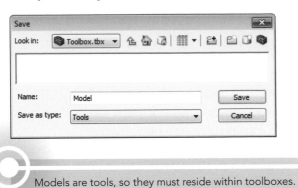

Models are tools, so they must reside within toolboxes.

17 Run the model and add the distance output to the map display.

On your own

If you want to see for yourself how environment settings can change an output, change the environment of the Euclidean Distance tool only (not of the whole geoprocessing session) back to the default, and then run the tool. Zoom to the full extent of the output layer. Can you see why this output is not ideal? When you're finished, repeat steps 14–16.

The final criterion is a high percentage of children under age five. You have demographic data for census block groups. To balance total population differences in block groups, a percentage calculation is preferred over an absolute count. If you examine the attribute table of the block groups, you will see fields called AGE_UNDER5 and POP2000, but there is no percentage field, so next, you will create one. This requires two separate processes: first, you will add the field, and then you will calculate its desired values.

18 From ArcToolbox, drag the Add Field and Calculate Field tools (Data Management Tools > Fields toolset) to the ModelBuilder window. Drop them below your existing processes. Notice when there are two elements with the same name, ModelBuilder appends a number to the element (2, 3, and so on) to differentiate between them.

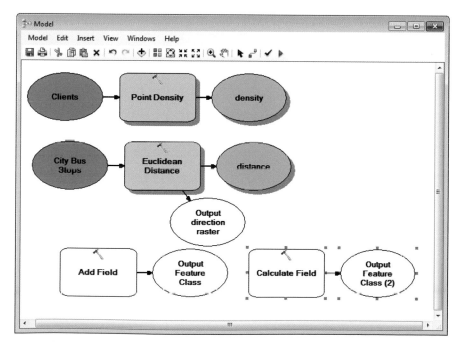

You can easily resize or rearrange model elements to fit the model window. Just click an element to select it. Drag it where you like or click one of the blue selection handles to resize it. Your model elements may differ in color and position from the exercise graphics somewhat. Just make sure the tool parameters (inputs and outputs) match.

19 **Open the Add Field tool and set the following parameters, and then click OK:**

Input Table: Study Area Block Groups
Field Name: PERC_UNDER5
Field Type: FLOAT
Field IsNullable (check box)

20 Open the Calculate Field tool. For Input Table, click the arrow and click the intermediate layer Study Area Block Groups (2), preceded by a blue icon 🔄 indicating intermediate data. *Intermediate data* means the layers that reside in the model, rather than those that reside in the map document.

21 For Field Name, click the arrow, scroll to the bottom of the list, and click PERC_UNDER5. (If you don't see this field name, you chose the wrong input.)

22 Click the Calculator button next to the Expression parameter. In the Field Calculator, build the following expression, and then click OK:
    [AGE_UNDER5] / [POP2000]

17

18

19

20a

23 Click OK to close the Calculate Field tool.

The field tools are connected and ready to run as a single operation in ArcMap. You will add a conversion process to the model before you run it. You must turn the block group layer into a raster so you can combine it with the other two suitability rasters.

24 Drag the Polygon to Raster tool (Conversion Tools > To Raster toolset) to the model and drop it to the right of the Calculate Field process. Open the tool dialog box and set the following parameters, and then click OK:
    Input features: Study Area Block Groups (3)
    Value field: PERC_UNDER5
    Output Raster Dataset: \Chapter20\MyData\children
    Priority field: NONE
    Cellsize: 100

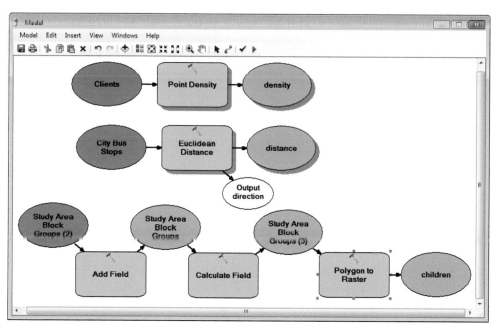

25 Run the model to execute the conversion process. Add the output (children) to the ArcMap display.

26 Change the symbology of the output layer using a stretched color ramp, with darker colors representing higher percentages.

On your own

Turn the three surfaces on and off. Can you determine areas that will satisfy all three criteria reasonably well?

27 Save your model. You have saved all the suitability surfaces to your MyData folder, so there is no need to save your map. If you are continuing to exercise 20b and you were successful creating the three suitability surfaces, you may continue using this map document. Otherwise, you can use the map document provided for you.

You have created three suitability surfaces to site a new location for the Santa Rosa medical clinic. In exercise 20b, you will reclassify and combine the surfaces for a final suitability layer.

1
2
3
4
5
6
7
8
9
10
11
12
13
14
15
16
17
18
19
20a

Exercise 20b
Combining raster surfaces

Before you combine raster surfaces, it's important to reclassify the cell values. *Reclassification* allows you to logically compare the inputs more easily. It's also helpful to standardize the extent of the inputs, although it's not always necessary. Raster calculations can be as simple as adding cell values together or as complex as doing mathematical equations.

In this exercise, you will finish the analysis project you began in exercise 20a. You have already created three suitability surfaces for the clinic placement study, so next, you will complete the steps to create the final output, in which the three surfaces are combined.

17
18
19
20b

1 In ArcMap, open the ex20b map document from your Chapter20\
Data folder. Or if you just completed exercise 20a, continue using the
same map document and zoom to the full extent of the map layers.

You should see the Santa Rosa map and three suitability surfaces.

2 Set the geoprocessing environment (Geoprocessing > Environments)
for Raster Analysis, set the cell size to 100, and set the mask to
Study Area Block Groups, if not already specified. Click OK.

3 Turn the suitability surfaces on and off. Notice the extent of the
layers.

On your own

Turn on the Effects toolbar and experiment with the Swipe tool ![icon] as another way to view multiple layers.

The distance raster covers a smaller area than the density and children rasters. That's because you entered a maximum calculation distance of a half mile from bus stops. Areas beyond a half mile are unsuitable locations for the new clinic.

Before you combine the three surfaces, it would be good to ensure that they cover the same area. In exercise 18b, you used the Clip tool to cut out features. With rasters, you can use a similar operation called extract.

4 If the model from exercise 20a is not open, expand the Catalog window in ArcMap and expand \Home\MyData\Toolbox.tbx. Right-click the model you created in exercise 20a and click Edit to open it in ModelBuilder. (If you didn't create a model, or if you aren't sure yours is correct, edit the model in \Chapter20\Results\Toolbox.tbx.)

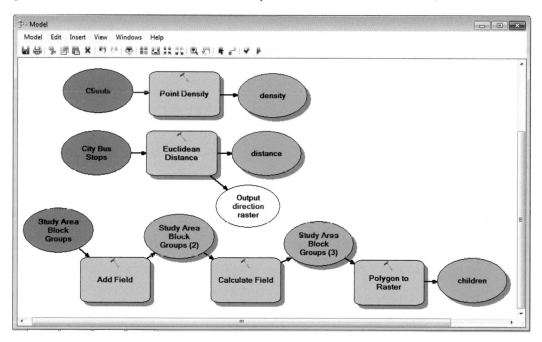

You will use the Extract by Mask tool twice — once, to reduce the extent of the density raster, and a second time, to reduce the extent of the children raster to match that of the distance raster. Then you will use the Reclassify tool on

17

18

19

20b

all three surfaces to give them a common scale, assigning the highest value to the most desirable characteristics. Finally, you will use a map algebra tool to add the three surfaces together. The output of this operation will be your final suitability map.

5 Drag the Extract by Mask tool from ArcToolbox (Spatial Analyst Tools > Extraction toolset) to the model and open the tool dialog box. Set the following parameters, and then click OK:

Input raster: density (the intermediate dataset preceded by the blue icon)

Input raster or feature mask data: distance (intermediate layer)

Output raster: \Chapter20\MyData\ext_density

6 Again, drag the Extract by Mask tool to the model and set the following parameters, and then click OK:

Input raster: children (intermediate)

Input raster or feature mask data: distance (intermediate layer)

Output raster: \Chapter20\MyData\ext_children

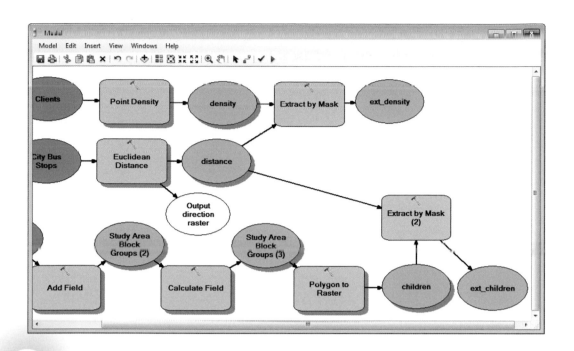

Your model will look somewhat different from the graphic, depending on how you have arranged it.

7 **Run the model, and then add the two new outputs to the map.** Your three suitability surfaces are now called distance, ext_density, and ext_children.

17

18

19

20b

8 Turn off the layers named children, density, and Study Area Block Groups. Zoom in slightly to see the new outputs.

Now you must reclassify the datasets so they can all be used together. So next, you will assign a common scale to each.

9 From ArcToolbox, drag the Reclassify tool (Spatial Analyst Tools > Reclass toolset) to the model three times and drop it there. (See facing page.)

20b

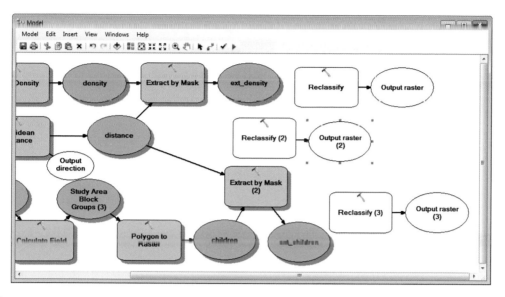

10 Open the first Reclassify tool dialog box. For Input raster, click the ext_density intermediate dataset.

11 Click the Classify button. Maintain the default classification method (Natural Breaks), but change the number of classes to 7 and click OK.

The output will have seven (7) classes, with the highest number (7) assigned to the areas that have the highest density of existing clients.

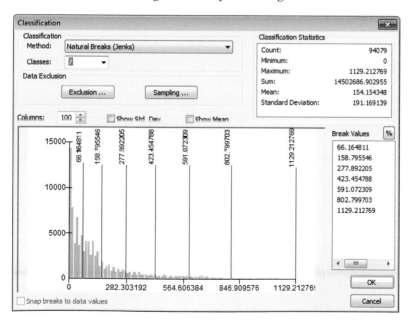

12 For Output raster, browse to your Chapter20\MyData folder and name the output recl_density. Click OK to close the tool dialog box.

Next, you will reclassify the distance raster. The process is similar, but this time you want to flip the output values so the highest number (7) is assigned to the lowest distance (because a shorter distance from bus stops is more desirable).

13 Open the Reclassify (2) tool dialog box. (See facing page.) Set the following parameters, and then click OK:
    Input raster: distance (intermediate layer)
    Classify: Natural Breaks, 7 classes
    Reverse the values (click Reverse New Values)
    Output raster: \Chapter20\MyData\recl_distance

17

18

19

20b

14 Open the Reclassify (3) tool dialog box. Set the following parameters, and then click OK:

Input raster: ext_children (intermediate layer)

Classify: Natural Breaks, 7 classes

Output raster: \Chapter20\MyData\recl_children

17

18

19

20b

The Auto Layout tool can be used to automatically organize your elements, so you'll use that next.

15 Click the Auto Layout tool ▦ .

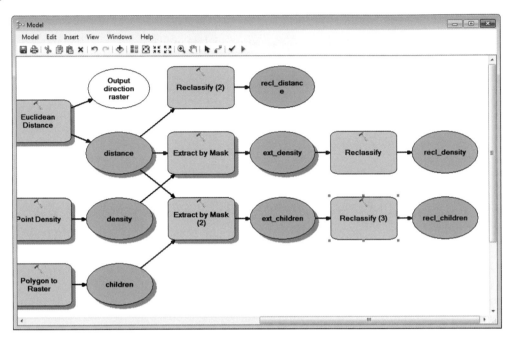

16 Run the model and add the results to the map.

17 From ArcToolbox, drag the Raster Calculator tool (Spatial Analyst Tools > Map Algebra toolset) to the model. Open the tool dialog box and use the list and operator buttons to add the values of recl_density, recl_distance, and recl_children (be sure to choose the intermediate data). The tool automatically creates a map algebra expression when you select the variables and operators. Save the output as Chapter20\MyData\final, and click OK. The model is complete.

18 Click the Auto Layout button.

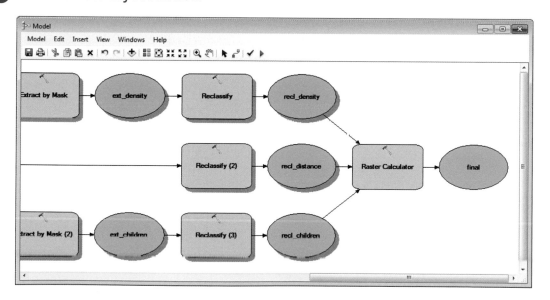

17

18

19

20b

19 To see the model in its entirety, expand the window and click the Full Extent button ⊡ .

For a weighted overlay, you could use a multiplier to apply a weight, or importance value, to the inputs. Or use a tool from the Spatial Analyst Tools > Overlay toolset, such as Weighted Overlay or Weighted Sum. For more information, in ArcGIS for Desktop Help, on the Contents tab, navigate to Geoprocessing > Tool reference > Spatial Analyst toolbox > Overlay toolset > An overview of the Overlay tools.

20 Save your model.

21 Run the final process and add the output to the map. Resymbolize the suitability surface using a color ramp for unique values. The most suitable areas for a new clinic (see facing page) have the highest values in this output (the highest is 21, because three (3) surfaces with seven (7) classes each were added together).

22 You have successfully completed a raster analysis project to determine site suitability for a new clinic that can serve more families. If you want to save your map, rename your map document and save to your MyData folder for chapter 20.

On your own

To run the entire analysis with a single click:

- Close the model and open 20a.mxd (the original map document without any processed layers)

- Edit the model in MyData\Toolbox.tbx. Click Model > Delete Intermediate Data. Then validate the model to clear all processes (Model > Validate).

- Run the model.

- Give yourself a pat on the back for creating and executing a sophisticated raster analysis.

---

23 Now that you have concluded the final chapter of the book, you may exit the application.

17

18

19

20b

Summary

If you want to combine datasets to create a new, meaningful output, you may want to take advantage of the powerful analysis tools offered by ArcGIS Spatial Analyst. Although a lot of analysis can be performed on vector data, overlaying multiple layers requires you to convert your data to the raster data model. Now you have some idea of how rasters work and how to create common analysis surfaces such as point density and Euclidean distance. Most important to remember, before you combine rasters, you must check the datasets to ensure they have the same cell size, extent, and classification scheme.

Those who have worked through this book from beginning to end have obtained a great deal of information and a plethora of skills in using ArcGIS for Desktop software, from basic map navigation to raster analysis, and everything in between. With this solid foundation, you are ready to study advanced GIS concepts and perhaps apply your knowledge to the real world. No matter what your field—forestry, health care, city management, infrastructure, defense, social services, marketing ... the list goes on—your newfound GIS skills can be used to spread geospatial information, further understanding and collaboration, and serve as a conduit for problem solving within your community.

Terms

cell

discrete data

continuous data

map algebra

Esri grid

NoData

environment settings

mask

intermediate data

reclassification

Appendix A
Data source credits

Chapter 1
No data.

Chapter 2
No data.

Chapter 3
\Chapter03\Data\World.gdb\Background from Esri Data & Maps 2010, courtesy of Esri.

\Chapter03\Data\World.gdb\Cities from Esri Data & Maps 2010, courtesy of ArcWorld.

\Chapter03\Data\World.gdb\Countries from Esri Data & Maps 2010, courtesy of ArcWorld, US Census International Division, CIA Factbook; The World Bank: Air Pollution in World Cities (PM10 Concentrations): David R. Wheeler, Uwe Deichmann , Kiran D. Pandey, Kirk E. Hamilton, Bart Ostro, and Katie Bolt.

\Chapter03\Data\World.gdb\Latlong from Esri Data & Maps 2010, courtesy of Esri.

Chapter 4
\Chapter04\Data\World.gdb\Background from Esri Data & Maps 2010, courtesy of Esri.

\Chapter04\Data\World.gdb\Cities from Esri Data & Maps 2010, courtesy of ArcWorld.

\Chapter04\Data\World.gdb\Countries from Esri Data & Maps 2010, courtesy of ArcWorld, US Census International Division, CIA Factbook; The World Bank: Air Pollution in World Cities (PM10 Concentrations): David R. Wheeler, Uwe Deichmann , Kiran D. Pandey, Kirk E. Hamilton, Bart Ostro, and Katie Bolt.

\Chapter04\Data\World.gdb\Latlong from Esri Data & Maps 2010, courtesy of Esri.

\Chapter04\Data\US_Cities_PM.shp from Esri Data & Maps 2010, courtesy of ArcWorld; The World Bank: Air Pollution in World Cities (PM10 Concentrations): David R. Wheeler, Uwe Deichmann , Kiran D. Pandey, Kirk E. Hamilton, Bart Ostro, and Katie Bolt.

Chapter 5
\Chapter05\Data\NanaimoBC.gdb\bike_routes, courtesy of City of Nanaimo, British Columbia, Canada.

\Chapter05\Data\NanaimoBC.gdb\intersections, courtesy of City of Nanaimo, British Columbia, Canada.

Chapter 6
\Chapter06\Data\Canada_Mexico.shp from Esri Data & Maps 2010, courtesy of ArcWorld Supplement.

\Chapter06\Data\Cities.shp from Esri Data & Maps 2010, courtesy of US Census Bureau.

\Chapter06\Data\ColoradoCounties.shp from Esri Data & Maps 2010, courtesy of Esri, derived from Tele Atlas, US Census Bureau, Esri (Popfield2010).

\Chapter06\Data\Lakes.shp from Esri Data & Maps 2010, courtesy of ArcWorld.

\Chapter06\Data\Latlong.shp from Esri Data & Maps 2010, courtesy of Esri.

\Chapter06\Data\USstates.shp from Esri Data & Maps 2010, courtesy of ArcUSA, US Census Bureau, Esri (Pop2010 field).

Chapter 7

\Chapter07\Data\Florida.gdb\Bathymetry from the Florida Geographic Data Library, courtesy of Florida Fish and Wildlife Conservation Commission, Fish and Wildlife Research Institute; National Oceanic and Atmospheric Administration.

\Chapter07\Data\Florida.gdb\Cities from the Florida Geographic Data Library, courtesy of National Atlas of the United States.

\Chapter07\Data\Florida.gdb\Coral from the Florida Geographic Data Library, courtesy of Florida Fish and Wildlife Conservation Commission, Fish and Wildlife Research Institute.

\Chapter07\Data\Florida.gdb\Counties from the Florida Geographic Data Library, courtesy of US Census Bureau.

\Chapter07\Data\Florida.gdb\Hwy1 from the Florida Geographic Data Library, courtesy of Florida Department of Transportation.

\Chapter07\Data\Florida.gdb\Lakes from the Florida Geographic Data Library, courtesy of Florida Department of Environmental Protection.

\Chapter07\Data\Florida.gdb\Rivers from the Florida Geographic Data Library, courtesy of Florida Department of Environmental Protection.

\Chapter07\Data\Florida.gdb\Shipwrecks from the Office of Coast Survey's Automated Wreck and Obstruction Information System (AWOIS), courtesy of National Oceanic and Atmospheric Administration.

\Chapter07\Data\Florida.gdb\Southeast from Esri Data & Maps 2010, courtesy of ArcUSA, US Census Bureau, Esri (Pop2010 field).

\Chapter07\Data\Florida.gdb\StParks from the Florida Geographic Data Library, courtesy of Florida Department of Environmental Protection.

\Chapter07\Data\florida_coast.tif made with Natural Earth.

\Chapter07\Data\shdrl from the Florida Geographic Data Library, courtesy of University of Florida GeoPlan Center.

\Chapter07\Data\usgsdem from the Florida Geographic Data Library, courtesy of US Geological Survey.

Chapter 8

\Chapter08\Data\TexasCounty.gdb\counties from Esri Data & Maps 2010, courtesy of Esri, derived from Tele Atlas, US Census Bureau, Esri (Pop2010 field). US Dept of Justice, Federal Bureau of Investigation. Uniform Crime Reporting Program Data [United States]: County-level Detailed Arrest and Offense Data, 2002 [Computer file]. ICPSR ed. Ann Arbor, MI: Inter-university Consortium for Political and Social Research [producer and distributor], 2004.

Chapter 9

\Chapter09\Data\Florida.gdb\Bathymetry from the Florida Geographic Data Library, courtesy of Florida Fish and Wildlife Conservation Commission, Fish and Wildlife Research Institute; National Oceanic and Atmospheric Administration.

\Chapter09\Data\Florida.gdb\Cities from the Florida Geographic Data Library, courtesy of National Atlas of the United States.

\Chapter09\Data\Florida.gdb\Coral from the Florida Geographic Data Library, courtesy of Florida Fish and Wildlife Conservation Commission, Fish and Wildlife Research Institute.

\Chapter09\Data\Florida.gdb\Counties from the Florida Geographic Data Library, courtesy of US Census Bureau.

\Chapter09\Data\Florida.gdb\Hwy1 from the Florida Geographic Data Library, courtesy of Florida Department of Transportation.

\Chapter09\Data\Florida.gdb\Lakes from the Florida Geographic Data Library, courtesy of Florida Department of Environmental Protection.

\Chapter09\Data\Florida.gdb\Rivers from the Florida Geographic Data Library, courtesy of Florida Department of Environmental Protection.

\Chapter09\Data\Florida.gdb\Shipwrecks from The Office of Coast Survey's Automated Wreck and Obstruction Information System (AWOIS), courtesy of National Oceanic and Atmospheric Administration.

\Chapter09\Data\Florida.gdb\Southeast from Esri Data & Maps 2010, courtesy of ArcUSA, US Census Bureau, Esri (Pop2010 field).

\Chapter09\Data\Florida.gdb\StParks from the Florida Geographic Data Library, courtesy of Florida Department of Environmental Protection.

\Chapter09\Data\florida_coast.tif made with Natural Earth.

Chapter 10

\Chapter10\Data\australia.shp from Esri Data & Maps 2010, courtesy of ArcWorld Supplement.

\Chapter10\Data\bioregions.shp courtesy of State of Queensland (Department of Environment and Resource Management) 2010.

\Chapter10\Data\biosubregions.shp courtesy of State of Queensland (Department of Environment and Resource Management) 2010.

\Chapter10\Data\country_background.shp from Esri Data & Maps 2010, courtesy ArcWorld Supplement.

\Chapter10\Data\OB_LR.tif made with Natural Earth.

Chapter 11

\Chapter11\Ex11aData\parcels, parcel data and imagery © 2011 County of Riley, Kansas, and its representatives. All rights reserved.

\Chapter11\Ex11aData\FireHydrants.shp courtesy of City of Manhattan, Kansas.

\Chapter11\Ex11aData\WaterValves.csv courtesy of City of Manhattan, Kansas.

\Chapter11\Ex11bData\CityData.gdb\FireHydrants courtesy of City of Manhattan, Kansas.

\Chapter11\Ex11bData\CityData.gdb\Parcels, parcel data, and imagery © 2011 County of Riley, Kansas, and its representatives. All rights reserved.

\Chapter11\Ex11bData\CityData.gdb\WaterValves courtesy of City of Manhattan, Kansas.

\Chapter11\Ex11cData\CityData.gdb\FireHydrants courtesy of City of Manhattan, Kansas.

\Chapter11\Ex11cData\CityData.gdb\Parcels, parcel data, and imagery © 2011 County of Riley, Kansas, and its representatives. All rights reserved.

\Chapter11\Ex11cData\CityData.gdb\WaterLines created by author.

\Chapter11\Ex11cData\CityData.gdb\WaterValves courtesy of City of Manhattan, Kansas.

Chapter 12

\Chapter12\Data\CityData.gdb\FireHydrants courtesy of City of Manhattan, Kansas.

\Chapter12\Data\CityData.gdb\Parcels, parcel data, and imagery © 2011 County of Riley, Kansas, and its representatives. All rights reserved.

\Chapter12\Data\CityData.gdb\WaterLines created by author.

\Chapter12\Data\CityData.gdb\WaterValves courtesy of City of Manhattan, Kansas.

\Chapter12\Data\Subdivision21.TIF, parcel data and imagery © 2011 County of Riley, Kansas, and its representatives. All rights reserved.

Chapter 13

\Chapter13\Data\PlanningData1.gdb\Parcels, parcel data, and imagery © 2011 County of Riley, Kansas, and its representatives. All rights reserved.

\Chapter13\Data\PlanningData2.gdb\Parcels, parcel data, and imagery © 2011 County of Riley, Kansas, and its representatives. All rights reserved.

\Chapter13\Data\PlanningData3.gdb\Parcels, parcel data, and imagery © 2011 County of Riley, Kansas, and its representatives. All rights reserved.

\Chapter13\Data\Subdivision21.tif, parcel data and imagery © 2011 County of Riley, Kansas, and its representatives. All rights reserved.

Chapter 14

\Chapter14\Ex14aData\Customers.csv courtesy of Esri.

\Chapter14\Ex14aData\interstates.shp from Esri Data & Maps 2010, courtesy of Tele Atlas.

\Chapter14\Ex14aData\streets.shp from Esri Data & Maps 2010, courtesy of Tele Atlas.

\Chapter14\Ex14aData\zipcodes.shp from Esri Data & Maps 2010, courtesy of Tele Atlas, US Census Bureau, Esri (Pop2010 field).

\Chapter14\Ex14bData\Customers.csv courtesy of Esri.

\Chapter14\Ex14bData\interstates.shp from Esri Data & Maps 2010, courtesy of Tele Atlas.

\Chapter14\Ex14bData\streets.shp from Esri Data & Maps 2010, courtesy of Tele Atlas.

\Chapter14\Ex14bData\zipcodes.shp from Esri Data & Maps 2010, courtesy of Tele Atlas, US Census Bureau, Esri (Pop2010 field).

\Chapter14\Ex14cData\Customers.csv courtesy of Esri.

\Chapter14\Ex14cData\interstates.shp from Esri Data & Maps 2010, courtesy of Tele Atlas.

\Chapter14\Ex14cData\streets.shp from Esri Data & Maps 2010, courtesy of Tele Atlas.

\Chapter14\Ex14cData\zipcodes.shp from Esri Data & Maps 2010, courtesy of Tele Atlas, US Census Bureau, Esri (Pop2010 field).

Chapter 15

\Chapter15\Data\cabbagetown.shp courtesy of Generation5, a Motricity Business © 2011 Motricity Canada Inc.

\Chapter15\Data\cabbagetown1.jpg photo taken by author.

\Chapter15\Data\cabbagetown2.jpg photo taken by author.

\Chapter15\Data\cabbagetown3.jpg photo taken by author.

\Chapter15\Data\cabbagetown4.jpg photo taken by author.

\Chapter15\Data\cabbagetown5.jpg photo taken by author.

\Chapter15\Data\cabbagetown6.jpg photo taken by author.

\Chapter15\Data\cabbagetown7.jpg photo taken by author.

\Chapter15\Data\cabbagetown8.jpg photo taken by author.

\Chapter15\Data\toronto_background.shp courtesy of Generation5, a Motricity Business © 2011 Motricity Canada Inc.

\Chapter15\Data\toronto_streets.shp courtesy of Open Data, Geospatial Competency Centre, City of Toronto (Ontario, Canada)

Chapter 16

\Chapter16\Data\Wisconsin.gdb\FoodDeserts from Esri Data & Maps 2010, courtesy of ArcUSA, US Census Bureau, Esri (Pop2010 field); US Department of Agriculture, Economic Research Service.

\Chapter16\Data\Wisconsin.gdb\USstates from Esri Data & Maps 2010, courtesy of Esri, derived from Tele Atlas, US Census, Esri (Pop2010 field).

\Chapter16\Data\Wisconsin.gdb\WisconsinCities from Esri Data & Maps 2010, courtesy of US Census Bureau.

\Chapter16\Data\Wisconsin.gdb\WisconsinCounties from Esri Data & Maps 2010, courtesy of Esri, derived from Tele Atlas, US Census Bureau, Esri (Pop2010 field).

\Chapter16\Data\Wisconsin.gdb\WisconsinTracts from Esri Data & Maps 2010, courtesy of Tele Atlas, US Census Bureau, Esri (Pop2010 field).

\Chapter16\Data\FoodDesert_Data.xls from Esri Data & Maps 2010, courtesy of ArcUSA, US Census Bureau, Esri (Pop2010 field); US Department of Agriculture, Economic Research Service.

Chapter 17

\Chapter17\Data\BayArea.shp courtesy of County of Sonoma, Information Systems Department (ISD)/GIS.

\Chapter17\Data\Cities.shp from Esri Data & Maps 2010, courtesy of National Atlas of the United States.

\Chapter17\Data\Guerneville.jpg courtesy of County of Sonoma, Information Systems Department (ISD)/GIS.

\Chapter17\Data\GuernevilleParcels.shp courtesy of County of Sonoma, Information Systems Department (ISD)/GIS.

\Chapter17\Data\RussianRiver.shp courtesy of County of Sonoma, Information Systems Department (ISD)/GIS.

\Chapter17\Data\RussianRiverFloodZone.shp courtesy of County of Sonoma, Information Systems Department (ISD)/GIS.

Chapter 18

\Chapter18\Data\OregonForest.gdb\curlook from the Oregon Spatial Data Library, courtesy of Oregon Department of Forestry.

\Chapter18\Data\OregonForest.gdb\forestland from the Oregon Spatial Data Library, courtesy of Oregon Department of Forestry.

\Chapter18\Data\OregonForest.gdb\fwz100k from the Oregon Spatial Data Library, courtesy of Oregon Department of Forestry.

\Chapter18\Data\OregonForest.gdb\populated_jurisdictions from the Oregon Spatial Data Library, courtesy of Oregon Department of Forestry.

Chapter 19

\Chapter19\Data\OregonForest.gdb\AreaAtRisk from the Oregon Spatial Data Library, courtesy of Oregon Department of Forestry.

\Chapter19\Data\OregonForest.gdb\CountyWarningAreas from the Oregon Spatial Data Library, courtesy of Oregon Department of Forestry.

\Chapter19\Data\OregonForest.gdb\Final from the Oregon Spatial Data Library, courtesy of Oregon Department of Forestry.

\Chapter19\Data\OregonForest.gdb\FinalDissolve from the Oregon Spatial Data Library, courtesy of Oregon Department of Forestry.

\Chapter19\Data\OregonForest.gdb\LookoutsBuffer from the Oregon Spatial Data Library, courtesy of Oregon Department of Forestry.

\Chapter19\Data\OregonForest.gdb\PopJursBuffer from the Oregon Spatial Data Library, courtesy of Oregon Department of Forestry.

\Chapter19\Data\OregonForest.gdb\PortlandForest from the Oregon Spatial Data Library, courtesy of Oregon Department of Forestry.

\Chapter19\Data\OregonForest.gdb\PortlandLookouts from the Oregon Spatial Data Library, courtesy of Oregon Department of Forestry.

\Chapter19\Data\OregonForest.gdb\PortlandPopJurs from the Oregon Spatial Data Library, courtesy of Oregon Department of Forestry.

Chapter 20

\Chapter20\Data\New_Clinic.gdb\Blockgroups from Esri Data & Maps 2010, courtesy of Tele Atlas, US Census Bureau, Esri (Pop2010 field).

\Chapter20\Data\New_Clinic.gdb\CityBusStops courtesy of County of Sonoma, Department of Health Services.

\Chapter20\Data\New_Clinic.gdb\ClientList courtesy of County of Sonoma, Department of Health Services.

\Chapter20\Data\New_Clinic.gdb\streets courtesy of County of Sonoma, Department of Health Services.

Appendix B
Data license agreement

Important: Read carefully before opening the sealed media package.

Environmental Systems Research Institute, Inc. (Esri), is willing to license the enclosed data and related materials to you only upon the condition that you accept all of the terms and conditions contained in this license agreement. Please read the terms and conditions carefully before opening the sealed media package. By opening the sealed media package, you are indicating your acceptance of the Esri License Agreement. If you do not agree to the terms and conditions as stated, then Esri is unwilling to license the data and related materials to you. In such event, you should return the media package with the seal unbroken and all other components to Esri.

Esri License Agreement

This is a license agreement, and not an agreement for sale, between you (Licensee) and Environmental Systems Research Institute, Inc. (Esri). This Esri License Agreement (Agreement) gives Licensee certain limited rights to use the data and related materials (Data and Related Materials). All rights not specifically granted in this Agreement are reserved to Esri and its Licensors.

Reservation of Ownership and Grant of License

Esri and its Licensors retain exclusive rights, title, and ownership to the copy of the Data and Related Materials licensed under this Agreement and, hereby, grant to Licensee a personal, nonexclusive, nontransferable, royalty-free, worldwide license to use the Data and Related Materials based on the terms and conditions of this Agreement. Licensee agrees to use reasonable effort to protect the Data and Related Materials from unauthorized use, reproduction, distribution, or publication.

Proprietary Rights and Copyright

Licensee acknowledges that the Data and Related Materials are proprietary and confidential property of Esri and its Licensors and are protected by United

States copyright laws and applicable international copyright treaties and/or conventions.

Permitted Uses

Licensee may install the Data and Related Materials onto permanent storage device(s) for Licensee's own internal use.

Licensee may make only one (1) copy of the original Data and Related Materials for archival purposes during the term of this Agreement unless the right to make additional copies is granted to Licensee in writing by Esri.

Licensee may internally use the Data and Related Materials provided by Esri for the stated purpose of GIS training and education.

Uses Not Permitted

Licensee shall not sell, rent, lease, sublicense, lend, assign, time-share, or transfer, in whole or in part, or provide unlicensed Third Parties access to the Data and Related Materials or portions of the Data and Related Materials, any updates, or Licensee's rights under this Agreement.

Licensee shall not remove or obscure any copyright or trademark notices of Esri or its Licensors.

Term and Termination

The license granted to Licensee by this Agreement shall commence upon the acceptance of this Agreement and shall continue until such time that Licensee elects in writing to discontinue use of the Data or Related Materials and terminates this Agreement. The Agreement shall automatically terminate without notice if Licensee fails to comply with any provision of this Agreement. Licensee shall then return to Esri the Data and Related Materials. The parties hereby agree that all provisions that operate to protect the rights of Esri and its Licensors shall remain in force should breach occur.

Disclaimer of Warranty

The Data and Related Materials contained herein are provided "as-is," without warranty of any kind, either express or implied, including, but not limited to, the implied warranties of merchantability, fitness for a particular purpose, or noninfringement. Esri does not warrant that the Data and Related Materials will meet Licensee's needs or expectations, that the use of the Data and Related Materials will be uninterrupted, or that all nonconformities, defects, or errors can or will be corrected. Esri is not inviting reliance on the Data or Related

Materials for commercial planning or analysis purposes, and Licensee should always check actual data.

Data Disclaimer

The Data used herein has been derived from actual spatial or tabular information. In some cases, Esri has manipulated and applied certain assumptions, analyses, and opinions to the Data solely for educational training purposes. Assumptions, analyses, opinions applied, and actual outcomes may vary. Again, Esri is not inviting reliance on this Data, and the Licensee should always verify actual Data and exercise their own professional judgment when interpreting any outcomes.

Limitation of Liability

Esri shall not be liable for direct, indirect, special, incidental, or consequential damages related to Licensee's use of the Data and Related Materials, even if Esri is advised of the possibility of such damage.

No Implied Waivers

No failure or delay by Esri or its Licensors in enforcing any right or remedy under this Agreement shall be construed as a waiver of any future or other exercise of such right or remedy by Esri or its Licensors.

Order for Precedence

Any conflict between the terms of this Agreement and any FAR, DFAR, purchase order, or other terms shall be resolved in favor of the terms expressed in this Agreement, subject to the government's minimum rights unless agreed otherwise.

Export Regulation

Licensee acknowledges that this Agreement and the performance thereof are subject to compliance with any and all applicable United States laws, regulations, or orders relating to the export of data thereto. Licensee agrees to comply with all laws, regulations, and orders of the United States in regard to any export of such technical data.

Severability

If any provision(s) of this Agreement shall be held to be invalid, illegal, or unenforceable by a court or other tribunal of competent jurisdiction, the validity, legality, and enforceability of the remaining provisions shall not in any way be affected or impaired thereby.

Governing Law

This Agreement, entered into in the County of San Bernardino, shall be construed and enforced in accordance with and be governed by the laws of the United States of America and the State of California without reference to conflict of laws principles. The parties hereby consent to the personal jurisdiction of the courts of this county and waive their rights to change venue.

Entire Agreement

The parties agree that this Agreement constitutes the sole and entire agreement of the parties as to the matter set forth herein and supersedes any previous agreements, understandings, and arrangements between the parties relating hereto.

Appendix C
Installing the data and software

Getting to Know ArcGIS for Desktop includes a DVD containing maps and exercise data. A free, fully functioning 180-day trial version of ArcGIS 10.1 for Desktop Advanced, can be downloaded at www.esri.com /gtkArcGISforDesktop10-1. You will find an authorization number printed on the inside back cover of this book. You will use this number when you are ready to install the software.

If you already have a licensed copy of ArcGIS 10.1 for Desktop installed on your computer (or have access to the software through a network), do not install the trial software. Use your licensed software to do the exercises in this book. If you have an older version of ArcGIS installed on your computer, you must uninstall it before you can install the software that is provided with this book.

.NET Framework 3.5 SP1 must be installed on your computer before you install ArcGIS 10.1 for Desktop. Some features of ArcGIS 10.1 for Desktop require Windows Internet Explorer version 8.0. If you do not have Internet Explorer version 8.0 or higher, you must install it before installing ArcGIS 10.1 for Desktop.

Installing the exercise data

Follow the steps below to install the exercise data.

1. Put the data DVD in your computer's DVD drive. A splash screen will appear.

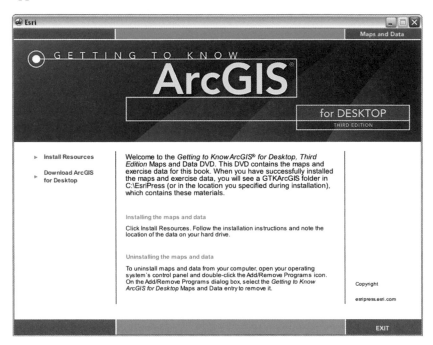

Note: If the splash screen does not appear, right-click the Start button on your computer and click My Computer. Navigate to the contents of the DVD drive and double-click Esri.exe.

2. Read the welcome, and then click the Install Resources link. This launches the InstallShield Wizard.

3. Click Next. Read and accept the license agreement terms, and then click Next.

4. Accept the default installation folder or click Browse and navigate to the drive or folder location where you want to install the data.

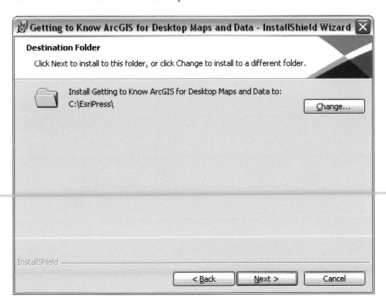

5. Click Next. The installation will take a few moments. When the installation is complete, you will see the following message.

6. Click Finish. The exercise data is installed on your computer in a folder called C:\EsriPress\GTKArcGIS.

Uninstalling the data and resources

To uninstall the data and resources from your computer, open your operating system's control panel and double-click the Add/Remove Programs icon. In the Add/Remove Programs dialog box, select the following entry and follow the prompts to remove it:

Getting to Know ArcGIS for Desktop Maps and Data

Installing the software

Note: If you already have a licensed copy of ArcGIS 10.1 for Desktop Advanced installed on your computer or have access to the software through a network, use it to do the exercises in this book. If you need a copy of ArcGIS 10.1 for Desktop Advanced, you can obtain a free trial version from Esri. This trial version is intended for educational purposes only and will expire 180 days after you install and register the software. The software cannot be reinstalled nor can the time limit be extended. It is recommended that you uninstall this software when it expires.

Follow these steps to obtain the 180-day free trial of ArcGIS 10.1 for Desktop Advanced:

1. Check the system requirements for ArcGIS to make sure your computer has the hardware and software required for the trial: www.esri.com/arcgis101sysreq

2. Uninstall any previous versions of ArcGIS Desktop that you already have on your computer. For best results, please download and use the ArcGIS Uninstall Utility from the download page.

3. Go to www.esri.com/gtkArcGISforDesktop10-1. Select the option that's right for you and click Next.

Note: You must have an Esri Global Account to receive your free trial software. Click "Create an Account" if you do not have one. When prompted, enter your 12 character authorization number (EVAxxxxxxxxx) printed on the inside back cover of the book, and then click Submit.

4. Install the Microsoft .NET Framework 3.5 Service Pack 1 from the Download Center on the Microsoft website.

5. Install ArcGIS for Desktop.

To download the software

- From the download page, click Download located to the right of ArcGIS for Desktop.
- Once downloaded, locate the EXE file to begin extracting the install files.
- Accept the default location or select a folder in your computer where the install files will be extracted.
- Go to the location of the extracted files and run setup.exe.

To install the software from a DVD

- Insert the DVD into your computer.
- The Quick Start Guide will run automatically. If not, go to My Computer and double-click your DVD drive to run setup.exe.
- To the right of ArcGIS for Desktop, click Setup. Click Next and accept the license agreement. Choose Complete Installation and click Next. Click Next to accept all default settings; then click Finish.

6. Authorize your trial software
- Once the ArcGIS Administrator Wizard window opens, select Advanced (ArcInfo) Single Use in the ArcGIS Desktop section; then click Authorize Now.
- Continue to click Next until you are prompted to enter your 12-character authorization number (EVAxxxxxxxxx), which you can find printed on the inside back cover of the book. Click Next.
- Select "I do not want to authorize any extensions at this time." The trial software automatically authorizes the extensions.
- On the next screen, verify that the extensions are listed on the left under Available Extensions. Click Next.
- After the authorization process is complete, click Finish.
- Click OK to close the ArcGIS Administrator.

7. To start ArcGIS 10.1 for Desktop, on the taskbar, click the Start button. On the Start menu, click All Programs > ArcGIS and choose ArcMap 10.1.
8. Enable the extensions.
- Start ArcMap 10.1. Go to Customize > Extensions and make sure each extension is checked.
- Click Close. The software is ready to use.

Assistance, FAQs, and support for your trial software are available on the online resources page at www.esri.com/evalhelp.

Uninstalling the software

To uninstall the software from your computer, use the Add or Remove Programs from Windows Control Panel. Select the following entry and follow the prompts to remove it:

ArcGIS 10.1 for Desktop

Glossary

This glossary contains definitions for the ArcGIS terms used throughout the book and noted in the list at the end of each chapter. A majority of these definitions can also be found at http://support.esri.com/en/knowledgebase/GISdictionary.

address locator A dataset in ArcGIS that stores the address attributes, associated indexes, and rules that define the process for translating nonspatial descriptions of places, such as street addresses, into spatial data that can be displayed as features on a map. An address locator contains a snapshot of the reference data used for geocoding and includes the parameters for standardizing addresses, searching for match locations, and creating output.

annotation In ArcGIS, text or graphics that can be individually selected, positioned, and modified. Annotation may be manually entered, or generated from labels. Annotation can be stored as features in a geodatabase or as map annotation in a data frame.

attribute domain In a geodatabase, a mechanism for enforcing data integrity. Attribute domains define what values are allowed in a field in a feature class or a nonspatial attribute table. If the features or nonspatial objects have been grouped into subtypes, different attribute domains can be assigned to each of the subtypes.

attribute join Appending the fields of one table to those of another through an attribute common to both tables. *See also* **spatial join** and **relate**.

attribute query A request for records of features in a table based on their attribute values.

attributes Nonspatial information about a geographic feature in a GIS, usually stored in a table and linked to the feature by a unique identifier. For example, attributes of a river might include its name, length, and sediment load at a gauging station. Attribute tables also store information about feature geometry, such as length and area.

basemap A map depicting background reference information, such as landforms, roads, landmarks, and political boundaries, onto which other thematic information is placed.

buffer A zone around a map feature measured in units of distance or time. A buffer is useful for proximity analysis.

cache In ArcGIS Online, a temporary storage of map tiles in the desktop computer's RAM, which may result in improved display performance.

cartography The art and science of experssing graphically, usually through maps, the natural and social features of the earth.

cell The smallest unit of information in raster data, usually square in shape. In a map or GIS dataset, each cell represents a portion of the earth, such as a square meter or square mile, and usually has an attribute value associated with it, such as soil type or vegetation class.

clip A command that extracts features from one feature class that reside entirely within a boundary defined by features in another feature class.

cloud computing A rapidly emerging, scalable technology that furnishes technological capabilities—commonly maintained off premises—that are delivered on demand as applications or services via the Internet.

continuous data Data such as elevation or temperature that varies without discrete steps. Since computers store data discretely, continuous data is usually represented by triangulated irregular networks (TINs), rasters, or contour lines, so that any location has either a specified value or one that can be derived.

coordinate system A reference framework consisting of a set of points, lines, and/or surfaces, and a set of rules, used to define the position of points in space in either two or three dimensions. The Cartesian coordinate system and the geographic coordinate system used on the earth's surface are common examples of coordinate systems.

data frame A map element that defines a geographic extent, a page extent, a coordinate system, and other display properties for one or more layers in ArcMap. A dataset can be represented in one or more data frames. In data view, only one data frame is displayed at a time; in layout view, all of a map's data frames are displayed at the same time.

data view An all-purpose view in ArcMap and ArcReader for exploring, displaying, and querying geographic data. This view hides all map elements, such as titles, north arrows, and scale bars.

datum The reference specifications of a measurement system, usually a system of coordinate positions on a surface (a horizontal datum) or heights above or below a surface (a vertical datum).

datum transformation A transformation definition that is used to transform the coordinates of a point defined in one datum to coordinates in a different datum.

decimal degrees Values of latitude and longitude expressed in decimal format rather than in degrees minutes seconds.

defined interval Also known as class interval. A set of categories for classification that divide the range of all values so that each piece of data is contained within a nonoverlapping category.

definition query In ArcMap, a request that examines feature or tabular attributes based on user-selected criteria and displays only those features or records that satisfy the criteria.

degrees-minutes-seconds The unit of measure for describing latitude and longitude. A degree is 1/360th of a circle. A degree is further divided into 60 minutes, and a minute is divided into 60 seconds.

digitizing The process of converting the geographic features on an analog map into digital format using a digitizing tablet, or digitizer, which is connected to a computer. Features on a paper map are traced with a digitizer puck, a device similar to a mouse, and the x,y coordinates of these features are automatically recorded and stored as spatial data. Digitizing also refers to creating new features with sketch tools in ArcMap.

discrete data Data that represents phenomena with distinct boundaries. Property lines and land-use areas are examples of discrete data.

display scale The scale at which data is rendered on a computer screen or on a printed map.

dissolve A geoprocessing command that removes boundaries between adjacent polygons that have the same value for a specified attribute.

dot density map A quantitative, thematic map on which dots of the same size are randomly placed in proportion to a numeric attribute associated with an area. Dot density maps convey the intensity of an attribute.

edge A line between two points that forms a boundary.

edit session In ArcMap, the environment in which spatial and attribute editing take place. In an edit session, a user can modify feature locations, geometry, or attributes. Modifications are not saved unless the user explicitly chooses to save them.

edit sketch In ArcMap, a temporary sketch that is used to perform a variety of tasks. Creating an edit sketch is a standard way to edit feature geometry.

endpoints The points that define the length of a line segment.

environment settings Settings that can apply to all tools within the application, all tools within a model or script, or to a particular process within a model or script. Environment settings include current workspace, output spatial reference, output spatial grids, cell size, and tile size. They are generally set before running tools.

equal interval A data classification method that divides a set of attribute values into groups that contain an equal range of values.

Esri grid An Esri data format for storing raster data that defines geographic space as an array of equally sized square cells arranged in rows and columns. Each cell stores a numeric value that represents a geographic attribute (such as elevation) for that unit of space. When the grid is drawn as a map, cells are assigned colors according to their numeric values. Each grid cell is referenced by its x,y coordinate location.

false easting The linear value added to all x-coordinates of a map projection so that none of the values in the geographic region being mapped are negative.

false northing The linear value added to all y-coordinates of a map projection so that none of the values in the geographic region being mapped are negative.

feature class In ArcGIS, a collection of geographic features with the same geometry type (such as point, line, or polygon), the same attributes, and the same spatial reference. Feature classes can be stored in a geodatabase, shapefile, or coverage, among other data formats. Feature classes allow homogeneous features to be grouped into a single unit for data storage purposes. For example, highways, primary roads, and secondary roads can be grouped into a line feature class named "roads." In a geodatabase, feature classes can also store annotation and dimensions.

feature dataset A collection of feature classes stored together that share the same spatial reference; that is, they share a coordinate system, and their features fall within a common geographic area. Feature classes with different geometry types may be stored in a feature dataset.

geocoding A GIS operation for converting street addresses into spatial data that can be displayed as features on a map, usually by referencing address information from a street segment data layer.

geodatabase A database or file structure used primarily to store, query, and manipulate spatial data. Geodatabases store geometry, a spatial reference system, attributes, and behavioral rules for data. Various types of geographic datasets can be collected within a geodatabase, including feature classes, attribute tables, raster datasets, network datasets, topologies, and many others. Geodatabases can be stored in IBM DB2, IBM Informix, Oracle, Microsoft Access, Microsoft SQL Server, and PostgreSQL relational database management systems or in a system of files, such as a file geodatabase.

geodatabase topology The arrangement that constrains how point, line, and polygon features share geometry. For example, street centerlines and census blocks share geometry, and adjacent soil polygons share geometry. Topology defines and enforces data integrity rules (for example, there should be no gaps between polygons). It supports topological relationship queries and navigation (for example, navigating feature adjacency or connectivity), works with sophisticated editing tools, and allows feature construction from unstructured geometry (for example, constructing polygons from lines).

geographic coordinate system A reference system that uses latitude and longitude to define the location of points on the surface of a sphere or spheroid. A geographic coordinate system definition includes a datum, prime meridian, and angular unit.

geoid A hypothetical surface representing the form the earth's oceans would take if there were no land and the water were free to respond to the earth's gravitational and centrifugal forces. The resulting geoid is irregular and varies from a perfect sphere by as much as 75 meters above and 100 meters below its surface.

geometrical interval A classification scheme where the class breaks are based on class intervals that are in a geometric series. The geometric coefficient in this classifier can change once (to its inverse) to optimize the class ranges.

geoprocessing A GIS operation used to manipulate GIS data. A typical geoprocessing operation takes an input dataset, performs an operation on that dataset, and returns the result of the operation as an output dataset. Common geoprocessing operations include geographic feature overlay, feature selection and analysis, topology processing, raster processing, and data conversion. Geoprocessing helps define, manage, and analyze the information used to form decisions.

graduated-color map A map on which a range of colors indicates a progression of numeric values. For example, increases in population density might be represented by the increased saturation of a single color, or temperature differences might be shown by a sequence of colors from blue to red.

graduated-symbol map A map with symbols that change in size according to the value of the attribute they represent. For example, denser populations might be represented by larger dots or larger rivers by thicker lines.

graticule A network of longitude and latitude lines on a map or chart that relates points on a map to their true location on the earth.

group In ArcGIS.com, a way to collaborate with other ArcGIS users and to exchange content related to a specific project or common activity.

hillshade Shadows drawn on a map to simulate the effect of the sun's rays over the varied terrain of the land.

histogram A graph showing the distribution of values in a set of data. Individual values are displayed along a horizontal axis, and the frequency of their occurrence is displayed along a vertical axis.

hyperlink A reference (link) from one point in an electronic document to another document or another location in the same document (the target).

Activating the link, usually by clicking it with the mouse, causes the browser to display the target of the link.

intermediate data Any data in a process that did not exist before the process existed and that will not be maintained after the process is executed.

intersect A geometric integration of spatial datasets that preserves features or portions of features that fall within areas common to all input datasets.

join table The table in a join operation, often a nonspatial table, that is appended to the target layer's attribute table.

label In ArcGIS, descriptive text, usually based on one or more feature attributes. Labels are placed dynamically on or near features based on user-defined rules and in response to changes in the map display. Labels cannot be individually selected and modified by the user. Label placement rules and display properties (such as font size and color) are defined for an entire layer.

latitude and longitude A reference system used to locate positions on the earth's surface. Distances east–west are measured with lines of longitude (also called meridians), which run north–south and converge at the North and South Poles. Distance measurements begin at the prime meridian and are measured positively 180 degrees to the east and negatively 180 degrees to the west. Distances north–south are measured with lines of latitude (also called parallels), which run east–west. Distance measurements begin at the equator and are measured positively 90 degrees to the north and negatively 90 degrees to the south.

layer In ArcGIS, a reference to a data source, such as a shapefile, coverage, geodatabase feature class, or raster, that defines how the data should be symbolized on a map. Layers can also define additional properties, such as which features from the data source are included. Layers can be stored in a map document (.mxd) or saved individually as a layer file (.lyr).

layer file A file with a .lyr extension that stores the path of a source dataset and other layer properties, including symbology.

layer package A file that includes both the layer properties and the source data referenced by the layer. Layer packages (.lpk) contain symbolization, labeling, table properties, and the data. They can be created in ArcMap, ArcGlobe, and ArcScene and can be viewed in any of these applications, as well

as in
ArcGIS Explorer.

layout The arrangement of elements on a map, including a title, legend, north arrow, scale bar, and geographic data.

layout view In ArcMap and ArcReader, a view that shows the virtual page upon which geographic data and map elements, such as titles, legends, and scale bars, are placed and arranged for printing.

legend The description of the types of features included in a map, usually displayed in the map layout. Legends often use graphics of symbols or examples of features from the map with a written description of what each symbol or graphic represents.

location query Also called a spatial query, a location query is a statement or logical expression that selects geographic features based on location or spatial relationship. For example, a spatial query might find which points are contained within a polygon or set of polygons, find features within a specified distance of a feature, or find features that are adjacent to each other.

manual classification The process of sorting or arranging entities into groups or categories; on a map, the process of representing members of a group by the same symbol, usually defined in a legend.

map algebra A language that defines a syntax for combining map themes by applying mathematical operations and analytical functions to create new map themes. In a map algebra expression, the operators are a combination of mathematical, logical, or Boolean operators (+, >, AND, tan, and so on) and spatial analysis functions (slope, shortest path, spline, and so on), and the operands are spatial data and numbers.

map extent The limit of the geographic area shown on a map, usually defined by a rectangle. In a dynamic map display, the map extent can be changed by zooming and panning.

map projection A method by which the curved surface of the earth is portrayed on a flat surface. This generally requires a systematic mathematical transformation of the earth's graticule of lines of longitude and latitude onto a plane. Some projections can be visualized as a transparent globe with a light bulb at its center (although not all projections emanate from the globe's center) casting lines of latitude and longitude onto a sheet of paper. Generally, the paper is either flat and placed tangent to the globe (a planar or azimuthal projection) or formed into a cone or cylinder and placed over the globe (a cylindrical or conical projection). Every map projection distorts distance, area, shape, or direction, or some combination thereof.

map topology A temporary set of topological relationships between coincident parts of simple features on a map, used to edit shared parts of multiple features.

map unit The ground unit of measurement—for example, feet, miles, meters, or kilometers—in which coordinates of spatial data are stored.

mashup In Internet mapping, the combination of content from more than one data source into one dynamic map service.

mask A means of identifying areas to be included in analysis. Such a mask is often referred to as an analysis mask and may be either a raster layer or a feature layer.

metadata Information that describes the content, quality, condition, origin, and other characteristics of data or other pieces of information. Metadata for spatial data may describe and document its subject matter; how, when, where, and by whom the data was collected; availability and distribution information; its projection, scale, resolution, and accuracy; and its reliability with regard to some standard. Metadata consists of properties and documentation. Properties are derived from the data source (for example, the coordinate system and projection of the data), and documentation is entered by a person (for example, keywords used to describe the data).

model In geoprocessing in ArcGIS, one process or a sequence of connected processes diagrammed in ModelBuilder.

multipart polygon In ArcGIS, a digital representation of a place or thing that has more than one part but is defined as one feature because it references one set of attributes. In a layer of states, for example, the state of Hawaii could be considered a multipart feature because its separate geometric parts are classified as a single state. A multipart feature can be a point, line, or polygon.

natural breaks A method of manual data classification that seeks to partition data into classes based on natural groups in the data distribution. Natural breaks occur in the histogram at the low points of valleys. Breaks are assigned in the order of the size of the valleys, with the largest valley assigned the first natural break.

neatline The border delineating and defining the extent of geographic data on a map. It demarcates map units so that, depending on the map projection, the neatline does not always have 90-degree corners. In a properly made map, it is the most accurate element of the data; other map features may be moved slightly or exaggerated for generalization or readability, but the neatline is never adjusted.

NoData In raster data, the absence of a recorded value. NoData does not equate to a zero (0) value. Although the measure of a particular attribute in a cell may be zero, a NoData value indicates that no measurements have been taken for that cell at all.

normalization The process of dividing one numeric attribute value by another to minimize differences in values based on the size of areas or the number of features in each area. For example, normalizing (dividing) total population by total area yields population per unit area, or density.

north arrow A map symbol that shows the direction of north on the map, thereby showing how the map is oriented.

on-the-fly projection ArcMap can display data stored in one projection as if it were in another projection. The new projection is used for display and query purposes only. The actual data is not altered.

operational layer Web mapping layers that users view and work with to perform their tasks.

operator The symbolic representation of a process or operation performed against one or more operands in an expression, such as + (plus, or addition) and > (greater than). When evaluated, operators return a value as their result. If multiple operators appear in an expression, they are evaluated in order of their operator precedence.

overlay In geoprocessing, the geometric intersection of multiple datasets to combine, erase, modify, or update features in a new output dataset.

projected coordinate system A reference system used to locate x, y, and z positions of point, line, and area features in two or three dimensions. A projected coordinate system is defined by a geographic coordinate system, a map projection, any parameters needed by the map projection, and a linear unit of measure.

proportional-symbol map A map that contains symbols whose sizes differ in relation to the phenomenon being mapped.

pyramid In raster datasets, a reduced-resolution layer that copies the original data in decreasing levels of resolution to enhance performance. The coarsest level of resolution is used to quickly draw the entire dataset. As the display zooms in, layers with finer resolution are drawn; drawing speed is maintained because fewer pixels are needed to represent the successively smaller areas.

quantile A data classification method that distributes a set of values into groups that contain an equal number of values.

query expression A type of expression that evaluates to a Boolean (true or false) value and is typically used to select those rows in a table whose values cause the expression to evaluate to true. Query expressions are generally part of a SQL statement.

raster A spatial data model that defines space as an array of equally sized cells arranged in rows and columns and composed of single or multiple bands. Each cell contains an attribute value and location coordinates. Unlike a vector structure, which stores coordinates explicitly, raster coordinates are contained in the ordering of the matrix. Groups of cells that share the same value represent the same type of geographic feature.

reclassification The process of taking input cell values and replacing them with new output cell values. Reclassification is often used to simplify or change the interpretation of raster data by changing a single value to a new value or grouping ranges of values into single values—for example, assigning a value of one (1) to cells that have values of 1 to 50, two (2) to cells that range from 51 to 100, and so on.

relate Establishing a connection between two tables using an attribute common to both.

scale bar A map element used to graphically represent the scale of a map. A scale bar is typically a line marked like a ruler in units proportional to the map's scale.

shapefile A vector data storage format for storing the location, shape, and attributes of geographic features. A shapefile is stored in a set of related files and contains one feature class.

shortcut (context) menu The menu of commands that is accessed by right-clicking an object.

snapping An automatic editing operation in which points or features within a specified distance (tolerance) of other points or features are moved to match or coincide exactly with each other's coordinates.

spatial join A type of table join operation in which fields from one layer's attribute table are appended to another layer's attribute table based on the relative locations of the features in the two layers.

spheroid When used to represent the earth, a three-dimensional shape obtained by rotating an ellipse about its minor axis, either with dimensions that approximate the earth as a whole or with a part that approximates the corresponding portion of the geoid.

standard deviation A data classification method that finds the mean value, and then places class breaks above and below the mean at intervals of either .25, .5, or 1 standard deviation until all the data values are contained within the classes. Values that are beyond three standard deviations from the mean are aggregated into two classes—greater than three standard deviations above the mean and less than three standard deviations below the mean.

standard meridian A great circle on the earth that passes through the poles, often used synonymously with longitude. Meridians run north–south between the poles. By convention, meridians are labeled with positive numbers that ascend as one moves eastward from the prime meridian and negative numbers as one moves westward from the prime meridian until the Eastern and Western Hemispheres meet at the 180-degree line. Meridians can also, however, be labeled with all positive or negative numbers, including positive numbers increasing westward from the prime meridian.

standard parallel The line of latitude in a conic or cylindrical projection in normal aspect where the projection surface touches the globe. A tangent conic or cylindrical projection has one standard parallel, while a secant conic or cylindrical projection has two. At the standard parallel, the projection shows no distortion.

style An organized collection of predefined colors, symbols, properties of symbols, and map elements. Styles promote standardization and consistency in mapping products.

symbology The set of conventions, rules, or encoding systems that define how geographic features are represented with symbols on a map. A characteristic of a map feature may influence the size, color, and shape of the symbol used.

target table The table to which new attributes are appended. Records in a target table are often referred to as target features.

tiles An internal subset of a spatial dataset (commonly a raster dataset) into a manageable rectangular set, or rows and columns of pixels, typically used to process or analyze a large raster dataset without consuming vast quantities of computer memory.

topography The study and mapping of land surfaces, including relief (relative positions and elevations) and the position of natural and constructed features.

union A topological overlay of two or more polygon spatial datasets that preserves the features that fall within the spatial extent of either input dataset—that is, all features from both datasets are retained and extracted into a new polygon dataset.

vector A coordinate-based data model that represents geographic features as points, lines, and polygons. Each point feature is represented as a single coordinate pair, and line and polygon features are represented as ordered lists of vertices. Attributes are associated with each vector feature, as opposed to a raster data model, which associates attributes with grid cells.

vertex (vertices) One of a set of ordered x,y coordinate pairs that defines the shape of a line or polygon feature.

Index